Sets, Logic, Computation

The Open Logic Project

Instigator

Richard Zach, *University of Calgary*

Editorial Board

Aldo Antonelli,[†] *University of California, Davis*
Andrew Arana, *Université de Lorraine*
Jeremy Avigad, *Carnegie Mellon University*
Tim Button, *University College London*
Walter Dean, *University of Warwick*
Gillian Russell, *Dianoia Institute of Philosophy*
Nicole Wyatt, *University of Calgary*
Audrey Yap, *University of Victoria*

Contributors

Samara Burns, *Columbia University*
Dana Hägg, *University of Calgary*
Zesen Qian, *Carnegie Mellon University*

Sets, Logic, Computation

An Open Introduction to Metalogic

Remixed by Richard Zach

FALL 2021

The Open Logic Project would like to acknowledge the generous support of the Taylor Institute of Teaching and Learning of the University of Calgary, and the Alberta Open Educational Resources (ABOER) Initiative, which is made possible through an investment from the Alberta government.

Cover illustrations by Matthew Leadbeater, used under a Creative Commons Attribution-NonCommercial 4.0 International License.

Typeset in Baskervald X and Nimbus Sans by LaTeX.

This version of *Sets, Logic, Computation* is revision 42ef903 (2021-08-02), with content generated from *Open Logic Text* revision ef13ef0 (2022-01-05). Free download at:
https://slc.openlogicproject.org/

 Sets, Logic, Computation by Richard Zach is licensed under a Creative Commons Attribution 4.0 International License. It is based on *The Open Logic Text* by the Open Logic Project, used under a Creative Commons Attribution 4.0 International License.

Contents

Preface		xiii
I	**Sets, Relations, Functions**	**1**
1	**Sets**	**2**
1.1	Extensionality	2
1.2	Subsets and Power Sets	4
1.3	Some Important Sets	5
1.4	Unions and Intersections	6
1.5	Pairs, Tuples, Cartesian Products	10
1.6	Russell's Paradox	12
	Summary	14
	Problems	14
2	**Relations**	**16**
2.1	Relations as Sets	16
2.2	Special Properties of Relations	18
2.3	Equivalence Relations	20
2.4	Orders	21
2.5	Graphs	24
2.6	Operations on Relations	26
	Summary	27
	Problems	28

3 Functions — 29
- 3.1 Basics — 29
- 3.2 Kinds of Functions — 32
- 3.3 Functions as Relations — 34
- 3.4 Inverses of Functions — 36
- 3.5 Composition of Functions — 38
- 3.6 Partial Functions — 39
- Summary — 41
- Problems — 41

4 The Size of Sets — 43
- 4.1 Introduction — 43
- 4.2 Enumerations and Countable Sets — 43
- 4.3 Cantor's Zig-Zag Method — 48
- 4.4 Pairing Functions and Codes — 50
- 4.5 An Alternative Pairing Function — 52
- 4.6 Uncountable Sets — 54
- 4.7 Reduction — 58
- 4.8 Equinumerosity — 59
- 4.9 Sets of Different Sizes, and Cantor's Theorem — 61
- 4.10 The Notion of Size, and Schröder-Bernstein — 63
- Summary — 64
- Problems — 65

II First-order Logic — 69

5 Introduction to First-Order Logic — 70
- 5.1 First-Order Logic — 70
- 5.2 Syntax — 72
- 5.3 Formulas — 73
- 5.4 Satisfaction — 75
- 5.5 Sentences — 77
- 5.6 Semantic Notions — 78
- 5.7 Substitution — 79
- 5.8 Models and Theories — 80
- 5.9 Soundness and Completeness — 82

6 Syntax of First-Order Logic — 84
- 6.1 Introduction — 84
- 6.2 First-Order Languages — 85
- 6.3 Terms and Formulas — 87
- 6.4 Unique Readability — 90
- 6.5 Main operator of a Formula — 94
- 6.6 Subformulas — 95
- 6.7 Free Variables and Sentences — 97
- 6.8 Substitution — 98
- Summary — 101
- Problems — 101

7 Semantics of First-Order Logic — 102
- 7.1 Introduction — 102
- 7.2 Structures for First-order Languages — 103
- 7.3 Covered Structures for First-order Languages — 105
- 7.4 Satisfaction of a Formula in a Structure — 106
- 7.5 Variable Assignments — 113
- 7.6 Extensionality — 116
- 7.7 Semantic Notions — 118
- Summary — 121
- Problems — 121

8 Theories and Their Models — 124
- 8.1 Introduction — 124
- 8.2 Expressing Properties of Structures — 126
- 8.3 Examples of First-Order Theories — 128
- 8.4 Expressing Relations in a Structure — 131
- 8.5 The Theory of Sets — 132
- 8.6 Expressing the Size of Structures — 136
- Summary — 137
- Problems — 138

9 Derivation Systems — 140
- 9.1 Introduction — 140
- 9.2 The Sequent Calculus — 142

	9.3	Natural Deduction	143
	9.4	Tableaux	145
	9.5	Axiomatic Derivations	147

10 The Sequent Calculus   150

10.1	Rules and Derivations	150
10.2	Propositional Rules	151
10.3	Quantifier Rules	152
10.4	Structural Rules	153
10.5	Derivations	154
10.6	Examples of Derivations	156
10.7	Derivations with Quantifiers	161
10.8	Proof-Theoretic Notions	162
10.9	Derivability and Consistency	165
10.10	Derivability and the Propositional Connectives	167
10.11	Derivability and the Quantifiers	168
10.12	Soundness	169
10.13	Derivations with Identity predicate	176
10.14	Soundness with Identity predicate	177
	Summary	178
	Problems	179

11 Natural Deduction   180

11.1	Rules and Derivations	180
11.2	Propositional Rules	181
11.3	Quantifier Rules	182
11.4	Derivations	184
11.5	Examples of Derivations	185
11.6	Derivations with Quantifiers	190
11.7	Proof-Theoretic Notions	194
11.8	Derivability and Consistency	196
11.9	Derivability and the Propositional Connectives	198
11.10	Derivability and the Quantifiers	200
11.11	Soundness	201
11.12	Derivations with Identity predicate	206
11.13	Soundness with Identity predicate	208

Summary 209
Problems 209

12 The Completeness Theorem 211
12.1 Introduction 211
12.2 Outline of the Proof 213
12.3 Complete Consistent Sets of Sentences 216
12.4 Henkin Expansion 217
12.5 Lindenbaum's Lemma 220
12.6 Construction of a Model 221
12.7 Identity 224
12.8 The Completeness Theorem 227
12.9 The Compactness Theorem 228
12.10 A Direct Proof of the Compactness Theorem . . 231
12.11 The Löwenheim-Skolem Theorem 232
Summary 234
Problems 235

13 Beyond First-order Logic 237
13.1 Overview 237
13.2 Many-Sorted Logic 238
13.3 Second-Order logic 240
13.4 Higher-Order logic 245
13.5 Intuitionistic Logic 248
13.6 Modal Logics 254
13.7 Other Logics 256

III Turing Machines 259

14 Turing Machine Computations 260
14.1 Introduction 260
14.2 Representing Turing Machines 263
14.3 Turing Machines 267
14.4 Configurations and Computations 269
14.5 Unary Representation of Numbers 271
14.6 Halting States 275

14.7	Disciplined Machines	276
14.8	Combining Turing Machines	277
14.9	Variants of Turing Machines	281
14.10	The Church-Turing Thesis	283
	Summary	284
	Problems	285

15 Undecidability — 288

15.1	Introduction	288
15.2	Enumerating Turing Machines	290
15.3	Universal Turing Machines	293
15.4	The Halting Problem	296
15.5	The Decision Problem	299
15.6	Representing Turing Machines	300
15.7	Verifying the Representation	304
15.8	The Decision Problem is Unsolvable	310
15.9	Trakthenbrot's Theorem	312
	Summary	316
	Problems	317

A Proofs — 321

A.1	Introduction	321
A.2	Starting a Proof	323
A.3	Using Definitions	323
A.4	Inference Patterns	326
A.5	An Example	334
A.6	Another Example	338
A.7	Proof by Contradiction	340
A.8	Reading Proofs	345
A.9	I Can't Do It!	347
A.10	Other Resources	349
	Problems	350

B Induction — 351

B.1	Introduction	351
B.2	Induction on $\mathbb{N}$	352

B.3	Strong Induction	355
B.4	Inductive Definitions	356
B.5	Structural Induction	359
B.6	Relations and Functions	361
Problems		365

C Biographies — 366
- C.1 Georg Cantor 366
- C.2 Alonzo Church 367
- C.3 Gerhard Gentzen 368
- C.4 Kurt Gödel 370
- C.5 Emmy Noether 372
- C.6 Bertrand Russell 374
- C.7 Alfred Tarski 376
- C.8 Alan Turing 377
- C.9 Ernst Zermelo 379

D The Greek Alphabet — 381

Glossary — 382

Photo Credits — 389

Bibliography — 391

About the Open Logic Project — 398

Thoralf Skolem
1887 - 1963

Preface

This book is an introduction to metalogic, aimed especially at students of computer science and philosophy. "Metalogic" is so-called because it is the discipline that studies logic itself. Logic proper is concerned with canons of valid inference, and its symbolic or formal version presents these canons using formal languages, such as those of propositional and first-order logic. Metalogic investigates the properties of these languages, and of the canons of correct inference that use them. It studies topics such as how to give precise meaning to the expressions of these formal languages, how to justify the canons of valid inference, what the properties of various derivation systems are, including their computational properties. These questions are important and interesting in their own right, because the languages and derivation systems investigated are applied in many different areas—in mathematics, philosophy, computer science, and linguistics, especially—but they also serve as examples of how to study formal systems in general. The logical languages we study here are not the only ones people are interested in. For instance, linguists and philosophers are interested in languages that are much more complicated than those of propositional and first-order logic, and computer scientists are interested in other *kinds* of languages altogether, such as programming languages. And the methods we discuss—how to give semantics for formal languages, how to prove results about formal languages, how to investigate the properties of formal languages—are applicable in those cases as well.

Like any discipline, metalogic both has a set of results or facts, and a store of methods and techniques, and this text covers both. Many students won't need to know all of the results we discuss outside of this course, but they will need and use the methods we use to establish them. The Löwenheim-Skolem theorem, say, does not often make an appearance in computer science, but the methods we use to prove it do. On the other hand, many of the results we discuss do have relevance for certain debates, say, in the philosophy of science and in metaphysics. Philosophy students may not need to be able to prove these results outside this course, but they do need to understand what the results are—and you really only *understand* these results if you have thought through the definitions and proofs needed to establish them. These are, in part, the reasons for why the results and the methods covered in this text are recommended study—in some cases even required—for students of computer science and philosophy.

The material is divided into three parts. Part I concerns itself with the theory of sets. Logic and metalogic is historically connected very closely to what's called the "foundations of mathematics." Mathematical foundations deal with how ultimately mathematical objects such as integers, rational, and real numbers, functions, spaces, etc., should be understood. Set theory provides one answer (there are others), and so set theory and logic have long been studied side-by-side. Sets, relations, and functions are also ubiquitous in any sort of formal investigation, not just in mathematics but also in computer science and in some of the more technical corners of philosophy. Certainly for the purposes of formulating and proving results about the semantics and proof theory of logic and the foundation of computability it is essential to have a terminology in which to do this. For instance, we will talk about sets of expressions, relations of consequence and provability, interpretations of predicate symbols (which turn out to be relations), computable functions, and various relations between and constructions using them. It will be good to have shorthand symbols for these, and think through the general properties of sets, relations, and functions. If you are not used to thinking mathematically and to formulating mathematical proofs, then think of the first part on set theory as a training ground: all the basic definitions will be given, and we'll give increasingly complicated proofs using them. Note that understanding these proofs—and being able to find and formulate them yourself—is perhaps more important than understanding the results, especially in the first part. If mathematical thinking is new to you, it is important that you think through the examples and problems.

In the first part we will establish one important result, however. This result—Cantor's theorem—relies on one of the most striking examples of conceptual analysis to be found anywhere in the sciences, namely, Cantor's analysis of infinity. Infinity has puzzled mathematicians and philosophers alike for centuries. Until Cantor, no-one knew how to properly think about it. Many people even considered it a mistake to think about it at all, and believed that the notion of an infinite collection itself was incoherent. Cantor made infinity into a subject we can coherently work with, and developed an entire theory of infinite collections—and infinite numbers with which we can measure the sizes of infinite collections. He showed that there are different levels of infinity. This theory of "transfinite" numbers is beautiful and intricate, and we won't get very far into it; but we will be able to show that there are different levels of infinity, specifically, that there are "countable" and "uncountable" levels of infinity. This result has important applications, but it is also really the kind of result that any self-respecting mathematician, computer scientist, and philosopher should know.

In part II, we turn to first-order logic. We will define the language of first-order logic and its semantics, i.e., what first-order structures are and when a sentence of first-order logic is true in a structure. This will enable us to do two important things: (1) We can define, with mathematical precision, when a sentence is a logical consequence of another. (2) We can also consider how the relations that make up a first-order structure are described—characterized—by the sentences that are true in them. This in particular leads us to a discussion of the axiomatic method, in which sentences of first-order languages are used to characterize certain kinds of structures. Proof theory will occupy us next, and we will consider the original version of the sequent calculus and natural deduction as defined in the 1930s by Gerhard Gentzen. (Your instructor may choose to cover only one, then any reference to "derivations" and "derivability" will mean whatever system they chose.) The semantic notion of consequence and the syntactic notion of derivability give us two completely different ways to make precise the idea that a sentence may follow from some others. The soundness and completeness theorems link these two characterization. In particular, we will prove Gödel's completeness theorem, which states that whenever a sentence is a semantic consequence of some others, then it is also derivable from them. An equivalent formulation is: if a collection of sentences is consistent—in the sense that nothing contradictory can be proved from them—then there is a structure that makes all of them true.

The second formulation of the completeness theorem is perhaps the more surprising. Around the time Gödel proved this result (in 1929), the German mathematician David Hilbert famously held the view that consistency (i.e., freedom from contradiction) is all that mathematical existence requires. In other words, whenever a mathematician can coherently describe a structure or class of structures, then they should be entitled to believe in the existence of such structures. At the time, many found this idea preposterous: just because you can describe a structure without contradicting yourself, it surely does not follow that such a structure actually exists. But that is exactly what Gödel's completeness theorem says. In addition to this paradoxical—and certainly philosophically intriguing—aspect, the completeness theorem also has two important applications which allow us to prove further results about the existence of structures which make given sentences true. These are the compactness and the Löwenheim-Skolem theorems.

In part III, we connect logic with computability. Again, there is a historical connection: David Hilbert had posed as a fundamental problem of logic to find a mechanical method which would decide, of a given sentence of logic, whether it has a proof. Such a method exists, of course, for propositional logic: one just has to check all truth tables, and since there are only finitely many of them, the method eventually yields a correct answer. Such a straightforward method is not possible for first-order logic, since the number of possible structures is infinite (and structures themselves may be infinite). Logicians were working to find a more ingenious methods for years. Alonzo Church and Alan Turing eventually established that there is no such method. In order to do this, it was necessary to first provide a precise definition of what a mechanical method is in general. If a decision procedure had been proposed, presumably it would have been recognized as an effective method. To prove that no effective method exists, you have to define "effective method" first and give an impossibility proof on the basis of that definition. This is what Turing did: he proposed the idea of a Turing machine[1] as a mathematical model of what a mechanical procedure can, in principle, do. This is another example of a *conceptual analysis* of an informal concept using mathematical machinery; and it is perhaps of the same order of importance for computer science as Cantor's analysis of infinity is for mathematics. Our last major undertaking will be the proof of two impossibility theorems: we will show that the so-called "halting problem" cannot be solved by Turing machines, and finally that Hilbert's "decision problem" (for logic) also cannot.

This text is mathematical, in the sense that we discuss mathematical definitions and prove our results mathematically. But it is not mathematical in the sense that you need extensive mathematical background knowledge. Nothing in this text requires knowledge of algebra, trigonometry, or calculus. We have made a special effort to also not require any familiarity with the way mathematics works: in fact, part of the point is to *develop* the kinds

---

[1] Turing of course did not call it that himself.

of reasoning and proof skills required to understand and prove our results. The organization of the text follows mathematical convention, for one reason: these conventions have been developed because clarity and precision are especially important, and so, e.g., it is critical to know when something is asserted as the conclusion of an argument, is offered as a reason for something else, or is intended to introduce new vocabulary. So we follow mathematical convention and label passages as "definitions" if they are used to introduce new terminology or symbols; and as "theorems," "propositions," "lemmas," or "corollaries" when we record a result or finding. Other than these conventions, we will use the methods of logical proof that may already be familiar from a first logic course, and we will also make extensive use of the method of *induction* to prove results. Two chapters of the appendix are devoted to these proof methods.

Notes for instructors The material in this book is suitable for a semester-long second course in formal logic. I cover it in 12 weeks in Logic II taught at the University of Calgary, although I don't cover everything in as much detail as there is in this book. For instance, I typically only talk about natural deduction, and leave out detailed proofs of completeness for identity. Students have taken Logic I, typically taught from *forall x: Calgary*, which uses the same natural deduction rules, except in Fitch format.

The most recent version of this book is available in PDF at slc.openlogicproject.org, but changes frequently. The CC BY license gives you the right to download and distribute the book yourself. In order to ensure that all your students have the same version of the book throughout the term you're using it, you should do so: upload the PDF you decide to use to your LMS rather than merely give your students the link. You are also free to have the PDFs printed by your bookstore, but some bookstores will be able to purchase and stock the softcover books available on Amazon.

The syntax, semantics, and proof systems for first-order logic are supported by Graham Leach-Krouses's free, online logic teaching software application *Carnap* (carnap.io). This allows for submission and automated marking of exercises such as natural deduction and sequent calculus derivations, giving structures for simple theories, and symbolization exercises. There is also a Turing machine simulator at turing.openlogicproject.org that can be used to illustrate the material in part III. The examples there are available pre-loaded in the simulator.

Georg Cantor
1845 - 1918

PART I

Sets, Relations, Functions

CHAPTER 1
Sets

1.1 Extensionality

A *set* is a collection of objects, considered as a single object. The objects making up the set are called *elements* or *members* of the set. If x is an element of a set a, we write $x \in a$; if not, we write $x \notin a$. The set which has no elements is called the *empty* set and denoted "$\emptyset$".

It does not matter how we *specify* the set, or how we *order* its elements, or indeed how *many times* we count its elements. All that matters are what its elements are. We codify this in the following principle.

Definition 1.1 (Extensionality). If A and B are sets, then $A = B$ iff every element of A is also an element of B, and vice versa.

Extensionality licenses some notation. In general, when we have some objects $a_1, \ldots, a_n$, then $\{a_1, \ldots, a_n\}$ is *the* set whose elements are $a_1, \ldots, a_n$. We emphasise the word "*the*", since extensionality tells us that there can be only *one* such set. Indeed, extensionality also licenses the following:

$$\{a, a, b\} = \{a, b\} = \{b, a\}.$$

1.1. EXTENSIONALITY

This delivers on the point that, when we consider sets, we don't care about the order of their elements, or how many times they are specified.

Example 1.2. Whenever you have a bunch of objects, you can collect them together in a set. The set of Richard's siblings, for instance, is a set that contains one person, and we could write it as $S = \{\text{Ruth}\}$. The set of positive integers less than 4 is $\{1,2,3\}$, but it can also be written as $\{3,2,1\}$ or even as $\{1,2,1,2,3\}$. These are all the same set, by extensionality. For every element of $\{1,2,3\}$ is also an element of $\{3,2,1\}$ (and of $\{1,2,1,2,3\}$), and vice versa.

Frequently we'll specify a set by some property that its elements share. We'll use the following shorthand notation for that: $\{x : \varphi(x)\}$, where the $\varphi(x)$ stands for the property that x has to have in order to be counted among the elements of the set.

Example 1.3. In our example, we could have specified S also as

$$S = \{x : x \text{ is a sibling of Richard}\}.$$

Example 1.4. A number is called *perfect* iff it is equal to the sum of its proper divisors (i.e., numbers that evenly divide it but aren't identical to the number). For instance, 6 is perfect because its proper divisors are 1, 2, and 3, and $6 = 1 + 2 + 3$. In fact, 6 is the only positive integer less than 10 that is perfect. So, using extensionality, we can say:

$$\{6\} = \{x : x \text{ is perfect and } 0 \leq x \leq 10\}$$

We read the notation on the right as "the set of x's such that x is perfect and $0 \leq x \leq 10$". The identity here confirms that, when we consider sets, we don't care about how they are specified. And, more generally, extensionality guarantees that there is always only one set of x's such that $\varphi(x)$. So, extensionality justifies calling $\{x : \varphi(x)\}$ *the* set of x's such that $\varphi(x)$.

Extensionality gives us a way for showing that sets are identical: to show that $A = B$, show that whenever $x \in A$ then also $x \in B$, and whenever $y \in B$ then also $y \in A$.

1.2 Subsets and Power Sets

We will often want to compare sets. And one obvious kind of comparison one might make is as follows: *everything in one set is in the other too*. This situation is sufficiently important for us to introduce some new notation.

Definition 1.5 (Subset). If every element of a set A is also an element of B, then we say that A is a *subset* of B, and write $A \subseteq B$. If A is not a subset of B we write $A \nsubseteq B$. If $A \subseteq B$ but $A \neq B$, we write $A \subsetneq B$ and say that A is a *proper subset* of B.

Example 1.6. Every set is a subset of itself, and $\emptyset$ is a subset of every set. The set of even numbers is a subset of the set of natural numbers. Also, $\{a,b\} \subseteq \{a,b,c\}$. But $\{a,b,e\}$ is not a subset of $\{a,b,c\}$.

Example 1.7. The number 2 is an element of the set of integers, whereas the set of even numbers is a subset of the set of integers. However, a set may happen to *both* be an element and a subset of some other set, e.g., $\{0\} \in \{0, \{0\}\}$ and also $\{0\} \subseteq \{0, \{0\}\}$.

Extensionality gives a criterion of identity for sets: $A = B$ iff every element of A is also an element of B and vice versa. The definition of "subset" defines $A \subseteq B$ precisely as the first half of this criterion: every element of A is also an element of B. Of course the definition also applies if we switch A and B: that is, $B \subseteq A$ iff every element of B is also an element of A. And that, in turn, is exactly the "vice versa" part of extensionality. In other words, extensionality entails that sets are equal iff they are subsets of one another.

Proposition 1.8. $A = B$ *iff both* $A \subseteq B$ *and* $B \subseteq A$.

Now is also a good opportunity to introduce some further bits of helpful notation. In defining when A is a subset of B we said that "every element of A is ...," and filled the "..." with "an element of B". But this is such a common *shape* of expression that it will be helpful to introduce some formal notation for it.

Definition 1.9. $(\forall x \in A)\varphi$ abbreviates $\forall x(x \in A \to \varphi)$. Similarly, $(\exists x \in A)\varphi$ abbreviates $\exists x(x \in A \land \varphi)$.

Using this notation, we can say that $A \subseteq B$ iff $(\forall x \in A) x \in B$.

Now we move on to considering a certain kind of set: the set of all subsets of a given set.

Definition 1.10 (Power Set). The set consisting of all subsets of a set A is called the *power set of* A, written $\wp(A)$.

$$\wp(A) = \{B : B \subseteq A\}$$

Example 1.11. What are all the possible subsets of $\{a,b,c\}$? They are: $\emptyset$, $\{a\}$, $\{b\}$, $\{c\}$, $\{a,b\}$, $\{a,c\}$, $\{b,c\}$, $\{a,b,c\}$. The set of all these subsets is $\wp(\{a,b,c\})$:

$$\wp(\{a,b,c\}) = \{\emptyset, \{a\}, \{b\}, \{c\}, \{a,b\}, \{b,c\}, \{a,c\}, \{a,b,c\}\}$$

1.3 Some Important Sets

Example 1.12. We will mostly be dealing with sets whose elements are mathematical objects. Four such sets are important enough to have specific names:

$\mathbb{N} = \{0, 1, 2, 3, \ldots\}$
 the set of natural numbers

$\mathbb{Z} = \{\ldots, -2, -1, 0, 1, 2, \ldots\}$
 the set of integers

$\mathbb{Q} = \{m/n : m, n \in \mathbb{Z} \text{ and } n \neq 0\}$
 the set of rationals

$\mathbb{R} = (-\infty, \infty)$
 the set of real numbers (the continuum)

These are all *infinite* sets, that is, they each have infinitely many elements.

As we move through these sets, we are adding *more* numbers to our stock. Indeed, it should be clear that $\mathbb{N} \subseteq \mathbb{Z} \subseteq \mathbb{Q} \subseteq \mathbb{R}$: after all, every natural number is an integer; every integer is a rational; and every rational is a real. Equally, it should be clear that $\mathbb{N} \subsetneq \mathbb{Z} \subsetneq \mathbb{Q}$, since -1 is an integer but not a natural number, and $1/2$ is rational but not integer. It is less obvious that $\mathbb{Q} \subsetneq \mathbb{R}$, i.e., that there are some real numbers which are not rational.

We'll sometimes also use the set of positive integers $\mathbb{Z}^+ = \{1, 2, 3, \dots\}$ and the set containing just the first two natural numbers $\mathbb{B} = \{0, 1\}$.

Example 1.13 (Strings). Another interesting example is the set A^* of *finite strings* over an alphabet A: any finite sequence of elements of A is a string over A. We include the *empty string* Λ among the strings over A, for every alphabet A. For instance,

$$\mathbb{B}^* = \{\Lambda, 0, 1, 00, 01, 10, 11,$$
$$000, 001, 010, 011, 100, 101, 110, 111, 0000, \dots\}.$$

If $x = x_1 \dots x_n \in A^*$ is a string consisting of n "letters" from A, then we say *length* of the string is n and write $\text{len}(x) = n$.

Example 1.14 (Infinite sequences). For any set A we may also consider the set A^ω of infinite sequences of elements of A. An infinite sequence $a_1 a_2 a_3 a_4 \dots$ consists of a one-way infinite list of objects, each one of which is an element of A.

1.4 Unions and Intersections

In section 1.1, we introduced definitions of sets by abstraction, i.e., definitions of the form $\{x : \varphi(x)\}$. Here, we invoke some property φ, and this property can mention sets we've already defined. So for instance, if A and B are sets, the set $\{x : x \in A \vee x \in B\}$ consists of all those objects which are elements of either

1.4. UNIONS AND INTERSECTIONS   7

Figure 1.1: The union $A \cup B$ of two sets is set of elements of A together with those of B.

A or B, i.e., it's the set that combines the elements of A and B. We can visualize this as in Figure 1.1, where the highlighted area indicates the elements of the two sets A and B together.

This operation on sets—combining them—is very useful and common, and so we give it a formal name and a symbol.

Definition 1.15 (Union). The *union* of two sets A and B, written $A \cup B$, is the set of all things which are elements of A, B, or both.
$$A \cup B = \{x : x \in A \lor x \in B\}$$

Example 1.16. Since the multiplicity of elements doesn't matter, the union of two sets which have an element in common contains that element only once, e.g., $\{a, b, c\} \cup \{a, 0, 1\} = \{a, b, c, 0, 1\}$.

The union of a set and one of its subsets is just the bigger set: $\{a, b, c\} \cup \{a\} = \{a, b, c\}$.

The union of a set with the empty set is identical to the set: $\{a, b, c\} \cup \emptyset = \{a, b, c\}$.

We can also consider a "dual" operation to union. This is the operation that forms the set of all elements that are elements of A and are also elements of B. This operation is called *intersection*, and can be depicted as in Figure 1.2.

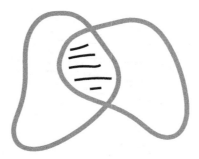

Figure 1.2: The intersection $A \cap B$ of two sets is the set of elements they have in common.

Definition 1.17 (Intersection). The *intersection* of two sets A and B, written $A \cap B$, is the set of all things which are elements of both A and B.

$$A \cap B = \{x : x \in A \wedge x \in B\}$$

Two sets are called *disjoint* if their intersection is empty. This means they have no elements in common.

Example 1.18. If two sets have no elements in common, their intersection is empty: $\{a,b,c\} \cap \{0,1\} = \emptyset$.

If two sets do have elements in common, their intersection is the set of all those: $\{a,b,c\} \cap \{a,b,d\} = \{a,b\}$.

The intersection of a set with one of its subsets is just the smaller set: $\{a,b,c\} \cap \{a,b\} = \{a,b\}$.

The intersection of any set with the empty set is empty: $\{a,b,c\} \cap \emptyset = \emptyset$.

We can also form the union or intersection of more than two sets. An elegant way of dealing with this in general is the following: suppose you collect all the sets you want to form the union (or intersection) of into a single set. Then we can define the union of all our original sets as the set of all objects which belong to at least one element of the set, and the intersection as the set of all objects which belong to every element of the set.

Definition 1.19. If A is a set of sets, then $\bigcup A$ is the set of elements of elements of A:

$$\bigcup A = \{x : x \text{ belongs to an element of } A\}, \text{ i.e.,}$$
$$= \{x : \text{there is a } B \in A \text{ so that } x \in B\}$$

Definition 1.20. If A is a set of sets, then $\bigcap A$ is the set of objects which all elements of A have in common:

$$\bigcap A = \{x : x \text{ belongs to every element of } A\}, \text{ i.e.,}$$
$$= \{x : \text{for all } B \in A, x \in B\}$$

Example 1.21. Suppose $A = \{\{a,b\},\{a,d,e\},\{a,d\}\}$. Then $\bigcup A = \{a,b,d,e\}$ and $\bigcap A = \{a\}$.

We could also do the same for a sequence of sets $A_1, A_2, \ldots$

$$\bigcup_i A_i = \{x : x \text{ belongs to one of the } A_i\}$$
$$\bigcap_i A_i = \{x : x \text{ belongs to every } A_i\}.$$

When we have an *index* of sets, i.e., some set I such that we are considering A_i for each $i \in I$, we may also use these abbreviations:

$$\bigcup_{i \in I} A_i = \bigcup \{A_i : i \in I\}$$
$$\bigcap_{i \in I} A_i = \bigcap \{A_i : i \in I\}$$

Finally, we may want to think about the set of all elements in A which are not in B. We can depict this as in Figure 1.3.

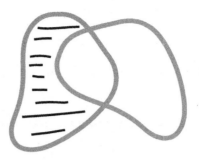

Figure 1.3: The difference $A \setminus B$ of two sets is the set of those elements of A which are not also elements of B.

Definition 1.22 (Difference). The *set difference* $A \setminus B$ is the set of all elements of A which are not also elements of B, i.e.,

$$A \setminus B = \{x : x \in A \text{ and } x \notin B\}.$$

1.5 Pairs, Tuples, Cartesian Products

It follows from extensionality that sets have no order to their elements. So if we want to represent order, we use *ordered pairs* $\langle x, y \rangle$. In an unordered pair $\{x, y\}$, the order does not matter: $\{x, y\} = \{y, x\}$. In an ordered pair, it does: if $x \neq y$, then $\langle x, y \rangle \neq \langle y, x \rangle$.

How should we think about ordered pairs in set theory? Crucially, we want to preserve the idea that ordered pairs are identical iff they share the same first element and share the same second element, i.e.:

$$\langle a, b \rangle = \langle c, d \rangle \text{ iff both } a = c \text{ and } b = d.$$

We can define ordered pairs in set theory using the Wiener-Kuratowski definition.

1.5. PAIRS, TUPLES, CARTESIAN PRODUCTS

Definition 1.23 (Ordered pair). $\langle a, b \rangle = \{\{a\}, \{a, b\}\}$.

Having fixed a definition of an ordered pair, we can use it to define further sets. For example, sometimes we also want ordered sequences of more than two objects, e.g., *triples* $\langle x, y, z \rangle$, *quadruples* $\langle x, y, z, u \rangle$, and so on. We can think of triples as special ordered pairs, where the first element is itself an ordered pair: $\langle x, y, z \rangle$ is $\langle \langle x, y \rangle, z \rangle$. The same is true for quadruples: $\langle x, y, z, u \rangle$ is $\langle \langle \langle x, y \rangle, z \rangle, u \rangle$, and so on. In general, we talk of *ordered n-tuples* $\langle x_1, \ldots, x_n \rangle$.

Certain sets of ordered pairs, or other ordered n-tuples, will be useful.

Definition 1.24 (Cartesian product). Given sets A and B, their *Cartesian product* $A \times B$ is defined by

$$A \times B = \{\langle x, y \rangle : x \in A \text{ and } y \in B\}.$$

Example 1.25. If $A = \{0, 1\}$, and $B = \{1, a, b\}$, then their product is

$$A \times B = \{\langle 0, 1 \rangle, \langle 0, a \rangle, \langle 0, b \rangle, \langle 1, 1 \rangle, \langle 1, a \rangle, \langle 1, b \rangle\}.$$

Example 1.26. If A is a set, the product of A with itself, $A \times A$, is also written A^2. It is the set of *all* pairs $\langle x, y \rangle$ with $x, y \in A$. The set of all triples $\langle x, y, z \rangle$ is A^3, and so on. We can give a recursive definition:

$$A^1 = A$$
$$A^{k+1} = A^k \times A$$

Proposition 1.27. *If A has n elements and B has m elements, then $A \times B$ has $n \cdot m$ elements.*

Proof. For every element x in A, there are m elements of the form $\langle x, y \rangle \in A \times B$. Let $B_x = \{\langle x, y \rangle : y \in B\}$. Since whenever $x_1 \neq x_2$, $\langle x_1, y \rangle \neq \langle x_2, y \rangle$, $B_{x_1} \cap B_{x_2} = \emptyset$. But if $A = \{x_1, \ldots, x_n\}$, then $A \times B = B_{x_1} \cup \cdots \cup B_{x_n}$, and so has $n \cdot m$ elements.

To visualize this, arrange the elements of $A \times B$ in a grid:

$$\begin{aligned} B_{x_1} &= \{\langle x_1, y_1 \rangle & \langle x_1, y_2 \rangle & \ldots & \langle x_1, y_m \rangle\} \\ B_{x_2} &= \{\langle x_2, y_1 \rangle & \langle x_2, y_2 \rangle & \ldots & \langle x_2, y_m \rangle\} \\ &\vdots & \vdots & & \\ B_{x_n} &= \{\langle x_n, y_1 \rangle & \langle x_n, y_2 \rangle & \ldots & \langle x_n, y_m \rangle\} \end{aligned}$$

Since the x_i are all different, and the y_j are all different, no two of the pairs in this grid are the same, and there are $n \cdot m$ of them. □

Example 1.28. If A is a set, a *word* over A is any sequence of elements of A. A sequence can be thought of as an n-tuple of elements of A. For instance, if $A = \{a, b, c\}$, then the sequence "*bac*" can be thought of as the triple $\langle b, a, c \rangle$. Words, i.e., sequences of symbols, are of crucial importance in computer science. By convention, we count elements of A as sequences of length 1, and $\emptyset$ as the sequence of length 0. The set of *all* words over A then is

$$A^* = \{\emptyset\} \cup A \cup A^2 \cup A^3 \cup \ldots$$

1.6 Russell's Paradox

Extensionality licenses the notation $\{x : \varphi(x)\}$, for *the* set of x's such that $\varphi(x)$. However, all that extensionality *really* licenses is the following thought. *If* there is a set whose members are all and only the φ's, *then* there is only one such set. Otherwise put: having fixed some φ, the set $\{x : \varphi(x)\}$ is unique, *if it exists*.

But this conditional is important! Crucially, not every property lends itself to *comprehension*. That is, some properties do *not* define sets. If they all did, then we would run into outright contradictions. The most famous example of this is Russell's Paradox.

1.6. RUSSELL'S PARADOX

Sets may be elements of other sets—for instance, the power set of a set A is made up of sets. And so it makes sense to ask or investigate whether a set is an element of another set. Can a set be a member of itself? Nothing about the idea of a set seems to rule this out. For instance, if *all* sets form a collection of objects, one might think that they can be collected into a single set—the set of all sets. And it, being a set, would be an element of the set of all sets.

Russell's Paradox arises when we consider the property of not having itself as an element, of being *non-self-membered*. What if we suppose that there is a set of all sets that do not have themselves as an element? Does

$$R = \{x : x \notin x\}$$

exist? It turns out that we can prove that it does not.

Theorem 1.29 (Russell's Paradox). *There is no set $R = \{x : x \notin x\}$.*

Proof. If $R = \{x : x \notin x\}$ exists, then $R \in R$ iff $R \notin R$, which is a contradiction. □

Let's run through this proof more slowly. If R exists, it makes sense to ask whether $R \in R$ or not. Suppose that indeed $R \in R$. Now, R was defined as the set of all sets that are not elements of themselves. So, if $R \in R$, then R does not itself have R's defining property. But only sets that have this property are in R, hence, R cannot be an element of R, i.e., $R \notin R$. But R can't both be and not be an element of R, so we have a contradiction.

Since the assumption that $R \in R$ leads to a contradiction, we have $R \notin R$. But this also leads to a contradiction! For if $R \notin R$, then R itself does have R's defining property, and so R would be an element of R just like all the other non-self-membered sets. And again, it can't both not be and be an element of R.

How do we set up a set theory which avoids falling into Russell's Paradox, i.e., which avoids making the *inconsistent* claim that

$R = \{x : x \notin x\}$ exists? Well, we would need to lay down axioms which give us very precise conditions for stating when sets exist (and when they don't).

The set theory sketched in this chapter doesn't do this. It's *genuinely naïve*. It tells you only that sets obey extensionality and that, if you have some sets, you can form their union, intersection, etc. It is possible to develop set theory more rigorously than this.

Summary

A **set** is a collection of objects, the elements of the set. We write $x \in A$ if x is an element of A. Sets are **extensional**—they are completely determined by their elements. Sets are specified by **listing** the elements explicitly or by giving a property the elements share (**abstraction**). Extensionality means that the order or way of listing or specifying the elements of a set doesn't matter. To prove that A and B are the same set ($A = B$) one has to prove that every element of X is an element of Y and vice versa.

Important sets include the natural ($\mathbb{N}$), integer ($\mathbb{Z}$), rational ($\mathbb{Q}$), and real ($\mathbb{R}$) numbers, but also **strings** (X^*) and infinite **sequences** (X^ω) of objects. A is a **subset** of B, $A \subseteq B$, if every element of A is also one of B. The collection of all subsets of a set B is itself a set, the **power set** $\wp(B)$ of B. We can form the **union** $A \cup B$ and **intersection** $A \cap B$ of sets. An **ordered pair** $\langle x, y \rangle$ consists of two objects x and y, but in that specific order. The pairs $\langle x, y \rangle$ and $\langle y, x \rangle$ are different pairs (unless $x = y$). The set of all pairs $\langle x, y \rangle$ where $x \in A$ and $y \in B$ is called the **Cartesian product** $A \times B$ of A and B. We write A^2 for $A \times A$; so for instance $\mathbb{N}^2$ is the set of pairs of natural numbers.

Problems

Problem 1.1. Prove that there is at most one empty set, i.e., show that if A and B are sets without elements, then $A = B$.

Problem 1.2. List all subsets of $\{a,b,c,d\}$.

Problem 1.3. Show that if A has n elements, then $\wp(A)$ has 2^n elements.

Problem 1.4. Prove that if $A \subseteq B$, then $A \cup B = B$.

Problem 1.5. Prove rigorously that if $A \subseteq B$, then $A \cap B = A$.

Problem 1.6. Show that if A is a set and $A \in B$, then $A \subseteq \bigcup B$.

Problem 1.7. Prove that if $A \subsetneq B$, then $B \setminus A \neq \emptyset$.

Problem 1.8. Using Definition 1.23, prove that $\langle a,b \rangle = \langle c,d \rangle$ iff both $a = c$ and $b = d$.

Problem 1.9. List all elements of $\{1,2,3\}^3$.

Problem 1.10. Show, by induction on k, that for all $k \geq 1$, if A has n elements, then A^k has n^k elements.

CHAPTER 2
Relations

2.1 Relations as Sets

In section 1.3, we mentioned some important sets: $\mathbb{N}, \mathbb{Z}, \mathbb{Q}, \mathbb{R}$. You will no doubt remember some interesting relations between the elements of some of these sets. For instance, each of these sets has a completely standard *order relation* on it. There is also the relation *is identical with* that every object bears to itself and to no other thing. There are many more interesting relations that we'll encounter, and even more possible relations. Before we review them, though, we will start by pointing out that we can look at relations as a special sort of set.

For this, recall two things from section 1.5. First, recall the notion of a *ordered pair*: given a and b, we can form $\langle a,b \rangle$. Importantly, the order of elements *does* matter here. So if $a \neq b$ then $\langle a,b \rangle \neq \langle b,a \rangle$. (Contrast this with unordered pairs, i.e., 2-element sets, where $\{a,b\} = \{b,a\}$.) Second, recall the notion of a *Cartesian product*: if A and B are sets, then we can form $A \times B$, the set of all pairs $\langle x,y \rangle$ with $x \in A$ and $y \in B$. In particular, $A^2 = A \times A$ is the set of all ordered pairs from A.

Now we will consider a particular relation on a set: the <-relation on the set $\mathbb{N}$ of natural numbers. Consider the set of all pairs of numbers $\langle n, m \rangle$ where $n < m$, i.e.,

$$R = \{\langle n, m \rangle : n, m \in \mathbb{N} \text{ and } n < m\}.$$

2.1. RELATIONS AS SETS

There is a close connection between n being less than m, and the pair $\langle n, m \rangle$ being a member of R, namely:

$$n < m \text{ iff } \langle n, m \rangle \in R.$$

Indeed, without any loss of information, we can consider the set R to *be* the $<$-relation on $\mathbb{N}$.

In the same way we can construct a subset of $\mathbb{N}^2$ for any relation between numbers. Conversely, given any set of pairs of numbers $S \subseteq \mathbb{N}^2$, there is a corresponding relation between numbers, namely, the relationship n bears to m if and only if $\langle n, m \rangle \in S$. This justifies the following definition:

Definition 2.1 (Binary relation). A *binary relation* on a set A is a subset of A^2. If $R \subseteq A^2$ is a binary relation on A and $x, y \in A$, we sometimes write Rxy (or xRy) for $\langle x, y \rangle \in R$.

Example 2.2. The set $\mathbb{N}^2$ of pairs of natural numbers can be listed in a 2-dimensional matrix like this:

$$\begin{array}{ccccc}
\langle \mathbf{0,0} \rangle & \langle 0,1 \rangle & \langle 0,2 \rangle & \langle 0,3 \rangle & \ldots \\
\langle 1,0 \rangle & \langle \mathbf{1,1} \rangle & \langle 1,2 \rangle & \langle 1,3 \rangle & \ldots \\
\langle 2,0 \rangle & \langle 2,1 \rangle & \langle \mathbf{2,2} \rangle & \langle 2,3 \rangle & \ldots \\
\langle 3,0 \rangle & \langle 3,1 \rangle & \langle 3,2 \rangle & \langle \mathbf{3,3} \rangle & \ldots \\
\vdots & \vdots & \vdots & \vdots & \ddots
\end{array}$$

We have put the diagonal, here, in bold, since the subset of $\mathbb{N}^2$ consisting of the pairs lying on the diagonal, i.e.,

$$\{\langle 0,0 \rangle, \langle 1,1 \rangle, \langle 2,2 \rangle, \ldots\},$$

is the *identity relation on* $\mathbb{N}$. (Since the identity relation is popular, let's define $\mathrm{Id}_A = \{\langle x, x \rangle : x \in X\}$ for any set A.) The subset of all pairs lying above the diagonal, i.e.,

$$L = \{\langle 0,1 \rangle, \langle 0,2 \rangle, \ldots, \langle 1,2 \rangle, \langle 1,3 \rangle, \ldots, \langle 2,3 \rangle, \langle 2,4 \rangle, \ldots\},$$

is the *less than* relation, i.e., Lnm iff $n < m$. The subset of pairs below the diagonal, i.e.,

$$G = \{\langle 1,0\rangle, \langle 2,0\rangle, \langle 2,1\rangle, \langle 3,0\rangle, \langle 3,1\rangle, \langle 3,2\rangle, \dots\},$$

is the *greater than* relation, i.e., Gnm iff $n > m$. The union of L with I, which we might call $K = L \cup I$, is the *less than or equal to* relation: Knm iff $n \le m$. Similarly, $H = G \cup I$ is the *greater than or equal to relation*. These relations L, G, K, and H are special kinds of relations called *orders*. L and G have the property that no number bears L or G to itself (i.e., for all n, neither Lnn nor Gnn). Relations with this property are called *irreflexive*, and, if they also happen to be orders, they are called *strict orders*.

Although orders and identity are important and natural relations, it should be emphasized that according to our definition *any* subset of A^2 is a relation on A, regardless of how unnatural or contrived it seems. In particular, $\emptyset$ is a relation on any set (the *empty relation*, which no pair of elements bears), and A^2 itself is a relation on A as well (one which every pair bears), called the *universal relation*. But also something like $E = \{\langle n,m\rangle : n > 5 \text{ or } m \times n \ge 34\}$ counts as a relation.

2.2 Special Properties of Relations

Some kinds of relations turn out to be so common that they have been given special names. For instance, $\le$ and $\subseteq$ both relate their respective domains (say, $\mathbb{N}$ in the case of $\le$ and $\wp(A)$ in the case of $\subseteq$) in similar ways. To get at exactly how these relations are similar, and how they differ, we categorize them according to some special properties that relations can have. It turns out that (combinations of) some of these special properties are especially important: orders and equivalence relations.

2.2. SPECIAL PROPERTIES OF RELATIONS

Definition 2.3 (Reflexivity). A relation $R \subseteq A^2$ is *reflexive* iff, for every $x \in A$, Rxx.

Definition 2.4 (Transitivity). A relation $R \subseteq A^2$ is *transitive* iff, whenever Rxy and Ryz, then also Rxz.

Definition 2.5 (Symmetry). A relation $R \subseteq A^2$ is *symmetric* iff, whenever Rxy, then also Ryx.

Definition 2.6 (Anti-symmetry). A relation $R \subseteq A^2$ is *anti-symmetric* iff, whenever both Rxy and Ryx, then $x = y$ (or, in other words: if $x \neq y$ then either $\neg Rxy$ or $\neg Ryx$).

In a symmetric relation, Rxy and Ryx always hold together, or neither holds. In an anti-symmetric relation, the only way for Rxy and Ryx to hold together is if $x = y$. Note that this does not *require* that Rxy and Ryx holds when $x = y$, only that it isn't ruled out. So an anti-symmetric relation can be reflexive, but it is not the case that every anti-symmetric relation is reflexive. Also note that being anti-symmetric and merely not being symmetric are different conditions. In fact, a relation can be both symmetric and anti-symmetric at the same time (e.g., the identity relation is).

Definition 2.7 (Connectivity). A relation $R \subseteq A^2$ is *connected* if for all $x, y \in A$, if $x \neq y$, then either Rxy or Ryx.

Definition 2.8 (Irreflexivity). A relation $R \subseteq A^2$ is called *irreflexive* if, for all $x \in A$, not Rxx.

Definition 2.9 (Asymmetry). A relation $R \subseteq A^2$ is called *asymmetric* if for no pair $x, y \in A$ we have both Rxy and Ryx.

Note that if $A \neq \emptyset$, then no irreflexive relation on A is reflexive and every asymmetric relation on A is also anti-symmetric. However, there are $R \subseteq A^2$ that are not reflexive and also not irreflexive, and there are anti-symmetric relations that are not asymmetric.

2.3 Equivalence Relations

The identity relation on a set is reflexive, symmetric, and transitive. Relations R that have all three of these properties are very common.

Definition 2.10 (Equivalence relation). A relation $R \subseteq A^2$ that is reflexive, symmetric, and transitive is called an *equivalence relation*. Elements x and y of A are said to be *R-equivalent* if Rxy.

Equivalence relations give rise to the notion of an *equivalence class*. An equivalence relation "chunks up" the domain into different partitions. Within each partition, all the objects are related to one another; and no objects from different partitions relate to one another. Sometimes, it's helpful just to talk about these partitions *directly*. To that end, we introduce a definition:

Definition 2.11. Let $R \subseteq A^2$ be an equivalence relation. For each $x \in A$, the *equivalence class* of x in A is the set $[x]_R = \{y \in A : Rxy\}$. The *quotient* of A under R is $A/R = \{[x]_R : x \in A\}$, i.e., the set of these equivalence classes.

The next result vindicates the definition of an equivalence class, in proving that the equivalence classes are indeed the partitions of A:

Proposition 2.12. *If $R \subseteq A^2$ is an equivalence relation, then Rxy iff $[x]_R = [y]_R$.*

Proof. For the left-to-right direction, suppose Rxy, and let $z \in [x]_R$. By definition, then, Rxz. Since R is an equivalence relation, Ryz. (Spelling this out: as Rxy and R is symmetric we have Ryx, and as Rxz and R is transitive we have Ryz.) So $z \in [y]_R$. Generalising, $[x]_R \subseteq [y]_R$. But exactly similarly, $[y]_R \subseteq [x]_R$. So $[x]_R = [y]_R$, by extensionality.

For the right-to-left direction, suppose $[x]_R = [y]_R$. Since R is reflexive, Ryy, so $y \in [y]_R$. Thus also $y \in [x]_R$ by the assumption that $[x]_R = [y]_R$. So Rxy. □

Example 2.13. A nice example of equivalence relations comes from modular arithmetic. For any a, b, and $n \in \mathbb{N}$, say that $a \equiv_n b$ iff dividing a by n gives the same remainder as dividing b by n. (Somewhat more symbolically: $a \equiv_n b$ iff, for some $k \in \mathbb{Z}$, $a - b = kn$.) Now, $\equiv_n$ is an equivalence relation, for any n. And there are exactly n distinct equivalence classes generated by $\equiv_n$; that is, $\mathbb{N}/_{\equiv_n}$ has n elements. These are: the set of numbers divisible by n without remainder, i.e., $[0]_{\equiv_n}$; the set of numbers divisible by n with remainder 1, i.e., $[1]_{\equiv_n}$; ...; and the set of numbers divisible by n with remainder $n - 1$, i.e., $[n-1]_{\equiv_n}$.

2.4 Orders

Many of our comparisons involve describing some objects as being "less than", "equal to", or "greater than" other objects, in a certain respect. These involve *order* relations. But there are different kinds of order relations. For instance, some require that any two objects be comparable, others don't. Some include identity (like ≤) and some exclude it (like <). It will help us to have a taxonomy here.

Definition 2.14 (Preorder). A relation which is both reflexive and transitive is called a *preorder*.

Definition 2.15 (Partial order). A preorder which is also anti-symmetric is called a *partial order*.

Definition 2.16 (Linear order). A partial order which is also connected is called a *total order* or *linear order*.

Example 2.17. Every linear order is also a partial order, and every partial order is also a preorder, but the converses don't hold. The universal relation on A is a preorder, since it is reflexive and transitive. But, if A has more than one element, the universal relation is not anti-symmetric, and so not a partial order.

Example 2.18. Consider the *no longer than* relation $\preccurlyeq$ on $\mathbb{B}^*$: $x \preccurlyeq y$ iff $\text{len}(x) \leq \text{len}(y)$. This is a preorder (reflexive and transitive), and even connected, but not a partial order, since it is not anti-symmetric. For instance, $01 \preccurlyeq 10$ and $10 \preccurlyeq 01$, but $01 \neq 10$.

Example 2.19. An important partial order is the relation $\subseteq$ on a set of sets. This is not in general a linear order, since if $a \neq b$ and we consider $\wp(\{a, b\}) = \{\emptyset, \{a\}, \{b\}, \{a, b\}\}$, we see that $\{a\} \nsubseteq \{b\}$ and $\{a\} \neq \{b\}$ and $\{b\} \nsubseteq \{a\}$.

Example 2.20. The relation of *divisibility without remainder* gives us a partial order which isn't a linear order. For integers n, m, we write $n \mid m$ to mean n (evenly) divides m, i.e., iff there is some integer k so that $m = kn$. On $\mathbb{N}$, this is a partial order, but not a linear order: for instance, $2 \nmid 3$ and also $3 \nmid 2$. Considered as a relation on $\mathbb{Z}$, divisibility is only a preorder since it is not anti-symmetric: $1 \mid -1$ and $-1 \mid 1$ but $1 \neq -1$.

2.4. ORDERS

Definition 2.21 (Strict order). A *strict order* is a relation which is irreflexive, asymmetric, and transitive.

Definition 2.22 (Strict linear order). A strict order which is also connected is called a *strict linear order*.

Example 2.23. $\leq$ is the linear order corresponding to the strict linear order $<$. $\subseteq$ is the partial order corresponding to the strict order $\subsetneq$.

Definition 2.24 (Total order). A strict order which is also connected is called a *total order*. This is also sometimes called a *strict linear order*.

Any strict order R on A can be turned into a partial order by adding the diagonal Id_A, i.e., adding all the pairs $\langle x,x \rangle$. (This is called the *reflexive closure* of R.) Conversely, starting from a partial order, one can get a strict order by removing Id_A. These next two results make this precise.

Proposition 2.25. *If R is a strict order on A, then $R^+ = R \cup \mathrm{Id}_A$ is a partial order. Moreover, if R is total, then R^+ is a linear order.*

Proof. Suppose R is a strict order, i.e., $R \subseteq A^2$ and R is irreflexive, asymmetric, and transitive. Let $R^+ = R \cup \mathrm{Id}_A$. We have to show that R^+ is reflexive, antisymmetric, and transitive.

R^+ is clearly reflexive, since $\langle x,x \rangle \in \mathrm{Id}_A \subseteq R^+$ for all $x \in A$.

To show R^+ is antisymmetric, suppose for reductio that R^+xy and R^+yx but $x \neq y$. Since $\langle x,y \rangle \in R \cup \mathrm{Id}_X$, but $\langle x,y \rangle \notin \mathrm{Id}_X$, we must have $\langle x,y \rangle \in R$, i.e., Rxy. Similarly, Ryx. But this contradicts the assumption that R is asymmetric.

To establish transitivity, suppose that R^+xy and R^+yz. If both $\langle x,y \rangle \in R$ and $\langle y,z \rangle \in R$, then $\langle x,z \rangle \in R$ since R is transitive. Otherwise, either $\langle x,y \rangle \in \mathrm{Id}_X$, i.e., $x = y$, or $\langle y,z \rangle \in \mathrm{Id}_X$, i.e.,

$y = z$. In the first case, we have that R^+yz by assumption, $x = y$, hence R^+xz. Similarly in the second case. In either case, R^+xz, thus, R^+ is also transitive.

Concerning the "moreover" clause, suppose R is a total order, i.e., that R is connected. So for all $x \neq y$, either Rxy or Ryx, i.e., either $\langle x,y \rangle \in R$ or $\langle y,x \rangle \in R$. Since $R \subseteq R^+$, this remains true of R^+, so R^+ is connected as well. □

Proposition 2.26. *If R is a partial order on X, then $R^- = R \setminus \mathrm{Id}_X$ is a strict order. Moreover, if R is linear, then R^- is total.*

Proof. This is left as an exercise. □

Example 2.27. $\leq$ is the linear order corresponding to the total order $<$. $\subseteq$ is the partial order corresponding to the strict order $\subsetneq$.

The following simple result which establishes that total orders satisfy an extensionality-like property:

Proposition 2.28. *If $<$ totally orders A, then:*

$$(\forall a, b \in A)((\forall x \in A)(x < a \leftrightarrow x < b) \rightarrow a = b)$$

Proof. Suppose $(\forall x \in A)(x < a \leftrightarrow x < b)$. If $a < b$, then $a < a$, contradicting the fact that $<$ is irreflexive; so $a \not< b$. Exactly similarly, $b \not< a$. So $a = b$, as $<$ is connected. □

2.5 Graphs

A *graph* is a diagram in which points—called "nodes" or "vertices" (plural of "vertex")—are connected by edges. Graphs are a ubiquitous tool in discrete mathematics and in computer science. They are incredibly useful for representing, and visualizing, relationships and structures, from concrete things like networks of various kinds to abstract structures such as the possible outcomes of decisions. There are many different kinds of graphs in the literature which differ, e.g., according to whether the edges are directed or not, have labels or not, whether there can be edges from a node to the same node, multiple edges between the same nodes, etc. *Directed graphs* have a special connection to relations.

Definition 2.29 (Directed graph). A *directed graph* $G = \langle V, E \rangle$ is a set of *vertices* V and a set of *edges* $E \subseteq V^2$.

According to our definition, a graph just is a set together with a relation on that set. Of course, when talking about graphs, it's only natural to expect that they are graphically represented: we can draw a graph by connecting two vertices v_1 and v_2 by an arrow iff $\langle v_1, v_2 \rangle \in E$. The only difference between a relation by itself and a graph is that a graph specifies the set of vertices, i.e., a graph may have isolated vertices. The important point, however, is that every relation R on a set X can be seen as a directed graph $\langle X, R \rangle$, and conversely, a directed graph $\langle V, E \rangle$ can be seen as a relation $E \subseteq V^2$ with the set V explicitly specified.

Example 2.30. The graph $\langle V, E \rangle$ with $V = \{1, 2, 3, 4\}$ and $E = \{\langle 1,1 \rangle, \langle 1,2 \rangle, \langle 1,3 \rangle, \langle 2,3 \rangle\}$ looks like this:

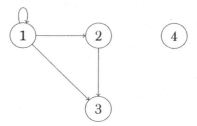

This is a different graph than $\langle V', E \rangle$ with $V' = \{1, 2, 3\}$, which looks like this:

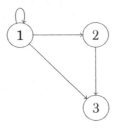

2.6 Operations on Relations

It is often useful to modify or combine relations. In Proposition 2.25, we considered the *union* of relations, which is just the union of two relations considered as sets of pairs. Similarly, in Proposition 2.26, we considered the relative difference of relations. Here are some other operations we can perform on relations.

Definition 2.31. Let R, S be relations, and A be any set.
The *inverse* of R is $R^{-1} = \{\langle y, x \rangle : \langle x, y \rangle \in R\}$.
The *relative product* of R and S is $(R \mid S) = \{\langle x, z \rangle : \exists y (Rxy \wedge Syz)\}$.
The *restriction* of R to A is $R\!\restriction_A = R \cap A^2$.
The *application* of R to A is $R[A] = \{y : (\exists x \in A) Rxy\}$

Example 2.32. Let $S \subseteq \mathbb{Z}^2$ be the successor relation on $\mathbb{Z}$, i.e., $S = \{\langle x, y \rangle \in \mathbb{Z}^2 : x + 1 = y\}$, so that Sxy iff $x + 1 = y$.
S^{-1} is the predecessor relation on $\mathbb{Z}$, i.e., $\{\langle x, y \rangle \in \mathbb{Z}^2 : x - 1 = y\}$.
$S \mid S$ is $\{\langle x, y \rangle \in \mathbb{Z}^2 : x + 2 = y\}$
$S\!\restriction_{\mathbb{N}}$ is the successor relation on $\mathbb{N}$.
$S[\{1, 2, 3\}]$ is $\{2, 3, 4\}$.

Definition 2.33 (Transitive closure). Let $R \subseteq A^2$ be a binary relation.

The *transitive closure* of R is $R^+ = \bigcup_{0<n\in\mathbb{N}} R^n$, where we recursively define $R^1 = R$ and $R^{n+1} = R^n \mid R$.

The *reflexive transitive closure* of R is $R^* = R^+ \cup \mathrm{Id}_A$.

Example 2.34. Take the successor relation $S \subseteq \mathbb{Z}^2$. S^2xy iff $x + 2 = y$, S^3xy iff $x + 3 = y$, etc. So S^+xy iff $x + n = y$ for some $n \geq 1$. In other words, S^+xy iff $x < y$, and S^*xy iff $x \leq y$.

Summary

A **relation** R on a set A is a way of relating elements of A. We write Rxy if the relation **holds** between x and y. Formally, we can consider R as the sets of pairs $\langle x,y \rangle \in A^2$ such that Rxy. Being less than, greater than, equal to, evenly dividing, being the same length as, a subset of, and the same size as are all important examples of relations (on sets of numbers, strings, or of sets). **Graphs** are a general way of visually representing relations. But a graph can also be seen as a binary relation (the **edge** relation) together with the underlying set of **vertices**.

Some relations share certain features which makes them especially interesting or useful. A relation R is **reflexive** if everything is R-related to itself; **symmetric**, if with Rxy also Ryx holds for any x and y; and **transitive** if Rxy and Ryz guarantees Rxz. Relations that have all three of these properties are **equivalence relations**. A relation is **anti-symmetric** if Rxy and Ryx guarantees $x = y$. **Partial orders** are those relations that are reflexive, anti-symmetric, and transitive. A **linear order** is any partial order which satisfies that for any x and y, either Rxy or $x = y$ or Ryx. (Generally, a relation with this property is **connected**).

Since relations are sets (of pairs), they can be operated on as sets (e.g., we can form the union and intersection of relations). We can also chain them together (**relative product** $R \mid S$). If we form the relative product of R with itself arbitrarily many times we get the **transitive closure** R^+ of R.

Problems

Problem 2.1. List the elements of the relation $\subseteq$ on the set $\wp(\{a,b,c\})$.

Problem 2.2. Give examples of relations that are (a) reflexive and symmetric but not transitive, (b) reflexive and anti-symmetric, (c) anti-symmetric, transitive, but not reflexive, and (d) reflexive, symmetric, and transitive. Do not use relations on numbers or sets.

Problem 2.3. Show that $\equiv_n$ is an equivalence relation, for any $n \in \mathbb{N}$, and that $\mathbb{N}/_{\equiv_n}$ has exactly n members.

Problem 2.4. Give a proof of Proposition 2.26.

Problem 2.5. Consider the less-than-or-equal-to relation $\leq$ on the set $\{1,2,3,4\}$ as a graph and draw the corresponding diagram.

Problem 2.6. Show that the transitive closure of R is in fact transitive.

CHAPTER 3
Functions

3.1 Basics

A *function* is a map which sends each element of a given set to a specific element in some (other) given set. For instance, the operation of adding 1 defines a function: each number n is mapped to a unique number $n + 1$.

More generally, functions may take pairs, triples, etc., as inputs and returns some kind of output. Many functions are familiar to us from basic arithmetic. For instance, addition and multiplication are functions. They take in two numbers and return a third.

In this mathematical, abstract sense, a function is a *black box*: what matters is only what output is paired with what input, not the method for calculating the output.

Definition 3.1 (Function). A *function* $f: A \to B$ is a mapping of each element of A to an element of B.

We call A the *domain* of f and B the *codomain* of f. The elements of A are called inputs or *arguments* of f, and the element of B that is paired with an argument x by f is called the *value of f* for argument x, written $f(x)$.

The *range* $\mathrm{ran}(f)$ of f is the subset of the codomain consisting of the values of f for some argument; $\mathrm{ran}(f) = \{f(x) : x \in A\}$.

Figure 3.1: A function is a mapping of each element of one set to an element of another. An arrow points from an argument in the domain to the corresponding value in the codomain.

The diagram in Figure 3.1 may help to think about functions. The ellipse on the left represents the function's *domain*; the ellipse on the right represents the function's *codomain*; and an arrow points from an *argument* in the domain to the corresponding *value* in the codomain.

Example 3.2. Multiplication takes pairs of natural numbers as inputs and maps them to natural numbers as outputs, so goes from $\mathbb{N} \times \mathbb{N}$ (the domain) to $\mathbb{N}$ (the codomain). As it turns out, the range is also $\mathbb{N}$, since every $n \in \mathbb{N}$ is $n \times 1$.

Example 3.3. Multiplication is a function because it pairs each input—each pair of natural numbers—with a single output: $\times \colon \mathbb{N}^2 \to \mathbb{N}$. By contrast, the square root operation applied to the domain $\mathbb{N}$ is not functional, since each positive integer n has two square roots: $\sqrt{n}$ and $-\sqrt{n}$. We can make it functional by only returning the positive square root: $\sqrt{} \colon \mathbb{N} \to \mathbb{R}$.

Example 3.4. The relation that pairs each student in a class with their final grade is a function—no student can get two different final grades in the same class. The relation that pairs each student in a class with their parents is not a function: students can have zero, or two, or more parents.

We can define functions by specifying in some precise way what the value of the function is for every possible argment. Different ways of doing this are by giving a formula, describing a method for computing the value, or listing the values for each argument. However functions are defined, we must make sure that for each argment we specify one, and only one, value.

Example 3.5. Let $f \colon \mathbb{N} \to \mathbb{N}$ be defined such that $f(x) = x + 1$. This is a definition that specifies f as a function which takes in natural numbers and outputs natural numbers. It tells us that, given a natural number x, f will output its successor $x + 1$. In this case, the codomain $\mathbb{N}$ is not the range of f, since the natural number 0 is not the successor of any natural number. The range of f is the set of all positive integers, $\mathbb{Z}^+$.

Example 3.6. Let $g \colon \mathbb{N} \to \mathbb{N}$ be defined such that $g(x) = x+2-1$. This tells us that g is a function which takes in natural numbers and outputs natural numbers. Given a natural number n, g will output the predecessor of the successor of the successor of x, i.e., $x + 1$.

We just considered two functions, f and g, with different *definitions*. However, these are the *same function*. After all, for any natural number n, we have that $f(n) = n + 1 = n + 2 - 1 = g(n)$. Otherwise put: our definitions for f and g specify the same mapping by means of different equations. Implicitly, then, we are relying upon a principle of extensionality for functions,

$$\text{if } \forall x\, f(x) = g(x), \text{ then } f = g$$

provided that f and g share the same domain and codomain.

Example 3.7. We can also define functions by cases. For instance, we could define $h \colon \mathbb{N} \to \mathbb{N}$ by

$$h(x) = \begin{cases} \frac{x}{2} & \text{if } x \text{ is even} \\ \frac{x+1}{2} & \text{if } x \text{ is odd.} \end{cases}$$

Since every natural number is either even or odd, the output of this function will always be a natural number. Just remember that if you define a function by cases, every possible input must fall into exactly one case. In some cases, this will require a proof that the cases are exhaustive and exclusive.

Figure 3.2: A surjective function has every element of the codomain as a value.

3.2 Kinds of Functions

It will be useful to introduce a kind of taxonomy for some of the kinds of functions which we encounter most frequently.

To start, we might want to consider functions which have the property that every member of the codomain is a value of the function. Such functions are called surjective, and can be pictured as in Figure 3.2.

Definition 3.8 (Surjective function). A function $f: A \to B$ is *surjective* iff B is also the range of f, i.e., for every $y \in B$ there is at least one $x \in A$ such that $f(x) = y$, or in symbols:

$$(\forall y \in B)(\exists x \in A) f(x) = y.$$

We call such a function a surjection from A to B.

If you want to show that f is a surjection, then you need to show that every object in f's codomain is the value of $f(x)$ for some input x.

Note that any function *induces* a surjection. After all, given a function $f: A \to B$, let $f': A \to \text{ran}(f)$ be defined by $f'(x) = f(x)$. Since $\text{ran}(f)$ is *defined* as $\{f(x) \in B : x \in A\}$, this function f' is guaranteed to be a surjection

Now, any function maps each possible input to a unique output. But there are also functions which never map different inputs to the same outputs. Such functions are called injective, and can be pictured as in Figure 3.3.

3.2. KINDS OF FUNCTIONS

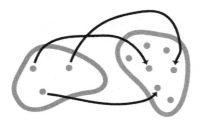

Figure 3.3: An injective function never maps two different arguments to the same value.

Definition 3.9 (Injective function). A function $f\colon A \to B$ is *injective* iff for each $y \in B$ there is at most one $x \in A$ such that $f(x) = y$. We call such a function an injection from A to B.

If you want to show that f is an injection, you need to show that for any elements x and y of f's domain, if $f(x) = f(y)$, then $x = y$.

Example 3.10. The constant function $f\colon \mathbb{N} \to \mathbb{N}$ given by $f(x) = 1$ is neither injective, nor surjective.

The identity function $f\colon \mathbb{N} \to \mathbb{N}$ given by $f(x) = x$ is both injective and surjective.

The successor function $f\colon \mathbb{N} \to \mathbb{N}$ given by $f(x) = x + 1$ is injective but not surjective.

The function $f\colon \mathbb{N} \to \mathbb{N}$ defined by:

$$f(x) = \begin{cases} \frac{x}{2} & \text{if } x \text{ is even} \\ \frac{x+1}{2} & \text{if } x \text{ is odd.} \end{cases}$$

is surjective, but not injective.

Often enough, we want to consider functions which are both injective and surjective. We call such functions bijective. They look like the function pictured in Figure 3.4. Bijections are also sometimes called *one-to-one correspondences*, since they uniquely pair elements of the codomain with elements of the domain.

Figure 3.4: A bijective function uniquely pairs the elements of the codomain with those of the domain.

Definition 3.11 (Bijection). A function $f: A \to B$ is *bijective* iff it is both surjective and injective. We call such a function a bijection from A to B (or between A and B).

3.3 Functions as Relations

A function which maps elements of A to elements of B obviously defines a relation between A and B, namely the relation which holds between x and y iff $f(x) = y$. In fact, we might even—if we are interested in reducing the building blocks of mathematics for instance—*identify* the function f with this relation, i.e., with a set of pairs. This then raises the question: which relations define functions in this way?

Definition 3.12 (Graph of a function). Let $f: A \to B$ be a function. The *graph* of f is the relation $R_f \subseteq A \times B$ defined by
$$R_f = \{\langle x, y \rangle : f(x) = y\}.$$

The graph of a function is uniquely determined, by extensionality. Moreover, extensionality (on sets) will immediate vindicate the implicit principle of extensionality for functions, whereby if f and g share a domain and codomain then they are identical if they agree on all values.

Similarly, if a relation is "functional", then it is the graph of a function.

3.3. FUNCTIONS AS RELATIONS

Proposition 3.13. *Let $R \subseteq A \times B$ be such that:*

1. *If Rxy and Rxz then $y = z$; and*

2. *for every $x \in A$ there is some $y \in B$ such that $\langle x, y \rangle \in R$.*

Then R is the graph of the function $f : A \to B$ defined by $f(x) = y$ iff Rxy.

Proof. Suppose there is a y such that Rxy. If there were another $z \neq y$ such that Rxz, the condition on R would be violated. Hence, if there is a y such that Rxy, this y is unique, and so f is well-defined. Obviously, $R_f = R$. □

Every function $f : A \to B$ has a graph, i.e., a relation on $A \times B$ defined by $f(x) = y$. On the other hand, every relation $R \subseteq A \times B$ with the properties given in Proposition 3.13 is the graph of a function $f : A \to B$. Because of this close connection between functions and their graphs, we can think of a function simply as its graph. In other words, functions can be identified with certain relations, i.e., with certain sets of tuples. We can now consider performing similar operations on functions as we performed on relations (see section 2.6). In particular:

Definition 3.14. Let $f : A \to B$ be a function with $C \subseteq A$.

The *restriction* of f to C is the function $f \restriction_C : C \to B$ defined by $(f \restriction_C)(x) = f(x)$ for all $x \in C$. In other words, $f \restriction_C = \{\langle x, y \rangle \in R_f : x \in C\}$.

The *application* of f to C is $f[C] = \{f(x) : x \in C\}$. We also call this the *image* of C under f.

It follows from these definition that $\operatorname{ran}(f) = f[\operatorname{dom}(f)]$, for any function f. These notions are exactly as one would expect, given the definitions in section 2.6 and our identification of functions with relations. But two other operations—inverses and relative products—require a little more detail. We will provide that in the section 3.4 and section 3.5.

3.4 Inverses of Functions

We think of functions as maps. An obvious question to ask about functions, then, is whether the mapping can be "reversed." For instance, the successor function $f(x) = x + 1$ can be reversed, in the sense that the function $g(y) = y - 1$ "undoes" what f does.

But we must be careful. Although the definition of g defines a function $\mathbb{Z} \to \mathbb{Z}$, it does not define a *function* $\mathbb{N} \to \mathbb{N}$, since $g(0) \notin \mathbb{N}$. So even in simple cases, it is not quite obvious whether a function can be reversed; it may depend on the domain and codomain.

This is made more precise by the notion of an inverse of a function.

Definition 3.15. A function $g\colon B \to A$ is an *inverse* of a function $f\colon A \to B$ if $f(g(y)) = y$ and $g(f(x)) = x$ for all $x \in A$ and $y \in B$.

If f has an inverse g, we often write f^{-1} instead of g.

Now we will determine when functions have inverses. A good candidate for an inverse of $f\colon A \to B$ is $g\colon B \to A$ "defined by"

$$g(y) = \text{"the"} \ x \text{ such that } f(x) = y.$$

But the scare quotes around "defined by" (and "the") suggest that this is not a definition. At least, it will not always work, with complete generality. For, in order for this definition to specify a function, there has to be one and only one x such that $f(x) = y$— the output of g has to be uniquely specified. Moreover, it has to be specified for every $y \in B$. If there are x_1 and $x_2 \in A$ with $x_1 \neq x_2$ but $f(x_1) = f(x_2)$, then $g(y)$ would not be uniquely specified for $y = f(x_1) = f(x_2)$. And if there is no x at all such that $f(x) = y$, then $g(y)$ is not specified at all. In other words, for g to be defined, f must be both injective and surjective.

Let's go slowly. We'll divide the question into two: Given a function $f\colon A \to B$, when is there a function $g\colon B \to A$ so that $g(f(x)) = x$? Such a g "undoes" what f does, and is called a *left inverse* of f. Secondly, when is there a function $h\colon B \to A$ so that

3.4. INVERSES OF FUNCTIONS

$f(h(y)) = y$? Such an h is called a *right inverse* of f—f "undoes" what h does.

Proposition 3.16. *If $f: A \to B$ is injective, then there is a left inverse $g: B \to A$ of f so that $g(f(x)) = x$ for all $x \in A$.*

Proof. Suppose that $f: A \to B$ is injective. Consider a $y \in B$. If $y \in \operatorname{ran}(f)$, there is an $x \in A$ so that $f(x) = y$. Because f is injective, there is only one such $x \in A$. Then we can define: $g(y) = x$, i.e., $g(y)$ is "the" $x \in A$ such that $f(x) = y$. If $y \notin \operatorname{ran}(f)$, we can map it to any $a \in A$. So, we can pick an $a \in A$ and define $g: B \to A$ by:

$$g(y) = \begin{cases} x & \text{if } f(x) = y \\ a & \text{if } y \notin \operatorname{ran}(f). \end{cases}$$

It is defined for all $y \in B$, since for each such $y \in \operatorname{ran}(f)$ there is exactly one $x \in A$ such that $f(x) = y$. By definition, if $y = f(x)$, then $g(y) = x$, i.e., $g(f(x)) = x$. □

Proposition 3.17. *If $f: A \to B$ is surjective, then there is a right inverse $h: B \to A$ of f so that $f(h(y)) = y$ for all $y \in B$.*

Proof. Suppose that $f: A \to B$ is surjective. Consider a $y \in B$. Since f is surjective, there is an $x_y \in A$ with $f(x_y) = y$. Then we can define: $h(y) = x_y$, i.e., for each $y \in B$ we choose some $x \in A$ so that $f(x) = y$; since f is surjective there is always at least one to choose from.[1] By definition, if $x = h(y)$, then $f(x) = y$, i.e., for any $y \in B$, $f(h(y)) = y$. □

---

[1] Since f is surjective, for every $y \in B$ the set $\{x : f(x) = y\}$ is nonempty. Our definition of h requires that we choose a single x from each of these sets. That this is always possible is actually not obvious—the possibility of making these choices is simply assumed as an axiom. In other words, this proposition assumes the so-called Axiom of Choice, an issue we will gloss over. However, in many specific cases, e.g., when $A = \mathbb{N}$ or is finite, or when f is bijective, the Axiom of Choice is not required. (In the particular case when f is bijective, for each $y \in B$ the set $\{x : f(x) = y\}$ has exactly one element, so that there is no choice to make.)

By combining the ideas in the previous proof, we now get that every bijection has an inverse, i.e., there is a single function which is both a left and right inverse of f.

Proposition 3.18. *If $f\colon A \to B$ is bijective, there is a function $f^{-1}\colon B \to A$ so that for all $x \in A$, $f^{-1}(f(x)) = x$ and for all $y \in B$, $f(f^{-1}(y)) = y$.*

Proof. Exercise. □

There is a slightly more general way to extract inverses. We saw in section 3.2 that every function f induces a surjection $f'\colon A \to \operatorname{ran}(f)$ by letting $f'(x) = f(x)$ for all $x \in A$. Clearly, if f is injective, then f' is bijective, so that it has a unique inverse by Proposition 3.18. By a very minor abuse of notation, we sometimes call the inverse of f' simply "the inverse of f."

Proposition 3.19. *Show that if $f\colon A \to B$ has a left inverse g and a right inverse h, then $h = g$.*

Proof. Exercise. □

Proposition 3.20. *Every function f has at most one inverse.*

Proof. Suppose g and h are both inverses of f. Then in particular g is a left inverse of f and h is a right inverse. By Proposition 3.19, $g = h$. □

3.5 Composition of Functions

We saw in section 3.4 that the inverse f^{-1} of a bijection f is itself a function. Another operation on functions is composition: we can define a new function by composing two functions, f and g, i.e., by first applying f and then g. Of course, this is only possible if the ranges and domains match, i.e., the range of f must be a subset of the domain of g. This operation on functions is the

3.6. PARTIAL FUNCTIONS

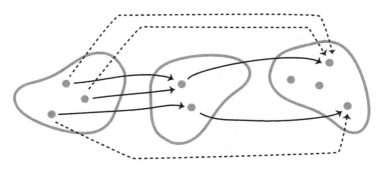

Figure 3.5: The composition $g \circ f$ of two functions f and g.

analogue of the operation of relative product on relations from section 2.6.

A diagram might help to explain the idea of composition. In Figure 3.5, we depict two functions $f\colon A \to B$ and $g\colon B \to C$ and their composition $(g \circ f)$. The function $(g \circ f)\colon A \to C$ pairs each element of A with an element of C. We specify which element of C an element of A is paired with as follows: given an input $x \in A$, first apply the function f to x, which will output some $f(x) = y \in B$, then apply the function g to y, which will output some $g(f(x)) = g(y) = z \in C$.

Definition 3.21 (Composition). Let $f\colon A \to B$ and $g\colon B \to C$ be functions. The *composition* of f with g is $g \circ f\colon A \to C$, where $(g \circ f)(x) = g(f(x))$.

Example 3.22. Consider the functions $f(x) = x + 1$, and $g(x) = 2x$. Since $(g \circ f)(x) = g(f(x))$, for each input x you must first take its successor, then multiply the result by two. So their composition is given by $(g \circ f)(x) = 2(x + 1)$.

3.6 Partial Functions

It is sometimes useful to relax the definition of function so that it is not required that the output of the function is defined for all possible inputs. Such mappings are called *partial functions*.

Definition 3.23. A *partial function* $f: A \twoheadrightarrow B$ is a mapping which assigns to every element of A at most one element of B. If f assigns an element of B to $x \in A$, we say $f(x)$ is *defined*, and otherwise *undefined*. If $f(x)$ is defined, we write $f(x) \downarrow$, otherwise $f(x) \uparrow$. The *domain* of a partial function f is the subset of A where it is defined, i.e., $\text{dom}(f) = \{x \in A : f(x) \downarrow\}$.

Example 3.24. Every function $f: A \to B$ is also a partial function. Partial functions that are defined everywhere on A—i.e., what we so far have simply called a function—are also called *total* functions.

Example 3.25. The partial function $f: \mathbb{R} \twoheadrightarrow \mathbb{R}$ given by $f(x) = 1/x$ is undefined for $x = 0$, and defined everywhere else.

Definition 3.26 (Graph of a partial function). Let $f: A \twoheadrightarrow B$ be a partial function. The *graph* of f is the relation $R_f \subseteq A \times B$ defined by
$$R_f = \{\langle x, y \rangle : f(x) = y\}.$$

Proposition 3.27. *Suppose $R \subseteq A \times B$ has the property that whenever Rxy and Rxy' then $y = y'$. Then R is the graph of the partial function $f: X \twoheadrightarrow Y$ defined by: if there is a y such that Rxy, then $f(x) = y$, otherwise $f(x) \uparrow$. If R is also* serial, *i.e., for each $x \in X$ there is a $y \in Y$ such that Rxy, then f is total.*

Proof. Suppose there is a y such that Rxy. If there were another $y' \neq y$ such that Rxy', the condition on R would be violated. Hence, if there is a y such that Rxy, that y is unique, and so f is well-defined. Obviously, $R_f = R$ and f is total if R is serial. □

Summary

A **function** $f: A \to B$ maps every element of the **domain** A to a unique element of the **codomain** B. If $x \in A$, we call the y that f maps x to the **value** $f(x)$ of f for **argument** x. If A is a set of pairs, we can think of the function f as taking two arguments. The **range** $\mathrm{ran}(f)$ of f is the subset of B that consists of all the values of f.

If $\mathrm{ran}(f) = B$ then f is called **surjective**. The value $f(x)$ is unique in that f maps x to only one $f(x)$, never more than one. If $f(x)$ is also unique in the sense that no two different arguments are mapped to the same value, f is called **injective**. Functions which are both injective and surjective are called **bijective**.

Bijective functions have a unique **inverse function** f^{-1}. Functions can also be chained together: the function $(g \circ f)$ is the **composition** of f with g. Compositions of injective functions are injective, and of surjective functions are surjective, and $(f^{-1} \circ f)$ is the identity function.

If we relax the requirement that f must have a value for every $x \in A$, we get the notion of a **partial functions**. If $f: A \twoheadrightarrow B$ is partial, we say $f(x)$ is **defined**, $f(x) \downarrow$ if f has a value for argument x, and otherwise we say that $f(x)$ is **undefined**, $f(x) \uparrow$. Any (partial) function f is associated with the **graph** R_f of f, the relation that holds iff $f(x) = y$.

Problems

Problem 3.1. Show that if $f: A \to B$ has a left inverse g, then f is injective.

Problem 3.2. Show that if $f: A \to B$ has a right inverse h, then f is surjective.

Problem 3.3. Prove Proposition 3.18. You have to define f^{-1}, show that it is a function, and show that it is an inverse of f, i.e., $f^{-1}(f(x)) = x$ and $f(f^{-1}(y)) = y$ for all $x \in A$ and $y \in B$.

Problem 3.4. Prove Proposition 3.19.

Problem 3.5. Show that if $f \colon A \to B$ and $g \colon B \to C$ are both injective, then $g \circ f \colon A \to C$ is injective.

Problem 3.6. Show that if $f \colon A \to B$ and $g \colon B \to C$ are both surjective, then $g \circ f \colon A \to C$ is surjective.

Problem 3.7. Suppose $f \colon A \to B$ and $g \colon B \to C$. Show that the graph of $g \circ f$ is $R_f \mid R_g$.

Problem 3.8. Given $f \colon A \twoheadrightarrow B$, define the partial function $g \colon B \twoheadrightarrow A$ by: for any $y \in B$, if there is a unique $x \in A$ such that $f(x) = y$, then $g(y) = x$; otherwise $g(y) \uparrow$. Show that if f is injective, then $g(f(x)) = x$ for all $x \in \mathrm{dom}(f)$, and $f(g(y)) = y$ for all $y \in \mathrm{ran}(f)$.

CHAPTER 4
The Size of Sets

4.1 Introduction

When Georg Cantor developed set theory in the 1870s, one of his aims was to make palatable the idea of an infinite collection—an actual infinity, as the medievals would say. A key part of this was his treatment of the *size* of different sets. If a, b and c are all distinct, then the set $\{a,b,c\}$ is intuitively *larger* than $\{a,b\}$. But what about infinite sets? Are they all as large as each other? It turns out that they are not.

The first important idea here is that of an enumeration. We can list every finite set by listing all its elements. For some infinite sets, we can also list all their elements if we allow the list itself to be infinite. Such sets are called countable. Cantor's surprising result, which we will fully understand by the end of this chapter, was that some infinite sets are not countable.

4.2 Enumerations and Countable Sets

We've already given examples of sets by listing their elements. Let's discuss in more general terms how and when we can list the elements of a set, even if that set is infinite.

Definition 4.1 (Enumeration, informally). Informally, an *enumeration* of a set A is a list (possibly infinite) of elements of A such that every element of A appears on the list at some finite position. If A has an enumeration, then A is said to be *countable*.

A couple of points about enumerations:

1. We count as enumerations only lists which have a beginning and in which every element other than the first has a single element immediately preceding it. In other words, there are only finitely many elements between the first element of the list and any other element. In particular, this means that every element of an enumeration has a finite position: the first element has position 1, the second position 2, etc.

2. We can have different enumerations of the same set A which differ by the order in which the elements appear: 4, 1, 25, 16, 9 enumerates the (set of the) first five square numbers just as well as 1, 4, 9, 16, 25 does.

3. Redundant enumerations are still enumerations: 1, 1, 2, 2, 3, 3, ... enumerates the same set as 1, 2, 3, ... does.

4. Order and redundancy *do* matter when we specify an enumeration: we can enumerate the positive integers beginning with 1, 2, 3, 1, ..., but the pattern is easier to see when enumerated in the standard way as 1, 2, 3, 4, ...

5. Enumerations must have a beginning: ..., 3, 2, 1 is not an enumeration of the positive integers because it has no first element. To see how this follows from the informal definition, ask yourself, "at what position in the list does the number 76 appear?"

6. The following is not an enumeration of the positive integers: 1, 3, 5, ..., 2, 4, 6, ... The problem is that the even

numbers occur at places $\infty+1$, $\infty+2$, $\infty+3$, rather than at finite positions.

7. The empty set is enumerable: it is enumerated by the empty list!

Proposition 4.2. *If A has an enumeration, it has an enumeration without repetitions.*

Proof. Suppose A has an enumeration $x_1, x_2, \ldots$ in which each x_i is an element of A. We can remove repetitions from an enumeration by removing repeated elements. For instance, we can turn the enumeration into a new one in which we list x_i if it is an element of A that is not among $x_1, \ldots, x_{i-1}$ or remove x_i from the list if it already appears among $x_1, \ldots, x_{i-1}$. □

The last argument shows that in order to get a good handle on enumerations and countable sets and to prove things about them, we need a more precise definition. The following provides it.

Definition 4.3 (Enumeration, formally). An *enumeration* of a set $A \neq \emptyset$ is any surjective function $f: \mathbb{Z}^+ \to A$.

Let's convince ourselves that the formal definition and the informal definition using a possibly infinite list are equivalent. First, any surjective function from $\mathbb{Z}^+$ to a set A enumerates A. Such a function determines an enumeration as defined informally above: the list $f(1), f(2), f(3), \ldots$. Since f is surjective, every element of A is guaranteed to be the value of $f(n)$ for some $n \in \mathbb{Z}^+$. Hence, every element of A appears at some finite position in the list. Since the function may not be injective, the list may be redundant, but that is acceptable (as noted above).

On the other hand, given a list that enumerates all elements of A, we can define a surjective function $f: \mathbb{Z}^+ \to A$ by letting $f(n)$ be the nth element of the list, or the final element of the list if there is no *n*th element. The only case where this does not produce a surjective function is when A is empty, and hence the list is empty. So, every non-empty list determines a surjective function $f: \mathbb{Z}^+ \to A$.

Definition 4.4. A set A is countable iff it is empty or has an enumeration.

Example 4.5. A function enumerating the positive integers ($\mathbb{Z}^+$) is simply the identity function given by $f(n) = n$. A function enumerating the natural numbers $\mathbb{N}$ is the function $g(n) = n - 1$.

Example 4.6. The functions $f: \mathbb{Z}^+ \to \mathbb{Z}^+$ and $g: \mathbb{Z}^+ \to \mathbb{Z}^+$ given by

$$f(n) = 2n \text{ and}$$
$$g(n) = 2n + 1$$

enumerate the even positive integers and the odd positive integers, respectively. However, neither function is an enumeration of $\mathbb{Z}^+$, since neither is surjective.

Example 4.7. The function $f(n) = (-1)^n \lceil \frac{(n-1)}{2} \rceil$ (where $\lceil x \rceil$ denotes the *ceiling* function, which rounds x up to the nearest integer) enumerates the set of integers $\mathbb{Z}$. Notice how f generates the values of $\mathbb{Z}$ by "hopping" back and forth between positive and negative integers:

$$\begin{array}{ccccccc} f(1) & f(2) & f(3) & f(4) & f(5) & f(6) & f(7) & \cdots \\ -\lceil \frac{0}{2} \rceil & \lceil \frac{1}{2} \rceil & -\lceil \frac{2}{2} \rceil & \lceil \frac{3}{2} \rceil & -\lceil \frac{4}{2} \rceil & \lceil \frac{5}{2} \rceil & -\lceil \frac{6}{2} \rceil & \cdots \\ 0 & 1 & -1 & 2 & -2 & 3 & & \cdots \end{array}$$

You can also think of f as defined by cases as follows:

$$f(n) = \begin{cases} 0 & \text{if } n = 1 \\ n/2 & \text{if } n \text{ is even} \\ -(n-1)/2 & \text{if } n \text{ is odd and } > 1 \end{cases}$$

4.2. ENUMERATIONS AND COUNTABLE SETS

Although it is perhaps more natural when listing the elements of a set to start counting from the 1st element, mathematicians like to use the natural numbers $\mathbb{N}$ for counting things. They talk about the 0th, 1st, 2nd, and so on, elements of a list. Correspondingly, we can define an enumeration as a surjective function from $\mathbb{N}$ to A. Of course, the two definitions are equivalent.

Proposition 4.8. *There is a surjection $f: \mathbb{Z}^+ \to A$ iff there is a surjection $g: \mathbb{N} \to A$.*

Proof. Given a surjection $f: \mathbb{Z}^+ \to A$, we can define $g(n) = f(n+1)$ for all $n \in \mathbb{N}$. It is easy to see that $g: \mathbb{N} \to A$ is surjective. Conversely, given a surjection $g: \mathbb{N} \to A$, define $f(n) = g(n-1)$. □

This gives us the following result:

Corollary 4.9. *A set A is countable iff it is empty or there is a surjective function $f: \mathbb{N} \to A$.*

We discussed above than an list of elements of a set A can be turned into a list without repetitions. This is also true for enumerations, but a bit harder to formulate and prove rigorously. Any function $f: \mathbb{Z}^+ \to A$ must be defined for all $n \in \mathbb{Z}^+$. If there are only finitely many elements in A then we clearly cannot have a function defined on the infinitely many elements of $\mathbb{Z}^+$ that takes as values all the elements of A but never takes the same value twice. In that case, i.e., in the case where the list without repetitions is finite, we must choose a different domain for f, one with only finitely many elements. Not having repetitions means that f must be injective. Since it is also surjective, we are looking for a bijection between some finite set $\{1,\ldots,n\}$ or $\mathbb{Z}^+$ and A.

Proposition 4.10. *If $f: \mathbb{Z}^+ \to A$ is surjective (i.e., an enumeration of A), there is a bijection $g: Z \to A$ where Z is either $\mathbb{Z}^+$ or $\{1,\ldots,n\}$ for some $n \in \mathbb{Z}^+$.*

Proof. We define the function g recursively: Let $g(1) = f(1)$. If $g(i)$ has already been defined, let $g(i+1)$ be the first value of $f(1)$, $f(2), \ldots$ not already among $g(1), \ldots, g(i)$, if there is one. If A has just n elements, then $g(1), \ldots, g(n)$ are all defined, and so we have defined a function $g \colon \{1, \ldots, n\} \to A$. If A has infinitely many elements, then for any i there must be an element of A in the enumeration $f(1), f(2), \ldots$, which is not already among $g(1), \ldots, g(i)$. In this case we have defined a funtion $g \colon \mathbb{Z}^+ \to A$.

The function g is surjective, since any element of A is among $f(1), f(2), \ldots$ (since f is surjective) and so will eventually be a value of $g(i)$ for some i. It is also injective, since if there were $j < i$ such that $g(j) = g(i)$, then $g(i)$ would already be among $g(1), \ldots, g(i-1)$, contrary to how we defined g. □

Corollary 4.11. *A set A is countable iff it is empty or there is a bijection $f \colon N \to A$ where either $N = \mathbb{N}$ or $N = \{0, \ldots, n\}$ for some $n \in \mathbb{N}$.*

Proof. A is countable iff A is empty or there is a surjective $f \colon \mathbb{Z}^+ \to A$. By Proposition 4.10, the latter holds iff there is a bijective function $f \colon Z \to A$ where $Z = \mathbb{Z}^+$ or $Z = \{1, \ldots, n\}$ for some $n \in \mathbb{Z}^+$. By the same argument as in the proof of Proposition 4.8, that in turn is the case iff there is a bijection $g \colon N \to A$ where either $N = \mathbb{N}$ or $N = \{0, \ldots, n-1\}$. □

4.3 Cantor's Zig-Zag Method

We've already considered some "easy" enumerations. Now we will consider something a bit harder. Consider the set of pairs of natural numbers, which we defined in section 1.5 thus:

$$\mathbb{N} \times \mathbb{N} = \{\langle n, m \rangle : n, m \in \mathbb{N}\}$$

4.3. CANTOR'S ZIG-ZAG METHOD

We can organize these ordered pairs into an *array*, like so:

	0	1	2	3	...
0	$\langle 0,0 \rangle$	$\langle 0,1 \rangle$	$\langle 0,2 \rangle$	$\langle 0,3 \rangle$	...
1	$\langle 1,0 \rangle$	$\langle 1,1 \rangle$	$\langle 1,2 \rangle$	$\langle 1,3 \rangle$	...
2	$\langle 2,0 \rangle$	$\langle 2,1 \rangle$	$\langle 2,2 \rangle$	$\langle 2,3 \rangle$	...
3	$\langle 3,0 \rangle$	$\langle 3,1 \rangle$	$\langle 3,2 \rangle$	$\langle 3,3 \rangle$	...
⋮	⋮	⋮	⋮	⋮	⋱

Clearly, every ordered pair in $\mathbb{N} \times \mathbb{N}$ will appear exactly once in the array. In particular, $\langle n, m \rangle$ will appear in the nth row and mth column. But how do we organize the elements of such an array into a "one-dimensional" list? The pattern in the array below demonstrates one way to do this (although of course there are many other options):

	0	1	2	3	4	...
0	0	1	3	6	10	...
1	2	4	7	11	...	...
2	5	8	12	...	...	...
3	9	13	...	...	...	...
4	14	...	...	...	...	...
⋮	⋮	⋮	⋮	⋮	...	⋱

This pattern is called *Cantor's zig-zag method*. It enumerates $\mathbb{N} \times \mathbb{N}$ as follows:

$$\langle 0,0 \rangle, \langle 0,1 \rangle, \langle 1,0 \rangle, \langle 0,2 \rangle, \langle 1,1 \rangle, \langle 2,0 \rangle, \langle 0,3 \rangle, \langle 1,2 \rangle, \langle 2,1 \rangle, \langle 3,0 \rangle, \ldots$$

And this establishes the following:

Proposition 4.12. $\mathbb{N} \times \mathbb{N}$ *is countable.*

Proof. Let $f : \mathbb{N} \to \mathbb{N} \times \mathbb{N}$ take each $k \in \mathbb{N}$ to the tuple $\langle n, m \rangle \in \mathbb{N} \times \mathbb{N}$ such that k is the value of the nth row and mth column in Cantor's zig-zag array. □

This technique also generalises rather nicely. For example, we can use it to enumerate the set of ordered triples of natural numbers, i.e.:

$$\mathbb{N} \times \mathbb{N} \times \mathbb{N} = \{\langle n,m,k \rangle : n,m,k \in \mathbb{N}\}$$

We think of $\mathbb{N} \times \mathbb{N} \times \mathbb{N}$ as the Cartesian product of $\mathbb{N} \times \mathbb{N}$ with $\mathbb{N}$, that is,

$$\mathbb{N}^3 = (\mathbb{N} \times \mathbb{N}) \times \mathbb{N} = \{\langle\langle n,m \rangle,k \rangle : n,m,k \in \mathbb{N}\}$$

and thus we can enumerate $\mathbb{N}^3$ with an array by labelling one axis with the enumeration of $\mathbb{N}$, and the other axis with the enumeration of $\mathbb{N}^2$:

	0	1	2	3	...
$\langle 0,0 \rangle$	$\langle 0,0,0 \rangle$	$\langle 0,0,1 \rangle$	$\langle 0,0,2 \rangle$	$\langle 0,0,3 \rangle$	...
$\langle 0,1 \rangle$	$\langle 0,1,0 \rangle$	$\langle 0,1,1 \rangle$	$\langle 0,1,2 \rangle$	$\langle 0,1,3 \rangle$	...
$\langle 1,0 \rangle$	$\langle 1,0,0 \rangle$	$\langle 1,0,1 \rangle$	$\langle 1,0,2 \rangle$	$\langle 1,0,3 \rangle$	...
$\langle 0,2 \rangle$	$\langle 0,2,0 \rangle$	$\langle 0,2,1 \rangle$	$\langle 0,2,2 \rangle$	$\langle 0,2,3 \rangle$	...
⋮	⋮	⋮	⋮	⋮	⋱

Thus, by using a method like Cantor's zig-zag method, we may similarly obtain an enumeration of $\mathbb{N}^3$. And we can keep going, obtaining enumerations of $\mathbb{N}^n$ for any natural number n. So, we have:

Proposition 4.13. *$\mathbb{N}^n$ is countable, for every $n \in \mathbb{N}$.*

4.4 Pairing Functions and Codes

Cantor's zig-zag method makes the enumerability of $\mathbb{N}^n$ visually evident. But let us focus on our array depicting $\mathbb{N}^2$. Following the zig-zag line in the array and counting the places, we can check that $\langle 1,2 \rangle$ is associated with the number 7. However, it would be nice if we could compute this more directly. That is, it would

4.4. PAIRING FUNCTIONS AND CODES

be nice to have to hand the *inverse* of the zig-zag enumeration, $g: \mathbb{N}^2 \to \mathbb{N}$, such that

$$g(\langle 0,0 \rangle) = 0, \; g(\langle 0,1 \rangle) = 1, \; g(\langle 1,0 \rangle) = 2, \; \ldots, g(\langle 1,2 \rangle) = 7, \; \ldots$$

This would enable us to calculate exactly where $\langle n, m \rangle$ will occur in our enumeration.

In fact, we can define g directly by making two observations. First: if the nth row and mth column contains value v, then the $(n+1)$st row and $(m-1)$st column contains value $v+1$. Second: the first row of our enumeration consists of the triangular numbers, starting with 0, 1, 3, 6, etc. The kth triangular number is the sum of the natural numbers $< k$, which can be computed as $k(k+1)/2$. Putting these two observations together, consider this function:

$$g(n,m) = \frac{(n+m+1)(n+m)}{2} + n$$

We often just write $g(n,m)$ rather that $g(\langle n,m \rangle)$, since it is easier on the eyes. This tells you first to determine the $(n+m)^{\text{th}}$ triangle number, and then add n to it. And it populates the array in exactly the way we would like. So in particular, the pair $\langle 1, 2 \rangle$ is sent to $\frac{4 \times 3}{2} + 1 = 7$.

This function g is the *inverse* of an enumeration of a set of pairs. Such functions are called *pairing functions*.

Definition 4.14 (Pairing function). A function $f: A \times B \to \mathbb{N}$ is an arithmetical *pairing function* if f is injective. We also say that f encodes $A \times B$, and that $f(x,y)$ is the *code* for $\langle x, y \rangle$.

We can use pairing functions to encode, e.g., pairs of natural numbers; or, in other words, we can represent each *pair* of elements using a *single* number. Using the inverse of the pairing function, we can *decode* the number, i.e., find out which pair it represents.

4.5 An Alternative Pairing Function

There are other enumerations of $\mathbb{N}^2$ that make it easier to figure out what their inverses are. Here is one. Instead of visualizing the enumeration in an array, start with the list of positive integers associated with (initially) empty spaces. Imagine filling these spaces successively with pairs $\langle n, m \rangle$ as follows. Starting with the pairs that have 0 in the first place (i.e., pairs $\langle 0, m \rangle$), put the first (i.e., $\langle 0,0 \rangle$) in the first empty place, then skip an empty space, put the second (i.e., $\langle 0,2 \rangle$) in the next empty place, skip one again, and so forth. The (incomplete) beginning of our enumeration now looks like this

1	2	3	4	5	6	7	8	9	10	...
$\langle 0,1 \rangle$		$\langle 0,2 \rangle$		$\langle 0,3 \rangle$		$\langle 0,4 \rangle$		$\langle 0,5 \rangle$		...

Repeat this with pairs $\langle 1, m \rangle$ for the place that still remain empty, again skipping every other empty place:

1	2	3	4	5	6	7	8	9	10	...
$\langle 0,0 \rangle$	$\langle 1,0 \rangle$	$\langle 0,1 \rangle$		$\langle 0,2 \rangle$	$\langle 1,1 \rangle$	$\langle 0,3 \rangle$		$\langle 0,4 \rangle$	$\langle 1,2 \rangle$	...

Enter pairs $\langle 2, m \rangle$, $\langle 2, m \rangle$, etc., in the same way. Our completed enumeration thus starts like this:

1	2	3	4	5	6	7	8	9	10	...
$\langle 0,0 \rangle$	$\langle 1,0 \rangle$	$\langle 0,1 \rangle$	$\langle 2,0 \rangle$	$\langle 0,2 \rangle$	$\langle 1,1 \rangle$	$\langle 0,3 \rangle$	$\langle 3,0 \rangle$	$\langle 0,4 \rangle$	$\langle 1,2 \rangle$	...

If we number the cells in the array above according to this enumeration, we will not find a neat zig-zag line, but this arrange-

4.5. AN ALTERNATIVE PAIRING FUNCTION

ment:

	0	1	2	3	4	5	...
0	1	3	5	7	9	11	...
1	2	6	10	14	18	...	...
2	4	12	20	28	...	...	...
3	8	24	40	...	...	...	...
4	16	48	...	...	...	...	...
5	32	...	...	...	...	...	...
⋮	⋮	⋮	⋮	⋮	⋮	⋮	⋱

We can see that the pairs in row 0 are in the odd numbered places of our enumeration, i.e., pair $\langle 0, m \rangle$ is in place $2m + 1$; pairs in the second row, $\langle 1, m \rangle$, are in places whose number is the double of an odd number, specifically, $2 \cdot (2m + 1)$; pairs in the third row, $\langle 2, m \rangle$, are in places whose number is four times an odd number, $4 \cdot (2m+1)$; and so on. The factors of $(2m+1)$ for each row, $1, 2, 4, 8, \ldots$, are exactly the powers of 2: $1 = 2^0$, $2 = 2^1$, $4 = 2^2$, $8 = 2^3$, $\ldots$ In fact, the relevant exponent is always the first member of the pair in question. Thus, for pair $\langle n, m \rangle$ the factor is 2^n. This gives us the general formula: $2^n \cdot (2m + 1)$. However, this is a mapping of pairs to *positive* integers, i.e., $\langle 0, 0 \rangle$ has position 1. If we want to begin at position 0 we must subtract 1 from the result. This gives us:

Example 4.15. The function $h \colon \mathbb{N}^2 \to \mathbb{N}$ given by

$$h(n, m) = 2^n(2m + 1) - 1$$

is a pairing function for the set of pairs of natural numbers $\mathbb{N}^2$.

Accordingly, in our second enumeration of $\mathbb{N}^2$, the pair $\langle 0, 0 \rangle$ has code $h(0,0) = 2^0(2 \cdot 0 + 1) - 1 = 0$; $\langle 1, 2 \rangle$ has code $2^1 \cdot (2 \cdot 2 + 1) - 1 = 2 \cdot 5 - 1 = 9$; $\langle 2, 6 \rangle$ has code $2^2 \cdot (2 \cdot 6 + 1) - 1 = 51$.

Sometimes it is enough to encode pairs of natural numbers $\mathbb{N}^2$ without requiring that the encoding is surjective. Such encodings have inverses that are only partial functions.

Example 4.16. The function $j\colon \mathbb{N}^2 \to \mathbb{N}^+$ given by

$$j(n,m) = 2^n 3^m$$

is an injective function $\mathbb{N}^2 \to \mathbb{N}$.

4.6 Uncountable Sets

Some sets, such as the set $\mathbb{Z}^+$ of positive integers, are infinite. So far we've seen examples of infinite sets which were all countable. However, there are also infinite sets which do not have this property. Such sets are called *uncountable*.

First of all, it is perhaps already surprising that there are uncountable sets. For any countable set A there is a surjective function $f\colon \mathbb{Z}^+ \to A$. If a set is uncountable there is no such function. That is, no function mapping the infinitely many elements of $\mathbb{Z}^+$ to A can exhaust all of A. So there are "more" elements of A than the infinitely many positive integers.

How would one prove that a set is uncountable? You have to show that no such surjective function can exist. Equivalently, you have to show that the elements of A cannot be enumerated in a one way infinite list. The best way to do this is to show that every list of elements of A must leave at least one element out; or that no function $f\colon \mathbb{Z}^+ \to A$ can be surjective. We can do this using Cantor's *diagonal method*. Given a list of elements of A, say, x_1, x_2, ..., we construct another element of A which, by its construction, cannot possibly be on that list.

Our first example is the set $\mathbb{B}^\omega$ of all infinite, non-gappy sequences of 0's and 1's.

Theorem 4.17. $\mathbb{B}^\omega$ *is uncountable.*

Proof. Suppose, by way of contradiction, that $\mathbb{B}^\omega$ is countable, i.e., suppose that there is a list s_1, s_2, s_3, s_4, ... of all elements of $\mathbb{B}^\omega$. Each of these s_i is itself an infinite sequence of 0's and 1's.

4.6. UNCOUNTABLE SETS

Let's call the j-th element of the i-th sequence in this list $s_i(j)$. Then the i-th sequence s_i is

$$s_i(1), s_i(2), s_i(3), \ldots$$

We may arrange this list, and the elements of each sequence s_i in it, in an array:

	1	2	3	4	...
1	$\mathbf{s_1(1)}$	$s_1(2)$	$s_1(3)$	$s_1(4)$	...
2	$s_2(1)$	$\mathbf{s_2(2)}$	$s_2(3)$	$s_2(4)$	...
3	$s_3(1)$	$s_3(2)$	$\mathbf{s_3(3)}$	$s_3(4)$	...
4	$s_4(1)$	$s_4(2)$	$s_4(3)$	$\mathbf{s_4(4)}$	...
⋮	⋮	⋮	⋮	⋮	⋱

The labels down the side give the number of the sequence in the list $s_1, s_2, \ldots$; the numbers across the top label the elements of the individual sequences. For instance, $s_1(1)$ is a name for whatever number, a 0 or a 1, is the first element in the sequence s_1, and so on.

Now we construct an infinite sequence, $\bar{s}$, of 0's and 1's which cannot possibly be on this list. The definition of $\bar{s}$ will depend on the list $s_1, s_2, \ldots$. Any infinite list of infinite sequences of 0's and 1's gives rise to an infinite sequence $\bar{s}$ which is guaranteed to not appear on the list.

To define $\bar{s}$, we specify what all its elements are, i.e., we specify $\bar{s}(n)$ for all $n \in \mathbb{Z}^+$. We do this by reading down the diagonal of the array above (hence the name "diagonal method") and then changing every 1 to a 0 and every 0 to a 1. More abstractly, we define $\bar{s}(n)$ to be 0 or 1 according to whether the n-th element of the diagonal, $s_n(n)$, is 1 or 0.

$$\bar{s}(n) = \begin{cases} 1 & \text{if } s_n(n) = 0 \\ 0 & \text{if } s_n(n) = 1. \end{cases}$$

If you like formulas better than definitions by cases, you could also define $\bar{s}(n) = 1 - s_n(n)$.

Clearly $\bar{s}$ is an infinite sequence of 0's and 1's, since it is just the mirror sequence to the sequence of 0's and 1's that appear on the diagonal of our array. So $\bar{s}$ is an element of $\mathbb{B}^\omega$. But it cannot be on the list $s_1, s_2, \ldots$ Why not?

It can't be the first sequence in the list, s_1, because it differs from s_1 in the first element. Whatever $s_1(1)$ is, we defined $\bar{s}(1)$ to be the opposite. It can't be the second sequence in the list, because $\bar{s}$ differs from s_2 in the second element: if $s_2(2)$ is 0, $\bar{s}(2)$ is 1, and vice versa. And so on.

More precisely: if $\bar{s}$ were on the list, there would be some k so that $\bar{s} = s_k$. Two sequences are identical iff they agree at every place, i.e., for any n, $\bar{s}(n) = s_k(n)$. So in particular, taking $n = k$ as a special case, $\bar{s}(k) = s_k(k)$ would have to hold. $s_k(k)$ is either 0 or 1. If it is 0 then $\bar{s}(k)$ must be 1—that's how we defined $\bar{s}$. But if $s_k(k) = 1$ then, again because of the way we defined $\bar{s}$, $\bar{s}(k) = 0$. In either case $\bar{s}(k) \neq s_k(k)$.

We started by assuming that there is a list of elements of $\mathbb{B}^\omega$, $s_1, s_2, \ldots$ From this list we constructed a sequence $\bar{s}$ which we proved cannot be on the list. But it definitely is a sequence of 0's and 1's if all the s_i are sequences of 0's and 1's, i.e., $\bar{s} \in \mathbb{B}^\omega$. This shows in particular that there can be no list of *all* elements of $\mathbb{B}^\omega$, since for any such list we could also construct a sequence $\bar{s}$ guaranteed to not be on the list, so the assumption that there is a list of all sequences in $\mathbb{B}^\omega$ leads to a contradiction. □

This proof method is called "diagonalization" because it uses the diagonal of the array to define $\bar{s}$. Diagonalization need not involve the presence of an array: we can show that sets are not countable by using a similar idea even when no array and no actual diagonal is involved.

Theorem 4.18. *$\wp(\mathbb{Z}^+)$ is not countable.*

Proof. We proceed in the same way, by showing that for every list of subsets of $\mathbb{Z}^+$ there is a subset of $\mathbb{Z}^+$ which cannot be on the

4.6. UNCOUNTABLE SETS

list. Suppose the following is a given list of subsets of $\mathbb{Z}^+$:

$$Z_1, Z_2, Z_3, \ldots$$

We now define a set $\overline{Z}$ such that for any $n \in \mathbb{Z}^+$, $n \in \overline{Z}$ iff $n \notin Z_n$:

$$\overline{Z} = \{n \in \mathbb{Z}^+ : n \notin Z_n\} \qquad \square$$

$\overline{Z}$ is clearly a set of positive integers, since by assumption each Z_n is, and thus $\overline{Z} \in \wp(\mathbb{Z}^+)$. But $\overline{Z}$ cannot be on the list. To show this, we'll establish that for each $k \in \mathbb{Z}^+$, $\overline{Z} \neq Z_k$.

So let $k \in \mathbb{Z}^+$ be arbitrary. We've defined $\overline{Z}$ so that for any $n \in \mathbb{Z}^+$, $n \in \overline{Z}$ iff $n \notin Z_n$. In particular, taking $n = k$, $k \in \overline{Z}$ iff $k \notin Z_k$. But this shows that $\overline{Z} \neq Z_k$, since k is an element of one but not the other, and so $\overline{Z}$ and Z_k have different elements. Since k was arbitrary, $\overline{Z}$ is not on the list $Z_1, Z_2, \ldots$

The preceding proof did not mention a diagonal, but you can think of it as involving a diagonal if you picture it this way: Imagine the sets $Z_1, Z_2, \ldots$, written in an array, where each element $j \in Z_i$ is listed in the j-th column. Say the first four sets on that list are $\{1, 2, 3, \ldots\}$, $\{2, 4, 6, \ldots\}$, $\{1, 2, 5\}$, and $\{3, 4, 5, \ldots\}$. Then the array would begin with

$$\begin{aligned}
Z_1 &= \{\mathbf{1},\ 2,\ 3,\ 4,\ 5,\ 6,\ \ldots\} \\
Z_2 &= \{\quad\ \mathbf{2},\quad\ 4,\quad\ \ 6,\ \ldots\} \\
Z_3 &= \{1,\ 2,\quad\quad\ 5\quad\quad\quad\ \} \\
Z_4 &= \{\quad\quad\ 3,\ \mathbf{4},\ 5,\ 6,\ \ldots\} \\
&\ \ \vdots \qquad\qquad\qquad \ddots
\end{aligned}$$

Then $\overline{Z}$ is the set obtained by going down the diagonal, leaving out any numbers that appear along the diagonal and include those j where the array has a gap in the j-th row/column. In the above case, we would leave out 1 and 2, include 3, leave out 4, etc.

4.7 Reduction

We showed $\wp(\mathbb{Z}^+)$ to be uncountable by a diagonalization argument. We already had a proof that $\mathbb{B}^\omega$, the set of all infinite sequences of 0s and 1s, is uncountable. Here's another way we can prove that $\wp(\mathbb{Z}^+)$ is uncountable: Show that *if $\wp(\mathbb{Z}^+)$ is countable then $\mathbb{B}^\omega$ is also countable.* Since we know $\mathbb{B}^\omega$ is not countable, $\wp(\mathbb{Z}^+)$ can't be either. This is called *reducing* one problem to another—in this case, we reduce the problem of enumerating $\mathbb{B}^\omega$ to the problem of enumerating $\wp(\mathbb{Z}^+)$. A solution to the latter—an enumeration of $\wp(\mathbb{Z}^+)$—would yield a solution to the former—an enumeration of $\mathbb{B}^\omega$.

How do we reduce the problem of enumerating a set B to that of enumerating a set A? We provide a way of turning an enumeration of A into an enumeration of B. The easiest way to do that is to define a surjective function $f: A \to B$. If $x_1, x_2, \ldots$ enumerates A, then $f(x_1), f(x_2), \ldots$ would enumerate B. In our case, we are looking for a surjective function $f: \wp(\mathbb{Z}^+) \to \mathbb{B}^\omega$.

Proof of Theorem 4.18 by reduction. Suppose that $\wp(\mathbb{Z}^+)$ were countable, and thus that there is an enumeration of it, Z_1, $Z_2, Z_3, \ldots$

Define the function $f: \wp(\mathbb{Z}^+) \to \mathbb{B}^\omega$ by letting $f(Z)$ be the sequence s_k such that $s_k(n) = 1$ iff $n \in Z$, and $s_k(n) = 0$ otherwise. This clearly defines a function, since whenever $Z \subseteq \mathbb{Z}^+$, any $n \in \mathbb{Z}^+$ either is an element of Z or isn't. For instance, the set $2\mathbb{Z}^+ = \{2, 4, 6, \ldots\}$ of positive even numbers gets mapped to the sequence $010101\ldots$, the empty set gets mapped to $0000\ldots$ and the set $\mathbb{Z}^+$ itself to $1111\ldots$.

It also is surjective: Every sequence of 0s and 1s corresponds to some set of positive integers, namely the one which has as its members those integers corresponding to the places where the sequence has 1s. More precisely, suppose $s \in \mathbb{B}^\omega$. Define $Z \subseteq \mathbb{Z}^+$ by:

$$Z = \{n \in \mathbb{Z}^+ : s(n) = 1\}$$

Then $f(Z) = s$, as can be verified by consulting the definition of f.

Now consider the list

$$f(Z_1), f(Z_2), f(Z_3), \ldots$$

Since f is surjective, every member of $\mathbb{B}^\omega$ must appear as a value of f for some argument, and so must appear on the list. This list must therefore enumerate all of $\mathbb{B}^\omega$.

So if $\wp(\mathbb{Z}^+)$ were countable, $\mathbb{B}^\omega$ would be countable. But $\mathbb{B}^\omega$ is uncountable (Theorem 4.17). Hence $\wp(\mathbb{Z}^+)$ is uncountable. □

It is easy to be confused about the direction the reduction goes in. For instance, a surjective function $g \colon \mathbb{B}^\omega \to \mathbb{B}$ does *not* establish that $\mathbb{B}$ is uncountable. (Consider $g \colon \mathbb{B}^\omega \to \mathbb{B}$ defined by $g(s) = s(1)$, the function that maps a sequence of 0's and 1's to its first element. It is surjective, because some sequences start with 0 and some start with 1. But $\mathbb{B}$ is finite.) Note also that the function f must be surjective, or otherwise the argument does not go through: $f(x_1), f(x_2), \ldots$ would then not be guaranteed to include all the elements of B. For instance,

$$h(n) = \underbrace{000\ldots 0}_{n \text{ 0's}}$$

defines a function $h \colon \mathbb{Z}^+ \to \mathbb{B}^\omega$, but $\mathbb{Z}^+$ is countable.

4.8 Equinumerosity

We have an intuitive notion of "size" of sets, which works fine for finite sets. But what about infinite sets? If we want to come up with a formal way of comparing the sizes of two sets of *any* size, it is a good idea to start by defining when sets are the same size. Here is Frege:

> If a waiter wants to be sure that he has laid exactly as many knives as plates on the table, he does not need

to count either of them, if he simply lays a knife to the right of each plate, so that every knife on the table lies to the right of some plate. The plates and knives are thus uniquely correlated to each other, and indeed through that same spatial relationship. (Frege, 1884, §70)

The insight of this passage can be brought out through a formal definition:

Definition 4.19. A is *equinumerous* with B, written $A \approx B$, iff there is a bijection $f: A \to B$.

Proposition 4.20. *Equinumerosity is an equivalence relation.*

Proof. We must show that equinumerosity is reflexive, symmetric, and transitive. Let A, B, and C be sets.
 Reflexivity. The identity map $\mathrm{Id}_A: A \to A$, where $\mathrm{Id}_A(x) = x$ for all $x \in A$, is a bijection. So $A \approx A$.
 Symmetry. Suppose $A \approx B$, i.e., there is a bijection $f: A \to B$. Since f is bijective, its inverse f^{-1} exists and is also bijective. Hence, $f^{-1}: B \to A$ is a bijection, so $B \approx A$.
 Transitivity. Suppose that $A \approx B$ and $B \approx C$, i.e., there are bijections $f: A \to B$ and $g: B \to C$. Then the composition $g \circ f: A \to C$ is bijective, so that $A \approx C$. □

Proposition 4.21. *If $A \approx B$, then A is countable if and only if B is.*

Proof. Suppose $A \approx B$, so there is some bijection $f: A \to B$, and suppose that A is countable. Then either $A = \emptyset$ or there is a surjective function $g: \mathbb{Z}^+ \to A$. If $A = \emptyset$, then $B = \emptyset$ also (otherwise there would be an element $y \in B$ but no $x \in A$ with $g(x) = y$). If, on the other hand, $g: \mathbb{Z}^+ \to A$ is surjective, then $f \circ g: \mathbb{Z}^+ \to B$ is surjective. To see this, let $y \in B$. Since f

is surjective, there is an $x \in A$ such that $f(x) = y$. Since g is surjective, there is an $n \in \mathbb{Z}^+$ such that $g(n) = x$. Hence,

$$(f \circ g)(n) = f(g(n)) = f(x) = y$$

and thus $f \circ g$ is surjective. We have that $f \circ g$ is an enumeration of B, and so B is countable.

If B is countable, we obtain that A is countable by repeating the argument with the bijection $f^{-1} \colon B \to A$ instead of f. □

4.9 Sets of Different Sizes, and Cantor's Theorem

We have offered a precise statement of the idea that two sets have the same size. We can also offer a precise statement of the idea that one set is smaller than another. Our definition of "is smaller than (or equinumerous)" will require, instead of a bijection between the sets, an injection from the first set to the second. If such a function exists, the size of the first set is less than or equal to the size of the second. Intuitively, an injection from one set to another guarantees that the range of the function has at least as many elements as the domain, since no two elements of the domain map to the same element of the range.

Definition 4.22. *A is no larger than B*, written $A \leq B$, iff there is an injection $f \colon A \to B$.

It is clear that this is a reflexive and transitive relation, but that it is not symmetric (this is left as an exercise). We can also introduce a notion, which states that one set is (strictly) smaller than another.

Definition 4.23. *A is smaller than B*, written $A < B$, iff there is an injection $f \colon A \to B$ but no bijection $g \colon A \to B$, i.e., $A \leq B$ and $A \not\approx B$.

It is clear that this relation is irreflexive and transitive. (This is left as an exercise.) Using this notation, we can say that a set A is countable iff $A \leq \mathbb{N}$, and that A is uncountable iff $\mathbb{N} < A$. This allows us to restate Theorem 4.18 as the observation that $\mathbb{Z}^+ < \wp(\mathbb{Z}^+)$. In fact, Cantor (1892) proved that this last point is *perfectly general*:

Theorem 4.24 (Cantor). *$A < \wp(A)$, for any set A.*

Proof. The map $f(x) = \{x\}$ is an injection $f: A \to \wp(A)$, since if $x \neq y$, then also $\{x\} \neq \{y\}$ by extensionality, and so $f(x) \neq f(y)$. So we have that $A \leq \wp(A)$.

We will now show that there cannot be a surjective function $g: A \to \wp(A)$, let alone a bijective one, and hence that $A \not\approx \wp(A)$. For suppose that $g: A \to \wp(A)$. Since g is total, every $x \in A$ is mapped to a subset $g(x) \subseteq A$. We can show that g cannot be surjective. To do this, we define a subset $\overline{A} \subseteq A$ which by definition cannot be in the range of g. Let

$$\overline{A} = \{x \in A : x \notin g(x)\}.$$

Since $g(x)$ is defined for all $x \in A$, $\overline{A}$ is clearly a well-defined subset of A. But, it cannot be in the range of g. Let $x \in A$ be arbitrary, we will show that $\overline{A} \neq g(x)$. If $x \in g(x)$, then it does not satisfy $x \notin g(x)$, and so by the definition of $\overline{A}$, we have $x \notin \overline{A}$. If $x \in \overline{A}$, it must satisfy the defining property of $\overline{A}$, i.e., $x \in A$ and $x \notin g(x)$. Since x was arbitrary, this shows that for each $x \in \overline{A}$, $x \in g(x)$ iff $x \notin \overline{A}$, and so $g(x) \neq \overline{A}$. In other words, $\overline{A}$ cannot be in the range of g, contradicting the assumption that g is surjective. □

It's instructive to compare the proof of Theorem 4.24 to that of Theorem 4.18. There we showed that for any list $Z_1, Z_2, \ldots,$ of subsets of $\mathbb{Z}^+$ one can construct a set $\overline{Z}$ of numbers guaranteed not to be on the list. It was guaranteed not to be on the list because, for every $n \in \mathbb{Z}^+$, $n \in Z_n$ iff $n \notin \overline{Z}$. This way, there is always some number that is an element of one of Z_n or $\overline{Z}$ but not the other. We follow the same idea here, except the indices n are now elements of A instead of $\mathbb{Z}^+$. The set $\overline{B}$ is defined so that it is different from $g(x)$ for each $x \in A$, because $x \in g(x)$ iff $x \notin \overline{B}$. Again, there is always an element of A which is an element of one of $g(x)$ and $\overline{B}$ but not the other. And just as $\overline{Z}$ therefore cannot be on the list $Z_1, Z_2, \ldots, \overline{B}$ cannot be in the range of g.

The proof is also worth comparing with the proof of Russell's Paradox, Theorem 1.29. Indeed, Cantor's Theorem was the inspiration for Russell's own paradox.

4.10 The Notion of Size, and Schröder-Bernstein

Here is an intuitive thought: if A is no larger than B and B is no larger than A, then A and B are equinumerous. To be honest, if this thought were *wrong*, then we could scarcely justify the thought that our defined notion of equinumerosity has anything to do with comparisons of "sizes" between sets! Fortunately, though, the intuitive thought is correct. This is justified by the Schröder-Bernstein Theorem.

Theorem 4.25 (Schröder-Bernstein). *If $A \leq B$ and $B \leq A$, then $A \approx B$.*

In other words, if there is an injection from A to B, and an injection from B to A, then there is a bijection from A to B.

This result, however, is really rather *difficult* to prove. Indeed, although Cantor stated the result, others proved it.[1] For now, you can (and must) take it on trust.

Fortunately, Schröder-Bernstein is *correct*, and it vindicates our thinking of the relations we defined, i.e., $A \approx B$ and $A \leq B$, as having something to do with "size". Moreover, Schröder-Bernstein is very *useful*. It can be difficult to think of a bijection between two equinumerous sets. The Schröder-Bernstein Theorem allows us

---
[1] For more on the history, see e.g., Potter (2004, pp. 165–6).

to break the comparison down into cases so we only have to think of an injection from the first to the second, and vice-versa.

Summary

The size of a set A can be measured by a natural number if the set is finite, and sizes can be compared by comparing these numbers. If sets are infinite, things are more complicated. The first level of infinity is that of **countably infinite** sets. A set A is countable if its elements can be arranged in an **enumeration**, a one-way infinite list, i.e., when there is a surjective function $f : \mathbb{Z}^+ \to A$. It is countably infinite if it is countable but not finite. Cantor's **zig-zag method** shows that the sets of pairs of elements of countably infinite sets is also countable; and this can be used to show that even the set of rational numbers $\mathbb{Q}$ is countable.

There are, however, infinite sets that are not countable: these sets are called **uncountable**. There are two ways of showing that a set is uncountable: directly, using a **diagonal argument**, or by **reduction**. To give a diagonal argument, we assume that the set A in question is countable, and use a hypothetical enumeration to define an element of A which, by the very way we define it, is guaranteed to be different from every element in the enumeration. So the enumeration can't be an enumeration of all of A after all, and we've shown that no enumeration of A can exist. A reduction shows that A is uncountable by associating every element of A with an element of some known uncountable set B in a surjective way. If this is possible, than a hypothetical enumeration of A would yield an enumeration of B. Since B is uncountable, no enumeration of A can exist.

In general, infinite sets can be compared sizewise: A and B are the same size, or **equinumerous**, if there is a bijection between them. We can also define that A is no larger than B ($A \preceq B$) if there is an injective function from A to B. By the Schröder-Bernstein Theorem, this in fact provides a sizewise order of infinite sets. Finally, **Cantor's theorem** says that for any

A, $A \prec \wp(A)$. This is a generalization of our result that $\wp(\mathbb{Z}^+)$ is uncountable, and shows that there are not just two, but infinitely many levels of infinity.

Problems

Problem 4.1. Define an enumeration of the positive squares 1, 4, 9, 16, ...

Problem 4.2. Show that if A and B are countable, so is $A \cup B$. To do this, suppose there are surjective functions $f \colon \mathbb{Z}^+ \to A$ and $g \colon \mathbb{Z}^+ \to B$, and define a surjective function $h \colon \mathbb{Z}^+ \to A \cup B$ and prove that it is surjective. Also consider the cases where A or $B = \emptyset$.

Problem 4.3. Show that if $B \subseteq A$ and A is countable, so is B. To do this, suppose there is a surjective function $f \colon \mathbb{Z}^+ \to A$. Define a surjective function $g \colon \mathbb{Z}^+ \to B$ and prove that it is surjective. What happens if $B = \emptyset$?

Problem 4.4. Show by induction on n that if $A_1, A_2, \ldots, A_n$ are all countable, so is $A_1 \cup \cdots \cup A_n$. You may assume the fact that if two sets A and B are countable, so is $A \cup B$.

Problem 4.5. According to Definition 4.4, a set A is enumerable iff $A = \emptyset$ or there is a surjective $f \colon \mathbb{Z}^+ \to A$. It is also possible to define "countable set" precisely by: a set is enumerable iff there is an injective function $g \colon A \to \mathbb{Z}^+$. Show that the definitions are equivalent, i.e., show that there is an injective function $g \colon A \to \mathbb{Z}^+$ iff either $A = \emptyset$ or there is a surjective $f \colon \mathbb{Z}^+ \to A$.

Problem 4.6. Show that $(\mathbb{Z}^+)^n$ is countable, for every $n \in \mathbb{N}$.

Problem 4.7. Show that $(\mathbb{Z}^+)^*$ is countable. You may assume problem 4.6.

Problem 4.8. Give an enumeration of the set of all non-negative rational numbers.

Problem 4.9. Show that $\mathbb{Q}$ is countable. Recall that any rational number can be written as a fraction z/m with $z \in \mathbb{Z}$, $m \in \mathbb{N}^+$.

Problem 4.10. Define an enumeration of $\mathbb{B}^*$.

Problem 4.11. Recall from your introductory logic course that each possible truth table expresses a truth function. In other words, the truth functions are all functions from $\mathbb{B}^k \to \mathbb{B}$ for some k. Prove that the set of all truth functions is enumerable.

Problem 4.12. Show that the set of all finite subsets of an arbitrary infinite countable set is countable.

Problem 4.13. A subset of $\mathbb{N}$ is said to be *cofinite* iff it is the complement of a finite set $\mathbb{N}$; that is, $A \subseteq \mathbb{N}$ is cofinite iff $\mathbb{N} \setminus A$ is finite. Let I be the set whose elements are exactly the finite and cofinite subsets of $\mathbb{N}$. Show that I is countable.

Problem 4.14. Show that the countable union of countable sets is countable. That is, whenever $A_1, A_2, \ldots$ are sets, and each A_i is countable, then the union $\bigcup_{i=1}^{\infty} A_i$ of all of them is also countable. [NB: this is hard!]

Problem 4.15. Let $f : A \times B \to \mathbb{N}$ be an arbitrary pairing function. Show that the inverse of f is an enumeration of $A \times B$.

Problem 4.16. Specify a function that encodes $\mathbb{N}^3$.

Problem 4.17. Show that $\wp(\mathbb{N})$ is uncountable by a diagonal argument.

Problem 4.18. Show that the set of functions $f : \mathbb{Z}^+ \to \mathbb{Z}^+$ is uncountable by an explicit diagonal argument. That is, show that if $f_1, f_2, \ldots,$ is a list of functions and each $f_i : \mathbb{Z}^+ \to \mathbb{Z}^+$, then there is some $\overline{f} : \mathbb{Z}^+ \to \mathbb{Z}^+$ not on this list.

4.10. THE NOTION OF SIZE, AND SCHRÖDER-BERNSTEIN

Problem 4.19. Show that if there is an injective function $g\colon B \to A$, and B is uncountable, then so is A. Do this by showing how you can use g to turn an enumeration of A into one of B.

Problem 4.20. Show that the set of all *sets of* pairs of positive integers is uncountable by a reduction argument.

Problem 4.21. Show that the set X of all functions $f\colon \mathbb{N} \to \mathbb{N}$ is uncountable by a reduction argument (Hint: give a surjective function from X to $\mathbb{B}^\omega$.)

Problem 4.22. Show that $\mathbb{N}^\omega$, the set of infinite sequences of natural numbers, is uncountable by a reduction argument.

Problem 4.23. Let P be the set of functions from the set of positive integers to the set $\{0\}$, and let Q be the set of *partial* functions from the set of positive integers to the set $\{0\}$. Show that P is countable and Q is not. (Hint: reduce the problem of enumerating $\mathbb{B}^\omega$ to enumerating Q).

Problem 4.24. Let S be the set of all surjective functions from the set of positive integers to the set $\{0,1\}$, i.e., S consists of all surjective $f\colon \mathbb{Z}^+ \to \mathbb{B}$. Show that S is uncountable.

Problem 4.25. Show that the set $\mathbb{R}$ of all real numbers is uncountable.

Problem 4.26. Show that if $A \approx C$ and $B \approx D$, and $A \cap B = C \cap D = \emptyset$, then $A \cup B \approx C \cup D$.

Problem 4.27. Show that if A is infinite and countable, then $A \approx \mathbb{N}$.

Problem 4.28. Show that there cannot be an injection $g\colon \wp(A) \to A$, for any set A. Hint: Suppose $g\colon \wp(A) \to A$ is injective. Consider $D = \{g(B) : B \subseteq A \text{ and } g(B) \notin B\}$. Let $x = g(D)$. Use the fact that g is injective to derive a contradiction.

Kurt Gödel
1906 - 1978

PART II

First-order Logic

CHAPTER 5

Introduction to First-Order Logic

5.1 First-Order Logic

You are probably familiar with first-order logic from your first introduction to formal logic.[1] You may know it as "quantificational logic" or "predicate logic." First-order logic, first of all, is a formal language. That means, it has a certain vocabulary, and its expressions are strings from this vocabulary. But not every string is permitted. There are different kinds of permitted expressions: terms, formulas, and sentences. We are mainly interested in sentences of first-order logic: they provide us with a formal analogue of sentences of English, and about them we can ask the questions a logician typically is interested in. For instance:

- Does B follow from A logically?

- Is A logically true, logically false, or contingent?

---

[1] In fact, we more or less assume you are! If you're not, you could review a more elementary textbook, such as *forall x* (Magnus et al., 2021).

- Are A and B equivalent?

These questions are primarily questions about the "meaning" of sentences of first-order logic. For instance, a philosopher would analyze the question of whether B follows logically from A as asking: is there a case where A is true but B is false (B doesn't follow from A), or does every case that makes A true also make B true (B does follow from A)? But we haven't been told yet what a "case" is—that is the job of *semantics*. The semantics of first-order logic provides a mathematically precise model of the philosopher's intuitive idea of "case," and also—and this is important—of what it is for a sentence A to be *true in* a case. We call the mathematically precise model that we will develop a structure. The relation which makes "true in" precise, is called the relation of *satisfaction*. So what we will define is "A is satisfied in M" (in symbols: $M \vDash A$) for sentences A and structures M. Once this is done, we can also give precise definitions of the other semantical terms such as "follows from" or "is logically true." These definitions will make it possible to settle, again with mathematical precision, whether, e.g., $\forall x\, (A(x) \to B(x)), \exists x\, A(x) \vDash \exists x\, B(x)$. The answer will, of course, be "yes." If you've already been trained to symbolize sentences of English in first-order logic, you will recognize this as, e.g., the symbolizations of, say, "All ants are insects, there are ants, therefore there are insects." That is obviously a valid argument, and so our mathematical model of "follows from" for our formal language should give the same answer.

Another topic you probably remember from your first introduction to formal logic is that there are *derivations*. If you have taken a first formal logic course, your instructor will have made you practice finding such derivations, perhaps even a derivation that shows that the above entailment holds. There are many different ways to give derivations: you may have done something called "natural deduction" or "truth trees," but there are many others. The purpose of derivation systems is to provide tools using which the logicians' questions above can be answered: e.g., a natural deduction derivation in which $\forall x\, (A(x) \to B(x))$ and

$\exists x\, A(x)$ are premises and $\exists x\, B(x)$ is the conclusion (last line) *verifies* that $\exists x\, B(x)$ logically follows from $\forall x\, (A(x) \to B(x))$ and $\exists x\, A(x)$.

But why is that? On the face of it, derivation systems have nothing to do with semantics: giving a formal derivation merely involves arranging symbols in certain rule-governed ways; they don't mention "cases" or "true in" at all. The connection between derivation systems and semantics has to be established by a metalogical investigation. What's needed is a mathematical proof, e.g., that a formal derivation of $\exists x\, B(x)$ from premises $\forall x\, (A(x) \to B(x))$ and $\exists x\, A(x)$ is possible, if, and only if, $\forall x\, (A(x) \to B(x))$ and $\exists x\, A(x)$ together entails $\exists x\, B(x)$. Before this can be done, however, a lot of painstaking work has to be carried out to get the definitions of syntax and semantics correct.

5.2 Syntax

We first must make precise what strings of symbols count as sentences of first-order logic. We'll do this later; for now we'll just proceed by example. The basic building blocks—the vocabulary—of first-order logic divides into two parts. The first part is the symbols we use to say specific things or to pick out specific things. We pick out things using constant symbols, and we say stuff about the things we pick out using predicate symbols. E.g, we might use a as a constant symbol to pick out a single thing, and then say something about it using the sentence $P(a)$. If you have meanings for "a" and "P" in mind, you can read $P(a)$ as a sentence of English (and you probably have done so when you first learned formal logic). Once you have such simple sentences of first-order logic, you can build more complex ones using the second part of the vocabulary: the logical symbols (connectives and quantifiers). So, for instance, we can form expressions like $(P(a) \wedge Q(b))$ or $\exists x\, P(x)$.

In order to provide the precise definitions of semantics and the rules of our derivation systems required for rigorous metalogical study, we first of all have to give a precise definition of what counts as a sentence of first-order logic. The basic idea is easy enough to understand: there are some simple sentences we can form from just predicate symbols and constant symbols, such as $P(a)$. And then from these we form more complex ones using the connectives and quantifiers. But what exactly are the rules by which we are allowed to form more complex sentences? These must be specified, otherwise we have not defined "sentence of first-order logic" precisely enough. There are a few issues. The first one is to get the right strings to count as sentences. The second one is to do this in such a way that we can give mathematical proofs about *all* sentences. Finally, we'll have to also give precise definitions of some rudimentary operations with sentences, such as "replace every x in A by b." The trouble is that the quantifiers and variables we have in first-order logic make it not entirely obvious how this should be done. E.g., should $\exists x\, P(a)$ count as a sentence? What about $\exists x\, \exists x\, P(x)$? What should the result of "replace x by b in $(P(x) \land \exists x\, P(x))$" be?

5.3 Formulas

Here is the approach we will use to rigorously specify sentences of first-order logic and to deal with the issues arising from the use of variables. We first define a *different* set of expressions: formulas. Once we've done that, we can consider the role variables play in them—and on the basis of some other ideas, namely those of "free" and "bound" variables, we can define what a sentence is (namely, a formula without free variables). We do this not just because it makes the definition of "sentence" more manageable, but also because it will be crucial to the way we define the semantic notion of satisfaction.

Let's define "formula" for a simple first-order language, one containing only a single predicate symbol P and a single constant symbol a, and only the logical symbols ¬, ∧, and ∃. Our full definitions will be much more general: we'll allow infinitely many predicate symbols and constant symbols. In fact, we will also consider function symbols which can be combined with constant symbols and variables to form "terms." For now, a and the variables will be our only terms. We do need infinitely many variables. We'll officially use the symbols $v_0, v_1, \ldots$, as variables.

Definition 5.1. The set of *formulas* Frm is defined as follows:

1. $P(a)$ and $P(v_i)$ are formulas ($i \in \mathbb{N}$).

2. If A is a formula, then $\neg A$ is formula.

3. If A and B are formulas, then $(A \wedge B)$ is a formula.

4. If A is a formula and x is a variable, then $\exists x\, A$ is a formula.

5. Nothing else is a formula.

(1) tell us that $P(a)$ and $P(v_i)$ are formulas, for any $i \in \mathbb{N}$. These are the so-called *atomic* formulas. They give us something to start from. The other clauses give us ways of forming new formulas from ones we have already formed. So for instance, we get that $\neg P(v_2)$ is a formula, since $P(v_2)$ is already a formula by (1), and then we get that $\exists v_2\, \neg P(v_2)$ is another formula, and so on. (5) tells us that *only* strings we can form in this way count as formulas. In particular, $\exists v_0\, P(a)$ and $\exists v_0\, \exists v_0\, P(a)$ *do* count as formulas, and $(\neg P(a))$ does not.

This way of defining formulas is called an *inductive definition*, and it allows us to prove things about formulas using a version of proof by induction called *structural induction*. These are discussed in a general way in appendix B.4 and appendix B.5, which you should review before delving into the proofs later on. Basically, the idea is that if you want to give a proof that something is true for all formulas you show first that it is true for the atomic formulas, and then that *if* it's true for any formula A (and B), it's *also* true for $\neg A$, $(A \wedge B)$, and $\exists x\, A$. For instance, this proves that it's true for $\exists v_2\, \neg P(v_2)$: from the first part you know that it's true for the atomic formula $P(v_2)$. Then you get that it's true for $\neg P(v_2)$ by the second part, and then again that it's true for $\exists v_2 \neg P(v_2)$ itself. Since all formulas are inductively generated from atomic formulas, this works for any of them.

5.4 Satisfaction

We can already skip ahead to the semantics of first-order logic once we know what formulas are: here, the basic definition is that of a structure. For our simple language, a structure M has just three components: a non-empty set $|M|$ called the *domain*, what a picks out in M, and what P is true of in M. The object picked out by a is denoted a^M and the set of things P is true of by P^M. A structure M consists of just these three things: $|M|$, $a^M \in |M|$ and $P^M \subseteq |M|$. The general case will be more complicated, since there will be many predicate symbols and constant symbols, the constant symbols can have more than one place, and there will also be function symbols.

This is enough to give a definition of satisfaction for formulas that don't contain variables. The idea is to give an inductive definition that mirrors the way we have defined formulas. We specify when an atomic formula is satisfied in M, and then when, e.g., $\neg A$ is satisfied in M on the basis of whether or not A is satisfied in M. E.g., we could define:

1. $P(a)$ is satisfied in M iff $a^M \in P^M$.

2. $\neg A$ is satisfied in M iff A is not satisfied in M.

3. $(A \wedge B)$ is satisfied in M iff A is satisfied in M, and B is satisfied in M as well.

Let's say that $|M| = \{0,1,2\}$, $a^M = 1$, and $P^M = \{1,2\}$. This definition would tell us that $P(a)$ is satisfied in M (since $a^M = 1 \in \{1,2\} = P^M$). It tells us further that $\neg P(a)$ is not satisfied in M, and that in turn that $\neg\neg P(a)$ is and $(\neg P(a) \wedge P(a))$ is not satisfied, and so on.

The trouble comes when we want to give a definition for the quantifiers: we'd like to say something like, "$\exists v_0\, P(v_0)$ is satisfied iff $P(v_0)$ is satisfied." But the structure M doesn't tell us what to do about variables. What we actually want to say is that $P(v_0)$ is satisfied *for some value of* v_0. To make this precise we need a way to assign elements of $|M|$ not just to a but also to v_0. To this end, we introduce variable *assignments*. A variable assignment is simply a function s that maps variables to elements of $|M|$ (in our example, to one of 1, 2, or 3). Since we don't know beforehand which variables might appear in a formula we can't limit which variables s assigns values to. The simple solution is to require that s assigns values to *all* variables v_0, v_1, ... We'll just use only the ones we need.

Instead of defining satisfaction of formulas just relative to a structure, we'll define it relative to a structure M *and* a variable assignment s, and write $M, s \vDash A$ for short. Our definition will now include an additional clause to deal with atomic formulas containing variables:

1. $M, s \vDash P(a)$ iff $a^M \in P^M$.

2. $M, s \vDash P(v_i)$ iff $s(v_i) \in P^M$.

3. $M, s \vDash \neg A$ iff not $M, s \vDash A$.

4. $M, s \vDash (A \wedge B)$ iff $M, s \vDash A$ and $M, s \vDash B$.

Ok, this solves one problem: we can now say when M satisfies $P(v_0)$ for the value $s(v_0)$. To get the definition right for $\exists v_0\, P(v_0)$ we have to do one more thing: We want to have that $M, s \vDash \exists v_0\, P(v_0)$ iff $M, s' \vDash P(v_0)$ for *some* way s' of assigning a value to v_0. But the value assigned to v_0 does not necessarily have to be the value that $s(v_0)$ picks out. We'll introduce a notation for that: if $m \in |M|$, then we let $s[m/v_0]$ be the assignment that is just like s (for all variables other than v_0), except to v_0 it assigns m. Now our definition can be:

5. $M, s \vDash \exists v_i\, A$ iff $M, s[m/v_i] \vDash A$ for some $m \in |M|$.

Does it work out? Let's say we let $s(v_i) = 0$ for all $i \in \mathbb{N}$. $M, s \vDash \exists v_0\, P(v_0)$ iff there is an $m \in |M|$ so that $M, s[m/v_0] \vDash P(v_0)$. And there is: we can choose $m = 1$ or $m = 2$. Note that this is true even if the value $s(v_0)$ assigned to v_0 by s itself—in this case, 0—doesn't do the job. We have $M, s[1/v_0] \vDash P(v_0)$ but not $M, s \vDash P(v_0)$.

If this looks confusing and cumbersome: it is. But the added complexity is required to give a precise, inductive definition of satisfaction for all formulas, and we need something like it to precisely define the semantic notions. There are other ways of doing it, but they are all equally (in)elegant.

5.5 Sentences

Ok, now we have a (sketch of a) definition of satisfaction ("true in") for structures and formulas. But it needs this additional bit—a variable assignment—and what we wanted is a definition of sentences. How do we get rid of assignments, and what are sentences?

You probably remember a discussion in your first introduction to formal logic about the relation between variables and quantifiers. A quantifier is always followed by a variable, and then in the part of the sentence to which that quantifier applies (its "scope"), we understand that the variable is "bound" by that quantifier. In formulas it was not required that every variable has a matching quantifier, and variables without matching quantifiers are "free" or "unbound." We will take sentences to be all those formulas that have no free variables.

Again, the intuitive idea of when an occurrence of a variable in a formula A is bound, which quantifier binds it, and when it is free, is not difficult to get. You may have learned a method for testing this, perhaps involving counting parentheses. We have to insist on a precise definition—and because we have defined formulas by induction, we can give a definition of the free and bound occurrences of a variable x in a formula A also by induction. E.g., it might look like this for our simplified language:

1. If A is atomic, all occurrences of x in it are free (that is, the occurrence of x in $P(x)$ is free).

2. If A is of the form $\neg B$, then an occurrence of x in $\neg B$ is free iff the corresponding occurrence of x is free in B (that is, the free occurrences of variables in B are exactly the corresponding occurrences in $\neg B$).

3. If A is of the form $(B \wedge C)$, then an occurrence of x in $(B \wedge C)$ is free iff the corresponding occurrence of x is free in B or in C.

4. If A is of the form $\exists x\, B$, then no occurrence of x in A is free; if it is of the form $\exists y\, B$ where y is a different variable than x, then an occurrence of x in $\exists y\, B$ is free iff the corresponding occurrence of x is free in B.

Once we have a precise definition of free and bound occurrences of variables, we can simply say: a sentence is any formula without free occurrences of variables.

5.6 Semantic Notions

We mentioned above that when we consider whether $M, s \vDash A$ holds, we (for convenience) let s assign values to all variables, but only the values it assigns to variables in A are used. In fact, it's only the values of *free* variables in A that matter. Of course, because we're careful, we are going to prove this fact. Since sentences have no free variables, s doesn't matter at all when it comes to whether or not they are satisfied in a structure. So, when A is a sentence we can define $M \vDash A$ to mean "$M, s \vDash A$ for all s," which as it happens is true iff $M, s \vDash A$ for at least one s. We need to introduce variable assignments to get a working definition of satisfaction for formulas, but for sentences, satisfaction is independent of the variable assignments.

5.7. SUBSTITUTION

Once we have a definition of "$M \vDash A$," we know what "case" and "true in" mean as far as sentences of first-order logic are concerned. On the basis of the definition of $M \vDash A$ for sentences we can then define the basic semantic notions of validity, entailment, and satisfiability. A sentence is valid, $\vDash A$, if every structure satisfies it. It is entailed by a set of sentences, $\Gamma \vDash A$, if every structure that satisfies all the sentences in Γ also satisfies A. And a set of sentences is satisfiable if some structure satisfies all sentences in it at the same time.

Because formulas are inductively defined, and satisfaction is in turn defined by induction on the structure of formulas, we can use induction to prove properties of our semantics and to relate the semantic notions defined. We'll collect and prove some of these properties, partly because they are individually interesting, but mainly because many of them will come in handy when we go on to investigate the relation between semantics and derivation systems. In order to do so, we'll also have to define (precisely, i.e., by induction) some syntactic notions and operations we haven't mentioned yet.

5.7 Substitution

We'll discuss an example to illustrate how things hang together, and how the development of syntax and semantics lays the foundation for our more advanced investigations later. Our derivation systems should let us derive $P(a)$ from $\forall v_0 \, P(v_0)$. Maybe we even want to state this as a rule of inference. However, to do so, we must be able to state it in the most general terms: not just for P, a, and v_0, but for any formula A, and term t, and variable x. (Recall that constant symbols are terms, but we'll consider also more complicated terms built from constant symbols and function symbols.) So we want to be able to say something like, "whenever you have derived $\forall x \, A(x)$ you are justified in inferring $A(t)$—the result of removing $\forall x$ and replacing x by t." But what exactly does "replacing x by t" mean? What is the relation between $A(x)$

and $A(t)$? Does this always work?

To make this precise, we define the operation of *substitution*. Substitution is actually tricky, because we can't just replace all x's in A by t, and not every t can be substituted for any x. We'll deal with this, again, using inductive definitions. But once this is done, specifying an inference rule as "infer $A(t)$ from $\forall x\, A(x)$" becomes a precise definition. Moreover, we'll be able to show that this is a good inference rule in the sense that $\forall x\, A(x)$ entails $A(t)$. But to prove this, we have to again prove something that may at first glance prompt you to ask "why are we doing this?" That $\forall x\, A(x)$ entails $A(t)$ relies on the fact that whether or not $M \vDash A(t)$ holds depends only on the value of the term t, i.e., if we let m be whatever element of $|M|$ is picked out by t, then $M, s \vDash A(t)$ iff $M, s[m/x] \vDash A(x)$. This holds even when t contains variables, but we'll have to be careful with how exactly we state the result.

5.8 Models and Theories

Once we've defined the syntax and semantics of first-order logic, we can get to work investigating the properties of structures, of the semantic notions, we can define derivation systems, and investigate those. For a set of sentences, we can ask: what structures make all the sentences in that set true? Given a set of sentences Γ, a structure M that satisfies them is called a *model of* Γ. We might start from Γ and try find its models—what do they look like? How big or small do they have to be? But we might also start with a single structure or collection of structures and ask: what sentences are true in them? Are there sentences that *characterize* these structures in the sense that they, and only they, are true in them? These kinds of questions are the domain of *model theory*. They also underlie the *axiomatic method*: describing a collection of structures by a set of sentences, the axioms of a theory. This is made possible by the observation that exactly those sentences entailed in first-order logic by the axioms are true in all models of the axioms.

As a very simple example, consider preorders. A preorder is a relation R on some set A which is both reflexive and transitive. A set A with a two-place relation $R \subseteq A \times A$ on it is exactly what we would need to give a structure for a first-order language with a single two-place relation symbol P: we would set $|M| = A$ and $P^M = R$. Since R is a preorder, it is reflexive and transitive, and we can find a set Γ of sentences of first-order logic that say this:

$$\forall v_0\, P(v_0, v_0)$$
$$\forall v_0 \forall v_1 \forall v_2\, ((P(v_0, v_1) \land P(v_1, v_2)) \to P(v_0, v_2))$$

These sentences are just the symbolizations of "for any x, Rxx" (R is reflexive) and "whenever Rxy and Ryz then also Rxz" (R is transitive). We see that a structure M is a model of these two sentences Γ iff R (i.e., P^M), is a preorder on A (i.e., $|M|$). In other words, the models of Γ are exactly the preorders. Any property of all preorders that can be expressed in the first-order language with just P as predicate symbol (like reflexivity and transitivity above), is entailed by the two sentences in Γ and vice versa. So anything we can prove about models of Γ we have proved about all preorders.

For any particular theory and class of models (such as Γ and all preorders), there will be interesting questions about what can be expressed in the corresponding first-order language, and what cannot be expressed. There are some properties of structures that are interesting for all languages and classes of models, namely those concerning the size of the domain. One can always express, for instance, that the domain contains exactly n elements, for any $n \in \mathbb{Z}^+$. One can also express, using a set of infinitely many sentences, that the domain is infinite. But one cannot express that the domain is finite, or that the domain is uncountable. These results about the limitations of first-order languages are consequences of the compactness and Löwenheim-Skolem theorems.

5.9 Soundness and Completeness

We'll also introduce derivation systems for first-order logic. There are many derivation systems that logicians have developed, but they all define the same derivability relation between sentences. We say that Γ *derives* A, $\Gamma \vdash A$, if there is a derivation of a certain precisely defined sort. Derivations are always finite arrangements of symbols—perhaps a list of sentences, or some more complicated structure. The purpose of derivation systems is to provide a tool to determine if a sentence is entailed by some set Γ. In order to serve that purpose, it must be true that $\Gamma \vDash A$ if, and only if, $\Gamma \vdash A$.

If $\Gamma \vdash A$ but not $\Gamma \vDash A$, our derivation system would be too strong, prove too much. The property that if $\Gamma \vdash A$ then $\Gamma \vDash A$ is called *soundness*, and it is a minimal requirement on any good derivation system. On the other hand, if $\Gamma \vDash A$ but not $\Gamma \vdash A$, then our derivation system is too weak, it doesn't prove enough. The property that if $\Gamma \vDash A$ then $\Gamma \vdash A$ is called *completeness*. Soundness is usually relatively easy to prove (by induction on the structure of derivations, which are inductively defined). Completeness is harder to prove.

Soundness and completeness have a number of important consequences. If a set of sentences Γ derives a contradiction (such as $A \wedge \neg A$) it is called *inconsistent*. Inconsistent Γ's cannot have any models, they are unsatisfiable. From completeness the converse follows: any Γ that is not inconsistent—or, as we will say, *consistent*—has a model. In fact, this is equivalent to completeness, and is the form of completeness we will actually prove. It is a deep and perhaps surprising result: just because you cannot prove $A \wedge \neg A$ from Γ guarantees that there is a structure that is as Γ describes it. So completeness gives an answer to the question: which sets of sentences have models? Answer: all and only consistent sets do.

The soundness and completeness theorems have two important consequences: the compactness and the Löwenheim-Skolem theorem. These are important results in the theory of models,

5.9. SOUNDNESS AND COMPLETENESS

and can be used to establish many interesting results. We've already mentioned two: first-order logic cannot express that the domain of a structure is finite or that it is uncountable.

Historically, all of this—how to define syntax and semantics of first-order logic, how to define good derivation systems, how to prove that they are sound and complete, getting clear about what can and cannot be expressed in first-order languages—took a long time to figure out and get right. We now know how to do it, but going through all the details can still be confusing and tedious. But it's also important, because the methods developed here for the formal language of first-order logic are applied all over the place in logic, computer science, and linguistics. So working through the details pays off in the long run.

CHAPTER 6
Syntax of First-Order Logic

6.1 Introduction

In order to develop the theory and metatheory of first-order logic, we must first define the syntax and semantics of its expressions. The expressions of first-order logic are terms and formulas. Terms are formed from variables, constant symbols, and function symbols. Formulas, in turn, are formed from predicate symbols together with terms (these form the smallest, "atomic" formulas), and then from atomic formulas we can form more complex ones using logical connectives and quantifiers. There are many different ways to set down the formation rules; we give just one possible one. Other systems will chose different symbols, will select different sets of connectives as primitive, will use parentheses differently (or even not at all, as in the case of so-called Polish notation). What all approaches have in common, though, is that the formation rules define the set of terms and formulas *inductively*. If done properly, every expression can result essentially in only one way according to the formation rules. The inductive definition resulting in expressions that are *uniquely readable* means we can give meanings to these expressions using the same method—inductive definition.

6.2 First-Order Languages

Expressions of first-order logic are built up from a basic vocabulary containing *variables, constant symbols, predicate symbols* and sometimes *function symbols*. From them, together with logical connectives, quantifiers, and punctuation symbols such as parentheses and commas, *terms* and *formulas* are formed.

Informally, predicate symbols are names for properties and relations, constant symbols are names for individual objects, and function symbols are names for mappings. These, except for the identity predicate =, are the *non-logical symbols* and together make up a language. Any first-order language $\mathscr{L}$ is determined by its non-logical symbols. In the most general case, $\mathscr{L}$ contains infinitely many symbols of each kind.

In the general case, we make use of the following symbols in first-order logic:

1. Logical symbols

    a) Logical connectives: $\neg$ (negation), $\wedge$ (conjunction), $\vee$ (disjunction), $\rightarrow$ (conditional), $\forall$ (universal quantifier), $\exists$ (existential quantifier).

    b) The propositional constant for falsity $\bot$.

    c) The two-place identity predicate =.

    d) A countably infinite set of variables: $v_0, v_1, v_2, \ldots$

2. Non-logical symbols, making up the *standard language* of first-order logic

    a) A countably infinite set of n-place predicate symbols for each $n > 0$: $A^n_0, A^n_1, A^n_2, \ldots$

b) A countably infinite set of constant symbols: c_0, c_1, c_2, ...

c) A countably infinite set of n-place function symbols for each $n > 0$: f_0^n, f_1^n, f_2^n, ...

3. Punctuation marks: (,), and the comma.

Most of our definitions and results will be formulated for the full standard language of first-order logic. However, depending on the application, we may also restrict the language to only a few predicate symbols, constant symbols, and function symbols.

Example 6.1. The language $\mathscr{L}_A$ of arithmetic contains a single two-place predicate symbol <, a single constant symbol 0, one one-place function symbol ′, and two two-place function symbols + and ×.

Example 6.2. The language of set theory $\mathscr{L}_Z$ contains only the single two-place predicate symbol ∈.

Example 6.3. The language of orders $\mathscr{L}_\leq$ contains only the two-place predicate symbol ≤.

Again, these are conventions: officially, these are just aliases, e.g., <, ∈, and ≤ are aliases for A_0^2, 0 for c_0, ′ for f_0^1, + for f_0^2, × for f_1^2.

In addition to the primitive connectives and quantifiers introduced above, we also use the following *defined* symbols: ↔ (biconditional), truth ⊤

A defined symbol is not officially part of the language, but is introduced as an informal abbreviation: it allows us to abbreviate formulas which would, if we only used primitive symbols, get quite long. This is obviously an advantage. The bigger advantage, however, is that proofs become shorter. If a symbol is primitive, it has to be treated separately in proofs. The more primitive symbols, therefore, the longer our proofs.

You may be familiar with different terminology and symbols than the ones we use above. Logic texts (and teachers) commonly use either ∼, ¬, and ! for "negation", ∧, ·, and & for "conjunction". Commonly used symbols for the "conditional" or "implication" are →, ⇒, and ⊃. Symbols for "biconditional," "bi-implication," or "(material) equivalence" are ↔, ⇔, and ≡. The ⊥ symbol is variously called "falsity," "falsum,", "absurdity,", or "bottom." The ⊤ symbol is variously called "truth," "verum,", or "top."

It is conventional to use lower case letters (e.g., a, b, c) from the beginning of the Latin alphabet for constant symbols (sometimes called names), and lower case letters from the end (e.g., x, y, z) for variables. Quantifiers combine with variables, e.g., x; notational variations include $\forall x$, $(\forall x)$, (x), Πx, $\bigwedge_x$ for the universal quantifier and $\exists x$, $(\exists x)$, (Ex), Σx, $\bigvee_x$ for the existential quantifier.

We might treat all the propositional operators and both quantifiers as primitive symbols of the language. We might instead choose a smaller stock of primitive symbols and treat the other logical operators as defined. "Truth functionally complete" sets of Boolean operators include $\{\neg, \vee\}$, $\{\neg, \wedge\}$, and $\{\neg, \rightarrow\}$—these can be combined with either quantifier for an expressively complete first-order language.

You may be familiar with two other logical operators: the Sheffer stroke | (named after Henry Sheffer), and Peirce's arrow ↓, also known as Quine's dagger. When given their usual readings of "nand" and "nor" (respectively), these operators are truth functionally complete by themselves.

6.3 Terms and Formulas

Once a first-order language $\mathscr{L}$ is given, we can define expressions built up from the basic vocabulary of $\mathscr{L}$. These include in particular *terms* and *formulas*.

Definition 6.4 (Terms). The set of *terms* $\text{Trm}(\mathcal{L})$ of $\mathcal{L}$ is defined inductively by:

1. Every variable is a term.

2. Every constant symbol of $\mathcal{L}$ is a term.

3. If f is an n-place function symbol and $t_1, \ldots, t_n$ are terms, then $f(t_1, \ldots, t_n)$ is a term.

4. Nothing else is a term.

A term containing no variables is a *closed term*.

The constant symbols appear in our specification of the language and the terms as a separate category of symbols, but they could instead have been included as zero-place function symbols. We could then do without the second clause in the definition of terms. We just have to understand $f(t_1, \ldots, t_n)$ as just f by itself if $n = 0$.

Definition 6.5 (Formula). The set of *formulas* $\text{Frm}(\mathcal{L})$ of the language $\mathcal{L}$ is defined inductively as follows:

1. $\bot$ is an atomic formula.

2. If R is an n-place predicate symbol of $\mathcal{L}$ and $t_1, \ldots, t_n$ are terms of $\mathcal{L}$, then $R(t_1, \ldots, t_n)$ is an atomic formula.

3. If t_1 and t_2 are terms of $\mathcal{L}$, then $=(t_1, t_2)$ is an atomic formula.

4. If A is a formula, then $\neg A$ is formula.

5. If A and B are formulas, then $(A \wedge B)$ is a formula.

6. If A and B are formulas, then $(A \vee B)$ is a formula.

7. If A and B are formulas, then $(A \to B)$ is a formula.

6.3. TERMS AND FORMULAS

8. If A is a formula and x is a variable, then $\forall x\, A$ is a formula.

9. If A is a formula and x is a variable, then $\exists x\, A$ is a formula.

10. Nothing else is a formula.

The definitions of the set of terms and that of formulas are *inductive definitions*. Essentially, we construct the set of formulas in infinitely many stages. In the initial stage, we pronounce all atomic formulas to be formulas; this corresponds to the first few cases of the definition, i.e., the cases for $\bot$, $R(t_1,\ldots,t_n)$ and $=(t_1,t_2)$. "Atomic formula" thus means any formula of this form.

The other cases of the definition give rules for constructing new formulas out of formulas already constructed. At the second stage, we can use them to construct formulas out of atomic formulas. At the third stage, we construct new formulas from the atomic formulas and those obtained in the second stage, and so on. A formula is anything that is eventually constructed at such a stage, and nothing else.

By convention, we write $=$ between its arguments and leave out the parentheses: $t_1 = t_2$ is an abbreviation for $=(t_1,t_2)$. Moreover, $\neg=(t_1,t_2)$ is abbreviated as $t_1 \neq t_2$. When writing a formula $(B * C)$ constructed from B, C using a two-place connective $*$, we will often leave out the outermost pair of parentheses and write simply $B * C$.

Some logic texts require that the variable x must occur in A in order for $\exists x\, A$ and $\forall x\, A$ to count as formulas. Nothing bad happens if you don't require this, and it makes things easier.

Definition 6.6. Formulas constructed using the defined operators are to be understood as follows:

1. $\top$ abbreviates $\neg\bot$.

2. $A \leftrightarrow B$ abbreviates $(A \to B) \land (B \to A)$.

If we work in a language for a specific application, we will often write two-place predicate symbols and function symbols between the respective terms, e.g., $t_1 < t_2$ and $(t_1 + t_2)$ in the language of arithmetic and $t_1 \in t_2$ in the language of set theory. The successor function in the language of arithmetic is even written conventionally *after* its argument: t'. Officially, however, these are just conventional abbreviations for $A_0^2(t_1, t_2)$, $f_0^2(t_1, t_2)$, $A_0^2(t_1, t_2)$ and $f_0^1(t)$, respectively.

Definition 6.7 (Syntactic identity). The symbol $\equiv$ expresses syntactic identity between strings of symbols, i.e., $A \equiv B$ iff A and B are strings of symbols of the same length and which contain the same symbol in each place.

The $\equiv$ symbol may be flanked by strings obtained by concatenation, e.g., $A \equiv (B \lor C)$ means: the string of symbols A is the same string as the one obtained by concatenating an opening parenthesis, the string B, the $\lor$ symbol, the string C, and a closing parenthesis, in this order. If this is the case, then we know that the first symbol of A is an opening parenthesis, A contains B as a substring (starting at the second symbol), that substring is followed by $\lor$, etc.

6.4 Unique Readability

The way we defined formulas guarantees that every formula has a *unique reading*, i.e., there is essentially only one way of constructing it according to our formation rules for formulas and only one way of "interpreting" it. If this were not so, we would have ambiguous formulas, i.e., formulas that have more than one reading or intepretation—and that is clearly something we want to avoid. But more importantly, without this property, most of the definitions and proofs we are going to give will not go through.

Perhaps the best way to make this clear is to see what would happen if we had given bad rules for forming formulas that would

6.4. UNIQUE READABILITY

not guarantee unique readability. For instance, we could have forgotten the parentheses in the formation rules for connectives, e.g., we might have allowed this:

If A and B are formulas, then so is $A \to B$.

Starting from an atomic formula D, this would allow us to form $D \to D$. From this, together with D, we would get $D \to D \to D$. But there are two ways to do this:

1. We take D to be A and $D \to D$ to be B.

2. We take A to be $D \to D$ and B is D.

Correspondingly, there are two ways to "read" the formula $D \to D \to D$. It is of the form $B \to C$ where B is D and C is $D \to D$, but *it is also* of the form $B \to C$ with B being $D \to D$ and C being D.

If this happens, our definitions will not always work. For instance, when we define the main operator of a formula, we say: in a formula of the form $B \to C$, the main operator is the indicated occurrence of $\to$. But if we can match the formula $D \to D \to D$ with $B \to C$ in the two different ways mentioned above, then in one case we get the first occurrence of $\to$ as the main operator, and in the second case the second occurrence. But we intend the main operator to be a *function* of the formula, i.e., every formula must have exactly one main operator occurrence.

Lemma 6.8. *The number of left and right parentheses in a formula A are equal.*

Proof. We prove this by induction on the way A is constructed. This requires two things: (a) We have to prove first that all atomic formulas have the property in question (the induction basis). (b) Then we have to prove that when we construct new formulas out of given formulas, the new formulas have the property provided the old ones do.

Let $l(A)$ be the number of left parentheses, and $r(A)$ the number of right parentheses in A, and $l(t)$ and $r(t)$ similarly the number of left and right parentheses in a term t. We leave the proof that for any term t, $l(t) = r(t)$ as an exercise.

1. $A \equiv \bot$: A has 0 left and 0 right parentheses.

2. $A \equiv R(t_1, \ldots, t_n)$: $l(A) = 1 + l(t_1) + \cdots + l(t_n) = 1 + r(t_1) + \cdots + r(t_n) = r(A)$. Here we make use of the fact, left as an exercise, that $l(t) = r(t)$ for any term t.

3. $A \equiv t_1 = t_2$: $l(A) = l(t_1) + l(t_2) = r(t_1) + r(t_2) = r(A)$.

4. $A \equiv \neg B$: By induction hypothesis, $l(B) = r(B)$. Thus $l(A) = l(B) = r(B) = r(A)$.

5. $A \equiv (B * C)$: By induction hypothesis, $l(B) = r(B)$ and $l(C) = r(C)$. Thus $l(A) = 1 + l(B) + l(C) = 1 + r(B) + r(C) = r(A)$.

6. $A \equiv \forall x\, B$: By induction hypothesis, $l(B) = r(B)$. Thus, $l(A) = l(B) = r(B) = r(A)$.

7. $A \equiv \exists x\, B$: Similarly. □

Definition 6.9 (Proper prefix). A string of symbols B is a *proper prefix* of a string of symbols A if concatenating B and a non-empty string of symbols yields A.

Lemma 6.10. *If A is a formula, and B is a proper prefix of A, then B is not a formula.*

Proof. Exercise. □

6.4. UNIQUE READABILITY

Proposition 6.11. *If A is an atomic formula, then it satisfies one, and only one of the following conditions.*

1. $A \equiv \bot$.

2. $A \equiv R(t_1, \ldots, t_n)$ *where R is an n-place predicate symbol, $t_1, \ldots, t_n$ are terms, and each of $R, t_1, \ldots, t_n$ is uniquely determined.*

3. $A \equiv t_1 = t_2$ *where t_1 and t_2 are uniquely determined terms.*

Proof. Exercise. □

Proposition 6.12 (Unique Readability). *Every formula satisfies one, and only one of the following conditions.*

1. *A is atomic.*

2. *A is of the form $\neg B$.*

3. *A is of the form $(B \land C)$.*

4. *A is of the form $(B \lor C)$.*

5. *A is of the form $(B \to C)$.*

6. *A is of the form $\forall x\, B$.*

7. *A is of the form $\exists x\, B$.*

Moreover, in each case B, or B and C, are uniquely determined. This means that, e.g., there are no different pairs B, C and B', C' so that A is both of the form $(B \to C)$ and $(B' \to C')$.

Proof. The formation rules require that if a formula is not atomic, it must start with an opening parenthesis (, ¬, or with a quantifier. On the other hand, every formula that starts with one of the following symbols must be atomic: a predicate symbol, a function symbol, a constant symbol, $\bot$.

So we really only have to show that if A is of the form $(B * C)$ and also of the form $(B' *' C')$, then $B \equiv B'$, $C \equiv C'$, and $* = *'$.

So suppose both $A \equiv (B * C)$ and $A \equiv (B' *' C')$. Then either $B \equiv B'$ or not. If it is, clearly $* = *'$ and $C \equiv C'$, since they then are substrings of A that begin in the same place and are of the same length. The other case is $B \not\equiv B'$. Since B and B' are both substrings of A that begin at the same place, one must be a proper prefix of the other. But this is impossible by Lemma 6.10. □

6.5 Main operator of a Formula

It is often useful to talk about the last operator used in constructing a formula A. This operator is called the *main operator* of A. Intuitively, it is the "outermost" operator of A. For example, the main operator of $\neg A$ is $\neg$, the main operator of $(A \vee B)$ is $\vee$, etc.

> **Definition 6.13 (Main operator).** The *main operator* of a formula A is defined as follows:
>
> 1. A is atomic: A has no main operator.
>
> 2. $A \equiv \neg B$: the main operator of A is $\neg$.
>
> 3. $A \equiv (B \wedge C)$: the main operator of A is $\wedge$.
>
> 4. $A \equiv (B \vee C)$: the main operator of A is $\vee$.
>
> 5. $A \equiv (B \to C)$: the main operator of A is $\to$.
>
> 6. $A \equiv \forall x\, B$: the main operator of A is $\forall$.
>
> 7. $A \equiv \exists x\, B$: the main operator of A is $\exists$.

In each case, we intend the specific indicated *occurrence* of the main operator in the formula. For instance, since the formula $((D \to E) \to (E \to D))$ is of the form $(B \to C)$ where B is $(D \to E)$ and C is $(E \to D)$, the second occurrence of $\to$ is the main operator.

This is a *recursive* definition of a function which maps all non-atomic formulas to their main operator occurrence. Because of the way formulas are defined inductively, every formula A satisfies one of the cases in Definition 6.13. This guarantees that for each non-atomic formula A a main operator exists. Because each formula satisfies only one of these conditions, and because the smaller formulas from which A is constructed are uniquely determined in each case, the main operator occurrence of A is unique, and so we have defined a function.

We call formulas by the following names depending on which symbol their main operator is:

Main operator	Type of formula	Example
none	atomic (formula)	$\bot$, $R(t_1,\ldots,t_n)$, $t_1 = t_2$
$\neg$	negation	$\neg A$
$\wedge$	conjunction	$(A \wedge B)$
$\vee$	disjunction	$(A \vee B)$
$\to$	conditional	$(A \to B)$
$\forall$	universal (formula)	$\forall x\, A$
$\exists$	existential (formula)	$\exists x\, A$

6.6 Subformulas

It is often useful to talk about the formulas that "make up" a given formula. We call these its *subformulas*. Any formula counts as a subformula of itself; a subformula of A other than A itself is a *proper subformula*.

Definition 6.14 (Immediate Subformula). If A is a formula, the *immediate subformulas* of A are defined inductively as follows:

1. Atomic formulas have no immediate subformulas.

2. $A \equiv \neg B$: The only immediate subformula of A is B.

3. $A \equiv (B * C)$: The immediate subformulas of A are B and

C (∗ is any one of the two-place connectives).

4. $A \equiv \forall x\, B$: The only immediate subformula of A is B.

5. $A \equiv \exists x\, B$: The only immediate subformula of A is B.

Definition 6.15 (Proper Subformula). If A is a formula, the *proper subformulas* of A are recursively as follows:

1. Atomic formulas have no proper subformulas.

2. $A \equiv \neg B$: The proper subformulas of A are B together with all proper subformulas of B.

3. $A \equiv (B * C)$: The proper subformulas of A are B, C, together with all proper subformulas of B and those of C.

4. $A \equiv \forall x\, B$: The proper subformulas of A are B together with all proper subformulas of B.

5. $A \equiv \exists x\, B$: The proper subformulas of A are B together with all proper subformulas of B.

Definition 6.16 (Subformula). The subformulas of A are A itself together with all its proper subformulas.

Note the subtle difference in how we have defined immediate subformulas and proper subformulas. In the first case, we have directly defined the immediate subformulas of a formula A for each possible form of A. It is an explicit definition by cases, and the cases mirror the inductive definition of the set of formulas. In the second case, we have also mirrored the way the set of all formulas is defined, but in each case we have also included the proper subformulas of the smaller formulas B, C in addition to these formulas themselves. This makes the definition *recursive*. In general, a definition of a function on an inductively defined set (in our case, formulas) is recursive if the cases in the definition of the function make use of the function itself. To be well defined, we must make sure, however, that we only ever use the values of the function for arguments that come "before" the one we are defining—in our case, when defining "proper subformula" for $(B * C)$ we only use the proper subformulas of the "earlier" formulas B and C.

6.7 Free Variables and Sentences

Definition 6.17 (Free occurrences of a variable). The *free* occurrences of a variable in a formula are defined inductively as follows:

1. A is atomic: all variable occurrences in A are free.

2. $A \equiv \neg B$: the free variable occurrences of A are exactly those of B.

3. $A \equiv (B * C)$: the free variable occurrences of A are those in B together with those in C.

4. $A \equiv \forall x\, B$: the free variable occurrences in A are all of those in B except for occurrences of x.

5. $A \equiv \exists x\, B$: the free variable occurrences in A are all of those in B except for occurrences of x.

Definition 6.18 (Bound Variables). An occurrence of a variable in a formula A is *bound* if it is not free.

Definition 6.19 (Scope). If $\forall x\, B$ is an occurrence of a subformula in a formula A, then the corresponding occurrence of B in A is called the *scope* of the corresponding occurrence of $\forall x$.

Similarly for $\exists x$.

If B is the scope of a quantifier occurrence $\forall x$ or $\exists x$ in A, then the free occurrences of x in B are bound in $\forall x\, B$ and $\exists x\, B$. We say that these occurrences are *bound by* the mentioned quantifier occurrence.

Example 6.20. Consider the following formula:

$$\exists v_0 \underbrace{A_0^2(v_0, v_1)}_{B}$$

B represents the scope of $\exists v_0$. The quantifier binds the occurence of v_0 in B, but does not bind the occurence of v_1. So v_1 is a free variable in this case.

We can now see how this might work in a more complicated formula A:

$$\forall v_0 \underbrace{(A_0^1(v_0) \to A_0^2(v_0, v_1))}_{B} \to \exists v_1 \underbrace{(A_1^2(v_0, v_1) \vee \forall v_0 \overbrace{\neg A_1^1(v_0)}^{D})}_{C}$$

B is the scope of the first $\forall v_0$, C is the scope of $\exists v_1$, and D is the scope of the second $\forall v_0$. The first $\forall v_0$ binds the occurrences of v_0 in B, $\exists v_1$ the occurrence of v_1 in C, and the second $\forall v_0$ binds the occurrence of v_0 in D. The first occurrence of v_1 and the fourth occurrence of v_0 are free in A. The last occurrence of v_0 is free in D, but bound in C and A.

Definition 6.21 (Sentence). A formula A is a *sentence* iff it contains no free occurrences of variables.

6.8 Substitution

6.8. SUBSTITUTION

Definition 6.22 (Substitution in a term). We define $s[t/x]$, the result of *substituting* t for every occurrence of x in s, recursively:

1. $s \equiv c$: $s[t/x]$ is just s.

2. $s \equiv y$: $s[t/x]$ is also just s, provided y is a variable and $y \not\equiv x$.

3. $s \equiv x$: $s[t/x]$ is t.

4. $s \equiv f(t_1, \ldots, t_n)$: $s[t/x]$ is $f(t_1[t/x], \ldots, t_n[t/x])$.

Definition 6.23. A term t is *free for* x in A if none of the free occurrences of x in A occur in the scope of a quantifier that binds a variable in t.

Example 6.24.

1. v_8 is free for v_1 in $\exists v_3 A_4^2(v_3, v_1)$

2. $f_1^2(v_1, v_2)$ is *not* free for v_0 in $\forall v_2 A_4^2(v_0, v_2)$

Definition 6.25 (Substitution in a formula). If A is a formula, x is a variable, and t is a term free for x in A, then $A[t/x]$ is the result of substituting t for all free occurrences of x in A.

1. $A \equiv \bot$: $A[t/x]$ is $\bot$.

2. $A \equiv P(t_1, \ldots, t_n)$: $A[t/x]$ is $P(t_1[t/x], \ldots, t_n[t/x])$.

3. $A \equiv t_1 = t_2$: $A[t/x]$ is $t_1[t/x] = t_2[t/x]$.

4. $A \equiv \neg B$: $A[t/x]$ is $\neg B[t/x]$.

5. $A \equiv (B \wedge C)$: $A[t/x]$ is $(B[t/x] \wedge C[t/x])$.

6. $A \equiv (B \vee C)$: $A[t/x]$ is $(B[t/x] \vee C[t/x])$.

7. $A \equiv (B \to C)$: $A[t/x]$ is $(B[t/x] \to C[t/x])$.

8. $A \equiv \forall y\, B$: $A[t/x]$ is $\forall y\, B[t/x]$, provided y is a variable other than x; otherwise $A[t/x]$ is just A.

9. $A \equiv \exists y\, B$: $A[t/x]$ is $\exists y\, B[t/x]$, provided y is a variable other than x; otherwise $A[t/x]$ is just A.

Note that substitution may be vacuous: If x does not occur in A at all, then $A[t/x]$ is just A.

The restriction that t must be free for x in A is necessary to exclude cases like the following. If $A \equiv \exists y\, x < y$ and $t \equiv y$, then $A[t/x]$ would be $\exists y\, y < y$. In this case the free variable y is "captured" by the quantifier $\exists y$ upon substitution, and that is undesirable. For instance, we would like it to be the case that whenever $\forall x\, B$ holds, so does $B[t/x]$. But consider $\forall x\, \exists y\, x < y$ (here B is $\exists y\, x < y$). It is sentence that is true about, e.g., the natural numbers: for every number x there is a number y greater than it. If we allowed y as a possible substitution for x, we would end up with $B[y/x] \equiv \exists y\, y < y$, which is false. We prevent this by requiring that none of the free variables in t would end up being bound by a quantifier in A.

We often use the following convention to avoid cumbersome notation: If A is a formula which may contain the variable x free, we also write $A(x)$ to indicate this. When it is clear which A and x we have in mind, and t is a term (assumed to be free for x in $A(x)$), then we write $A(t)$ as short for $A[t/x]$. So for instance, we might say, "we call $A(t)$ an instance of $\forall x\, A(x)$." By this we mean that if A is any formula, x a variable, and t a term that's free for x in A, then $A[t/x]$ is an instance of $\forall x\, A$.

Summary

A **first-order language** consists of **constant**, **function**, and **predicate** symbols. Function and constant symbols take a specified number of arguments. In the **language of arithmetic**, e.g., we have a single constant symbol 0, one 1-place function symbol ′, two 2-place function symbols + and ×, and one 2-place predicate symbol <. From **variables** and constant and function symbols we form the **terms** of a language. From the terms of a language together with its predicate symbols, as well as the **identity symbol** =, we form the **atomic formulas**. And in turn from them, using the logical connectives ¬, ∨, ∧, →, ↔ and the quantifiers ∀ and ∃ we form its formulas. Since we are careful to always include necessary parentheses in the process of forming terms and formulas, there is always exactly one way of reading a formula. This makes it possible to define things by induction on the structure of formulas.

Occurrences of variables in formulas are sometimes governed by a corresponding quantifier: if a variable occurs in the **scope** of a quantifier it is considered **bound**, otherwise **free**. These concepts all have inductive definitions, and we also inductively define the operation of **substitution** of a term for a variable in a formula. Formulas without free variable occurrences are called **sentences**.

Problems

Problem 6.1. Prove Lemma 6.10.

Problem 6.2. Prove Proposition 6.11 (Hint: Formulate and prove a version of Lemma 6.10 for terms.)

Problem 6.3. Give an inductive definition of the bound variable occurrences along the lines of Definition 6.17.

CHAPTER 7

Semantics of First-Order Logic

7.1 Introduction

Giving the meaning of expressions is the domain of semantics. The central concept in semantics is that of satisfaction in a structure. A structure gives meaning to the building blocks of the language: a domain is a non-empty set of objects. The quantifiers are interpreted as ranging over this domain, constant symbols are assigned elements in the domain, function symbols are assigned functions from the domain to itself, and predicate symbols are assigned relations on the domain. The domain together with assignments to the basic vocabulary constitutes a structure. Variables may appear in formulas, and in order to give a semantics, we also have to assign elements of the domain to them—this is a variable assignment. The satisfaction relation, finally, brings these together. A formula may be satisfied in a structure M relative to a variable assignment s, written as $M, s \vDash A$. This relation is also defined by induction on the structure of A, using the truth tables for the logical connectives to define, say, satisfaction of $(A \wedge B)$ in terms of satisfaction (or not) of A and B. It then turns out that the variable assignment is irrelevant if the formula A is a sentence, i.e., has no free variables, and so we can talk of sentences being simply satisfied (or not) in structures.

On the basis of the satisfaction relation $M \vDash A$ for sentences we can then define the basic semantic notions of validity, entailment, and satisfiability. A sentence is valid, $\vDash A$, if every structure satisfies it. It is entailed by a set of sentences, $\Gamma \vDash A$, if every structure that satisfies all the sentences in Γ also satisfies A. And a set of sentences is satisfiable if some structure satisfies all sentences in it at the same time. Because formulas are inductively defined, and satisfaction is in turn defined by induction on the structure of formulas, we can use induction to prove properties of our semantics and to relate the semantic notions defined.

7.2 Structures for First-order Languages

First-order languages are, by themselves, *uninterpreted:* the constant symbols, function symbols, and predicate symbols have no specific meaning attached to them. Meanings are given by specifying a *structure*. It specifies the *domain*, i.e., the objects which the constant symbols pick out, the function symbols operate on, and the quantifiers range over. In addition, it specifies which constant symbols pick out which objects, how a function symbol maps objects to objects, and which objects the predicate symbols apply to. Structures are the basis for *semantic* notions in logic, e.g., the notion of consequence, validity, satisfiablity. They are variously called "structures," "interpretations," or "models" in the literature.

Definition 7.1 (Structures). A *structure* M, for a language $\mathscr{L}$ of first-order logic consists of the following elements:

1. *Domain:* a non-empty set, $|M|$

2. *Interpretation of constant symbols:* for each constant symbol c of $\mathscr{L}$, an element $c^M \in |M|$

3. *Interpretation of predicate symbols:* for each n-place predicate symbol R of $\mathscr{L}$ (other than =), an n-place relation $R^M \subseteq |M|^n$

4. *Interpretation of function symbols:* for each n-place function symbol f of $\mathscr{L}$, an n-place function $f^M \colon |M|^n \to |M|$

Example 7.2. A structure M for the language of arithmetic consists of a set, an element of $|M|$, 0^M, as interpretation of the constant symbol 0, a one-place function $\prime^M \colon |M| \to |M|$, two two-place functions $+^M$ and $\times^M$, both $|M|^2 \to |M|$, and a two-place relation $<^M \subseteq |M|^2$.

An obvious example of such a structure is the following:

1. $|N| = \mathbb{N}$

2. $0^N = 0$

3. $\prime^N(n) = n + 1$ for all $n \in \mathbb{N}$

4. $+^N(n, m) = n + m$ for all $n, m \in \mathbb{N}$

5. $\times^N(n, m) = n \cdot m$ for all $n, m \in \mathbb{N}$

6. $<^N = \{\langle n, m \rangle : n \in \mathbb{N}, m \in \mathbb{N}, n < m\}$

The structure N for $\mathscr{L}_A$ so defined is called the *standard model of arithmetic*, because it interprets the non-logical constants of $\mathscr{L}_A$ exactly how you would expect.

However, there are many other possible structures for $\mathscr{L}_A$. For instance, we might take as the domain the set $\mathbb{Z}$ of integers instead of $\mathbb{N}$, and define the interpretations of 0, $\prime$, +, ×, < accordingly. But we can also define structures for $\mathscr{L}_A$ which have nothing even remotely to do with numbers.

Example 7.3. A structure M for the language $\mathscr{L}_Z$ of set theory requires just a set and a single-two place relation. So technically, e.g., the set of people plus the relation "x is older than y" could be used as a structure for $\mathscr{L}_Z$, as well as $\mathbb{N}$ together with $n \geq m$ for $n, m \in \mathbb{N}$.

A particularly interesting structure for $\mathscr{L}_Z$ in which the elements of the domain are actually sets, and the interpretation of $\in$ actually is the relation "x is an element of y" is the structure *HF* of *hereditarily finite sets*:

1. $|HF| = \emptyset \cup \wp(\emptyset) \cup \wp(\wp(\emptyset)) \cup \wp(\wp(\wp(\emptyset))) \cup \ldots;$
2. $\in^{HF} = \{\langle x, y \rangle : x, y \in |HF|, x \in y\}.$

The stipulations we make as to what counts as a structure impact our logic. For example, the choice to prevent empty domains ensures, given the usual account of satisfaction (or truth) for quantified sentences, that $\exists x\, (A(x) \vee \neg A(x))$ is valid—that is, a logical truth. And the stipulation that all constant symbols must refer to an object in the domain ensures that the existential generalization is a sound pattern of inference: $A(a)$, therefore $\exists x\, A(x)$. If we allowed names to refer outside the domain, or to not refer, then we would be on our way to a *free logic*, in which existential generalization requires an additional premise: $A(a)$ and $\exists x\, x = a$, therefore $\exists x\, A(x)$.

7.3 Covered Structures for First-order Languages

Recall that a term is *closed* if it contains no variables.

Definition 7.4 (Value of closed terms). If t is a closed term of the language $\mathscr{L}$ and M is a structure for $\mathscr{L}$, the *value* $\text{Val}^M(t)$ is defined as follows:

1. If t is just the constant symbol c, then $\text{Val}^M(c) = c^M$.

2. If t is of the form $f(t_1,\ldots,t_n)$, then

$$\mathrm{Val}^M(t) = f^M(\mathrm{Val}^M(t_1),\ldots,\mathrm{Val}^M(t_n)).$$

Definition 7.5 (Covered structure). A structure is *covered* if every element of the domain is the value of some closed term.

Example 7.6. Let $\mathscr{L}$ be the language with constant symbols *zero*, *one*, *two*, ..., the binary predicate symbol $<$, and the binary function symbols $+$ and $\times$. Then a structure M for $\mathscr{L}$ is the one with domain $|M| = \{0,1,2,\ldots\}$ and assignments $zero^M = 0$, $one^M = 1$, $two^M = 2$, and so forth. For the binary relation symbol $<$, the set $<^M$ is the set of all pairs $\langle c_1, c_2\rangle \in |M|^2$ such that c_1 is less than c_2: for example, $\langle 1,3\rangle \in <^M$ but $\langle 2,2\rangle \notin <^M$. For the binary function symbol $+$, define $+^M$ in the usual way—for example, $+^M(2,3)$ maps to 5, and similarly for the binary function symbol $\times$. Hence, the value of *four* is just 4, and the value of $\times(two,+(three,zero))$ (or in infix notation, $two \times (three + zero)$) is

$$\begin{aligned}
\mathrm{Val}^M(\times(two,+(three,zero)) &= \\
&= \times^M(\mathrm{Val}^M(two),\mathrm{Val}^M(+(three,zero))) \\
&= \times^M(\mathrm{Val}^M(two),+^M(\mathrm{Val}^M(three),\mathrm{Val}^M(zero))) \\
&= \times^M(two^M,+^M(three^M,zero^M)) \\
&= \times^M(2,+^M(3,0)) \\
&= \times^M(2,3) \\
&= 6
\end{aligned}$$

7.4 Satisfaction of a Formula in a Structure

The basic notion that relates expressions such as terms and formulas, on the one hand, and structures on the other, are those of *value* of a term and *satisfaction* of a formula. Informally, the

7.4. SATISFACTION OF A FORMULA IN A STRUCTURE

value of a term is an element of a structure—if the term is just a constant, its value is the object assigned to the constant by the structure, and if it is built up using function symbols, the value is computed from the values of constants and the functions assigned to the functions in the term. A formula is *satisfied* in a structure if the interpretation given to the predicates makes the formula true in the domain of the structure. This notion of satisfaction is specified inductively: the specification of the structure directly states when atomic formulas are satisfied, and we define when a complex formula is satisfied depending on the main connective or quantifier and whether or not the immediate subformulas are satisfied.

The case of the quantifiers here is a bit tricky, as the immediate subformula of a quantified formula has a free variable, and structures don't specify the values of variables. In order to deal with this difficulty, we also introduce *variable assignments* and define satisfaction not with respect to a structure alone, but with respect to a structure plus a variable assignment.

Definition 7.7 (Variable Assignment). A *variable assignment* s for a structure M is a function which maps each variable to an element of $|M|$, i.e., $s\colon \mathrm{Var} \to |M|$.

A structure assigns a value to each constant symbol, and a variable assignment to each variable. But we want to use terms built up from them to also name elements of the domain. For this we define the value of terms inductively. For constant symbols and variables the value is just as the structure or the variable assignment specifies it; for more complex terms it is computed recursively using the functions the structure assigns to the function symbols.

Definition 7.8 (Value of Terms). If t is a term of the language $\mathcal{L}$, M is a structure for $\mathcal{L}$, and s is a variable assignment for M, the *value* $\mathrm{Val}_s^M(t)$ is defined as follows:

1. $t \equiv c$: $\operatorname{Val}_s^M(t) = c^M$.

2. $t \equiv x$: $\operatorname{Val}_s^M(t) = s(x)$.

3. $t \equiv f(t_1, \ldots, t_n)$:

$$\operatorname{Val}_s^M(t) = f^M(\operatorname{Val}_s^M(t_1), \ldots, \operatorname{Val}_s^M(t_n)).$$

Definition 7.9 (x-Variant). If s is a variable assignment for a structure M, then any variable assignment s' for M which differs from s at most in what it assigns to x is called an *x-variant* of s. If s' is an x-variant of s we write $s' \sim_x s$.

Note that an x-variant of an assignment s does not *have* to assign something different to x. In fact, every assignment counts as an x-variant of itself.

Definition 7.10. If s is a variable assignment for a structure M and $m \in |M|$, then the assignment $s[m/x]$ is the variable assignment defined by

$$s[m/y] = \begin{cases} m & \text{if } y \equiv x \\ s(y) & \text{otherwise.} \end{cases}$$

In other words, $s[m/x]$ is the particular x-variant of s which assigns the domain element m to x, and assigns the same things to variables other than x that s does.

Definition 7.11 (Satisfaction). Satisfaction of a formula A in a structure M relative to a variable assignment s, in symbols: $M, s \vDash A$, is defined recursively as follows. (We write $M, s \nvDash A$ to mean "not $M, s \vDash A$.")

1. $A \equiv \bot$: $M, s \nvDash A$.

2. $A \equiv R(t_1, \ldots, t_n)$: $M, s \vDash A$ iff $\langle \operatorname{Val}_s^M(t_1), \ldots, \operatorname{Val}_s^M(t_n) \rangle \in$

7.4. SATISFACTION OF A FORMULA IN A STRUCTURE

R^M.

3. $A \equiv t_1 = t_2$: $M, s \vDash A$ iff $\mathrm{Val}^M_s(t_1) = \mathrm{Val}^M_s(t_2)$.

4. $A \equiv \neg B$: $M, s \vDash A$ iff $M, s \nvDash B$.

5. $A \equiv (B \wedge C)$: $M, s \vDash A$ iff $M, s \vDash B$ and $M, s \vDash C$.

6. $A \equiv (B \vee C)$: $M, s \vDash A$ iff $M, s \vDash B$ or $M, s \vDash C$ (or both).

7. $A \equiv (B \rightarrow C)$: $M, s \vDash A$ iff $M, s \nvDash B$ or $M, s \vDash C$ (or both).

8. $A \equiv \forall x\, B$: $M, s \vDash A$ iff for every element $m \in |M|$, $M, s[m/x] \vDash B$.

9. $A \equiv \exists x\, B$: $M, s \vDash A$ iff for at least one element $m \in |M|$, $M, s[m/x] \vDash B$.

The variable assignments are important in the last two clauses. We cannot define satisfaction of $\forall x\, B(x)$ by "for all $m \in |M|$, $M \vDash B(m)$." We cannot define satisfaction of $\exists x\, B(x)$ by "for at least one $m \in |M|$, $M \vDash B(m)$." The reason is that if $m \in |M|$, it is not symbol of the language, and so $B(a)$ is not a formula (that is, $B[m/x]$ is undefined). We also cannot assume that we have constant symbols or terms available that name every element of M, since there is nothing in the definition of structures that requires it. In the standard language, the set of constant symbols is countably infinite, so if $|M|$ is not countable there aren't even enough constant symbols to name every object.

We solve this problem by introducing variable assignments, which allow us to link variables directly with elements of the domain. Then instead of saying that, e.g., $\exists x\, B(x)$ is satisfied in M iff for at least one $m \in |M|$, we say it is satisfied in M *relative to s* iff $B(x)$ is satisfied relative to $s[m/x]$ for at least one $m \in |M|$.

Example 7.12. Let $\mathscr{L} = \{a, b, f, R\}$ where a and b are constant symbols, f is a two-place function symbol, and R is a two-place predicate symbol. Consider the structure M defined by:

1. $|M| = \{1,2,3,4\}$
2. $a^M = 1$
3. $b^M = 2$
4. $f^M(x,y) = x+y$ if $x+y \le 3$ and $= 3$ otherwise.
5. $R^M = \{\langle 1,1\rangle, \langle 1,2\rangle, \langle 2,3\rangle, \langle 2,4\rangle\}$

The function $s(x) = 1$ that assigns $1 \in |M|$ to every variable is a variable assignment for M.

Then
$$\mathrm{Val}_s^M(f(a,b)) = f^M(\mathrm{Val}_s^M(a), \mathrm{Val}_s^M(b)).$$

Since a and b are constant symbols, $\mathrm{Val}_s^M(a) = a^M = 1$ and $\mathrm{Val}_s^M(b) = b^M = 2$. So
$$\mathrm{Val}_s^M(f(a,b)) = f^M(1,2) = 1+2 = 3.$$

To compute the value of $f(f(a,b),a)$ we have to consider
$$\mathrm{Val}_s^M(f(f(a,b),a)) = f^M(\mathrm{Val}_s^M(f(a,b)), \mathrm{Val}_s^M(a)) = f^M(3,1) = 3,$$

since $3+1 > 3$. Since $s(x) = 1$ and $\mathrm{Val}_s^M(x) = s(x)$, we also have
$$\mathrm{Val}_s^M(f(f(a,b),x)) = f^M(\mathrm{Val}_s^M(f(a,b)), \mathrm{Val}_s^M(x)) = f^M(3,1) = 3,$$

An atomic formula $R(t_1, t_2)$ is satisfied if the tuple of values of its arguments, i.e., $\langle \mathrm{Val}_s^M(t_1), \mathrm{Val}_s^M(t_2)\rangle$, is an element of R^M. So, e.g., we have $M, s \models R(b, f(a,b))$ since $\langle \mathrm{Val}^M(b), \mathrm{Val}^M(f(a,b))\rangle = \langle 2,3\rangle \in R^M$, but $M, s \not\models R(x, f(a,b))$ since $\langle 1,3\rangle \notin R^M[s]$.

To determine if a non-atomic formula A is satisfied, you apply the clauses in the inductive definition that applies to the main connective. For instance, the main connective in $R(a,a) \to (R(b,x) \vee R(x,b))$ is the $\to$, and

$$M, s \models R(a,a) \to (R(b,x) \vee R(x,b)) \text{ iff}$$
$$M, s \not\models R(a,a) \text{ or } M, s \models R(b,x) \vee R(x,b)$$

Since $M, s \vDash R(a,a)$ (because $\langle 1,1 \rangle \in R^M$) we can't yet determine the answer and must first figure out if $M, s \vDash R(b,x) \vee R(x,b)$:

$$M, s \vDash R(b,x) \vee R(x,b) \text{ iff}$$
$$M, s \vDash R(b,x) \text{ or } M, s \vDash R(x,b)$$

And this is the case, since $M, s \vDash R(x,b)$ (because $\langle 1,2 \rangle \in R^M$).

Recall that an x-variant of s is a variable assignment that differs from s at most in what it assigns to x. For every element of $|M|$, there is an x-variant of s:

$$s_1 = s[1/x], \quad\quad s_2 = s[2/x],$$
$$s_3 = s[3/x], \quad\quad s_4 = s[4/x].$$

So, e.g., $s_2(x) = 2$ and $s_2(y) = s(y) = 1$ for all variables y other than x. These are all the x-variants of s for the structure M, since $|M| = \{1, 2, 3, 4\}$. Note, in particular, that $s_1 = s$ (s is always an x-variant of itself).

To determine if an existentially quantified formula $\exists x\, A(x)$ is satisfied, we have to determine if $M, s[m/x] \vDash A(x)$ for at least one $m \in |M|$. So,

$$M, s \vDash \exists x\, (R(b,x) \vee R(x,b)),$$

since $M, s[1/x] \vDash R(b,x) \vee R(x,b)$ ($s[3/x]$ would also fit the bill). But,

$$M, s \nvDash \exists x\, (R(b,x) \wedge R(x,b))$$

since, whichever $m \in |M|$ we pick, $M, s[m/x] \nvDash R(b,x) \wedge R(x,b)$.

To determine if a universally quantified formula $\forall x\, A(x)$ is satisfied, we have to determine if $M, s[m/x] \vDash A(x)$ for all $m \in |M|$. So,

$$M, s \vDash \forall x\, (R(x,a) \to R(a,x)),$$

since $M, s[m/x] \vDash R(x,a) \to R(a,x)$ for all $m \in |M|$. For $m = 1$, we have $M, s[1/x] \vDash R(a,x)$ so the consequent is true; for $m = 2$,

3, and 4, we have $M, s[m/x] \nvDash R(x, a)$, so the antecedent is false. But,
$$M, s \nvDash \forall x \, (R(a, x) \rightarrow R(x, a))$$
since $M, s[2/x] \nvDash R(a, x) \rightarrow R(x, a)$ (because $M, s[2/x] \vDash R(a, x)$ and $M, s[2/x] \nvDash R(x, a)$).

For a more complicated case, consider
$$\forall x \, (R(a, x) \rightarrow \exists y \, R(x, y)).$$
Since $M, s[3/x] \nvDash R(a, x)$ and $M, s[4/x] \nvDash R(a, x)$, the interesting cases where we have to worry about the consequent of the conditional are only $m = 1$ and $= 2$. Does $M, s[1/x] \vDash \exists y \, R(x, y)$ hold? It does if there is at least one $n \in |M|$ so that $M, s[1/x][n/y] \vDash R(x, y)$. In fact, if we take $n = 1$, we have $s[1/x][n/y] = s[1/y] = s$. Since $s(x) = 1$, $s(y) = 1$, and $\langle 1, 1 \rangle \in R^M$, the answer is yes.

To determine if $M, s[2/x] \vDash \exists y \, R(x, y)$, we have to look at the variable assignments $s[2/x][n/y]$. Here, for $n = 1$, this assignment is $s_2 = s[2/x]$, which does not satisfy $R(x, y)$ ($s_2(x) = 2$, $s_2(y) = 1$, and $\langle 2, 1 \rangle \notin R^M$). However, consider $s[2/x][3/y] = s_2[3/y]$. $M, s_2[3/y] \vDash R(x, y)$ since $\langle 2, 3 \rangle \in R^M$, and so $M, s_2 \vDash \exists y \, R(x, y)$.

So, for all $n \in |M|$, either $M, s[m/x] \nvDash R(a, x)$ (if $m = 3, 4$) or $M, s[m/x] \vDash \exists y \, R(x, y)$ (if $m = 1, 2$), and so
$$M, s \vDash \forall x \, (R(a, x) \rightarrow \exists y \, R(x, y)).$$

On the other hand,
$$M, s \nvDash \exists x \, (R(a, x) \wedge \forall y \, R(x, y)).$$

We have $M, s[m/x] \vDash R(a, x)$ only for $m = 1$ and $m = 2$. But for both of these values of m, there is in turn an $n \in |M|$, namely $n = 4$, so that $M, s[m/x][n/y] \nvDash R(x, y)$ and so $M, s[m/x] \nvDash \forall y \, R(x, y)$ for $m = 1$ and $m = 2$. In sum, there is no $m \in |M|$ such that $M, s[m/x] \vDash R(a, x) \wedge \forall y \, R(x, y)$.

7.5 Variable Assignments

A variable assignment s provides a value for *every* variable—and there are infinitely many of them. This is of course not necessary. We require variable assignments to assign values to all variables simply because it makes things a lot easier. The value of a term t, and whether or not a formula A is satisfied in a structure with respect to s, only depend on the assignments s makes to the variables in t and the free variables of A. This is the content of the next two propositions. To make the idea of "depends on" precise, we show that any two variable assignments that agree on all the variables in t give the same value, and that A is satisfied relative to one iff it is satisfied relative to the other if two variable assignments agree on all free variables of A.

Proposition 7.13. *If the variables in a term t are among $x_1, \ldots, x_n$, and $s_1(x_i) = s_2(x_i)$ for $i = 1, \ldots, n$, then $\operatorname{Val}_{s_1}^M(t) = \operatorname{Val}_{s_2}^M(t)$.*

Proof. By induction on the complexity of t. For the base case, t can be a constant symbol or one of the variables $x_1, \ldots, x_n$. If $t = c$, then $\operatorname{Val}_{s_1}^M(t) = c^M = \operatorname{Val}_{s_2}^M(t)$. If $t = x_i$, $s_1(x_i) = s_2(x_i)$ by the hypothesis of the proposition, and so $\operatorname{Val}_{s_1}^M(t) = s_1(x_i) = s_2(x_i) = \operatorname{Val}_{s_2}^M(t)$.

For the inductive step, assume that $t = f(t_1, \ldots, t_k)$ and that the claim holds for $t_1, \ldots, t_k$. Then

$$\operatorname{Val}_{s_1}^M(t) = \operatorname{Val}_{s_1}^M(f(t_1, \ldots, t_k)) =$$
$$= f^M(\operatorname{Val}_{s_1}^M(t_1), \ldots, \operatorname{Val}_{s_1}^M(t_k))$$

For $j = 1, \ldots, k$, the variables of t_j are among $x_1, \ldots, x_n$. By induction hypothesis, $\operatorname{Val}_{s_1}^M(t_j) = \operatorname{Val}_{s_2}^M(t_j)$. So,

$$\operatorname{Val}_{s_1}^M(t) = \operatorname{Val}_{s_2}^M(f(t_1, \ldots, t_k)) =$$
$$= f^M(\operatorname{Val}_{s_1}^M(t_1), \ldots, \operatorname{Val}_{s_1}^M(t_k)) =$$
$$= f^M(\operatorname{Val}_{s_2}^M(t_1), \ldots, \operatorname{Val}_{s_2}^M(t_k)) =$$
$$= \operatorname{Val}_{s_2}^M(f(t_1, \ldots, t_k)) = \operatorname{Val}_{s_2}^M(t). \qquad \square$$

Proposition 7.14. *If the free variables in A are among x_1, ..., x_n, and $s_1(x_i) = s_2(x_i)$ for $i = 1$, ..., n, then $M, s_1 \vDash A$ iff $M, s_2 \vDash A$.*

Proof. We use induction on the complexity of A. For the base case, where A is atomic, A can be: $\bot$, $R(t_1, \ldots, t_k)$ for a k-place predicate R and terms t_1, ..., t_k, or $t_1 = t_2$ for terms t_1 and t_2.

1. $A \equiv \bot$: both $M, s_1 \nvDash A$ and $M, s_2 \nvDash A$.

2. $A \equiv R(t_1, \ldots, t_k)$: let $M, s_1 \vDash A$. Then
$$\langle \text{Val}^M_{s_1}(t_1), \ldots, \text{Val}^M_{s_1}(t_k) \rangle \in R^M.$$
   For $i = 1$, ..., k, $\text{Val}^M_{s_1}(t_i) = \text{Val}^M_{s_2}(t_i)$ by Proposition 7.13. So we also have $\langle \text{Val}^M_{s_2}(t_i), \ldots, \text{Val}^M_{s_2}(t_k) \rangle \in R^M$.

3. $A \equiv t_1 = t_2$: suppose $M, s_1 \vDash A$. Then $\text{Val}^M_{s_1}(t_1) = \text{Val}^M_{s_1}(t_2)$. So,

$$\begin{aligned}
\text{Val}^M_{s_2}(t_1) &= \text{Val}^M_{s_1}(t_1) && \text{(by Proposition 7.13)} \\
&= \text{Val}^M_{s_1}(t_2) && \text{(since } M, s_1 \vDash t_1 = t_2\text{)} \\
&= \text{Val}^M_{s_2}(t_2) && \text{(by Proposition 7.13)},
\end{aligned}$$

so $M, s_2 \vDash t_1 = t_2$.

Now assume $M, s_1 \vDash B$ iff $M, s_2 \vDash B$ for all formulas B less complex than A. The induction step proceeds by cases determined by the main operator of A. In each case, we only demonstrate the forward direction of the biconditional; the proof of the reverse direction is symmetrical. In all cases except those for the quantifiers, we apply the induction hypothesis to sub-formulas B of A. The free variables of B are among those of A. Thus, if s_1 and s_2 agree on the free variables of A, they also agree on those of B, and the induction hypothesis applies to B.

1. $A \equiv \neg B$: if $M, s_1 \vDash A$, then $M, s_1 \nvDash B$, so by the induction hypothesis, $M, s_2 \nvDash B$, hence $M, s_2 \vDash A$.

7.5. VARIABLE ASSIGNMENTS

2. $A \equiv B \wedge C$: exercise.

3. $A \equiv B \vee C$: if $M, s_1 \vDash A$, then $M, s_1 \vDash B$ or $M, s_1 \vDash C$. By induction hypothesis, $M, s_2 \vDash B$ or $M, s_2 \vDash C$, so $M, s_2 \vDash A$.

4. $A \equiv B \to C$: exercise.

5. $A \equiv \exists x\, B$: if $M, s_1 \vDash A$, there is an $m \in |M|$ so that $M, s_1[m/x] \vDash B$. Let $s_1' = s_1[m/x]$ and $s_2' = s_2[m/x]$. The free variables of B are among $x_1, \ldots, x_n$, and x. $s_1'(x_i) = s_2'(x_i)$, since s_1' and s_2' are x-variants of s_1 and s_2, respectively, and by hypothesis $s_1(x_i) = s_2(x_i)$. $s_1'(x) = s_2'(x) = m$ by the way we have defined s_1' and s_2'. Then the induction hypothesis applies to B and s_1', s_2', so $M, s_2' \vDash B$. Hence, since $s_2' = s_2[m/x]$, there is an $m \in |M|$ such that $M, s_2[m/x] \vDash B$, and so $M, s_2 \vDash A$.

6. $A \equiv \forall x\, B$: exercise.

By induction, we get that $M, s_1 \vDash A$ iff $M, s_2 \vDash A$ whenever the free variables in A are among $x_1, \ldots, x_n$ and $s_1(x_i) = s_2(x_i)$ for $i = 1, \ldots, n$. □

Sentences have no free variables, so any two variable assignments assign the same things to all the (zero) free variables of any sentence. The proposition just proved then means that whether or not a sentence is satisfied in a structure relative to a variable assignment is completely independent of the assignment. We'll record this fact. It justifies the definition of satisfaction of a sentence in a structure (without mentioning a variable assignment) that follows.

Corollary 7.15. *If A is a sentence and s a variable assignment, then $M, s \vDash A$ iff $M, s' \vDash A$ for every variable assignment s'.*

Proof. Let s' be any variable assignment. Since A is a sentence, it has no free variables, and so every variable assignment s' trivially assigns the same things to all free variables of A as does s. So the condition of Proposition 7.14 is satisfied, and we have $M, s \vDash A$ iff $M, s' \vDash A$. □

Definition 7.16. If A is a sentence, we say that a structure M *satisfies* A, $M \vDash A$, iff $M, s \vDash A$ for all variable assignments s.

If $M \vDash A$, we also simply say that A *is true in* M.

Proposition 7.17. *Let M be a structure, A be a sentence, and s a variable assignment. $M \vDash A$ iff $M, s \vDash A$.*

Proof. Exercise. □

Proposition 7.18. *Suppose $A(x)$ only contains x free, and M is a structure. Then:*

1. *$M \vDash \exists x\, A(x)$ iff $M, s \vDash A(x)$ for at least one variable assignment s.*

2. *$M \vDash \forall x\, A(x)$ iff $M, s \vDash A(x)$ for all variable assignments s.*

Proof. Exercise. □

7.6 Extensionality

Extensionality, sometimes called relevance, can be expressed informally as follows: the only factors that bears upon the satisfaction of formula A in a structure M relative to a variable assignment s, are the size of the domain and the assignments made by M and s to the elements of the language that actually appear in A.

One immediate consequence of extensionality is that where two structures M and M' agree on all the elements of the language appearing in a sentence A and have the same domain, M and M' must also agree on whether or not A itself is true.

7.6. EXTENSIONALITY

Proposition 7.19 (Extensionality). *Let A be a formula, and M_1 and M_2 be structures with $|M_1| = |M_2|$, and s a variable assignment on $|M_1| = |M_2|$. If $c^{M_1} = c^{M_2}$, $R^{M_1} = R^{M_2}$, and $f^{M_1} = f^{M_2}$ for every constant symbol c, relation symbol R, and function symbol f occurring in A, then $M_1, s \vDash A$ iff $M_2, s \vDash A$.*

Proof. First prove (by induction on t) that for every term, $\mathrm{Val}_s^{M_1}(t) = \mathrm{Val}_s^{M_2}(t)$. Then prove the proposition by induction on A, making use of the claim just proved for the induction basis (where A is atomic). □

Corollary 7.20 (Extensionality for Sentences). *Let A be a sentence and M_1, M_2 as in Proposition 7.19. Then $M_1 \vDash A$ iff $M_2 \vDash A$.*

Proof. Follows from Proposition 7.19 by Corollary 7.15. □

Moreover, the value of a term, and whether or not a structure satisfies a formula, only depends on the values of its subterms.

Proposition 7.21. *Let M be a structure, t and t' terms, and s a variable assignment. Then $\mathrm{Val}_s^M(t[t'/x]) = \mathrm{Val}_{s[\mathrm{Val}_s^M(t')/x]}^M(t)$.*

Proof. By induction on t.

1. If t is a constant, say, $t \equiv c$, then $t[t'/x] = c$, and $\mathrm{Val}_s^M(c) = c^M = \mathrm{Val}_{s[\mathrm{Val}_s^M(t')/x]}^M(c)$.

2. If t is a variable other than x, say, $t \equiv y$, then $t[t'/x] = y$, and $\mathrm{Val}_s^M(y) = \mathrm{Val}_{s[\mathrm{Val}_s^M(t')/x]}^M(y)$ since $s \sim_x s[\mathrm{Val}_s^M(t')/x]$.

3. If $t \equiv x$, then $t[t'/x] = t'$. But $\mathrm{Val}_{s[\mathrm{Val}_s^M(t')/x]}^M(x) = \mathrm{Val}_s^M(t')$ by definition of $s[\mathrm{Val}_s^M(t')/x]$.

4. If $t \equiv f(t_1, \ldots, t_n)$ then we have:

$\mathrm{Val}_s^M(t[t'/x]) =$
$= \mathrm{Val}_s^M(f(t_1[t'/x], \ldots, t_n[t'/x]))$
   by definition of $t[t'/x]$
$= f^M(\mathrm{Val}_s^M(t_1[t'/x]), \ldots, \mathrm{Val}_s^M(t_n[t'/x]))$
   by definition of $\mathrm{Val}_s^M(f(\ldots))$
$= f^M(\mathrm{Val}_{s[\mathrm{Val}_s^M(t')/x]}^M(t_1), \ldots, \mathrm{Val}_{s[\mathrm{Val}_s^M(t')/x]}^M(t_n))$
   by induction hypothesis
$= \mathrm{Val}_{s[\mathrm{Val}_s^M(t')/x]}^M(t)$ by definition of $\mathrm{Val}_{s[\mathrm{Val}_s^M(t')/x]}^M(f(\ldots))$ □

Proposition 7.22. *Let M be a structure, A a formula, t' a term, and s a variable assignment. Then $M, s \vDash A[t'/x]$ iff $M, s[\mathrm{Val}_s^M(t')/x] \vDash A$.*

Proof. Exercise. □

The point of Propositions 7.21 and 7.22 is the following. Suppose we have a term t or a formula A and some term t', and we want to know the value of $t[t'/x]$ or whether or not $A[t'/x]$ is satisfied in a structure M relative to a variable assignment s. Then we can either perform the substitution first and then consider the value or satisfaction relative to M and s, or we can first determine the value $m = \mathrm{Val}_s^M(t')$ of t' in M relative to s, change the variable assignment to $s[m/x]$ and then consider the value of t in M and $s[m/x]$, or whether $M, s[m/x] \vDash A$. Propositions 7.21 and 7.22 guarantee that the answer will be the same, whichever way we do it.

7.7 Semantic Notions

Give the definition of structures for first-order languages, we can define some basic semantic properties of and relationships between sentences. The simplest of these is the notion of *validity*

7.7. SEMANTIC NOTIONS

of a sentence. A sentence is valid if it is satisfied in every structure. Valid sentences are those that are satisfied regardless of how the non-logical symbols in it are interpreted. Valid sentences are therefore also called *logical truths*—they are true, i.e., satisfied, in any structure and hence their truth depends only on the logical symbols occurring in them and their syntactic structure, but not on the non-logical symbols or their interpretation.

Definition 7.23 (Validity). A sentence A is *valid*, $\vDash A$, iff $M \vDash A$ for every structure M.

Definition 7.24 (Entailment). A set of sentences Γ *entails* a sentence A, $\Gamma \vDash A$, iff for every structure M with $M \vDash \Gamma$, $M \vDash A$.

Definition 7.25 (Satisfiability). A set of sentences Γ is *satisfiable* if $M \vDash \Gamma$ for some structure M. If Γ is not satisfiable it is called *unsatisfiable*.

Proposition 7.26. *A sentence A is valid iff $\Gamma \vDash A$ for every set of sentences Γ.*

Proof. For the forward direction, let A be valid, and let Γ be a set of sentences. Let M be a structure so that $M \vDash \Gamma$. Since A is valid, $M \vDash A$, hence $\Gamma \vDash A$.

For the contrapositive of the reverse direction, let A be invalid, so there is a structure M with $M \nvDash A$. When $\Gamma = \{\top\}$, since $\top$ is valid, $M \vDash \Gamma$. Hence, there is a structure M so that $M \vDash \Gamma$ but $M \nvDash A$, hence Γ does not entail A. □

Proposition 7.27. $\Gamma \vDash A$ iff $\Gamma \cup \{\neg A\}$ is unsatisfiable.

Proof. For the forward direction, suppose $\Gamma \vDash A$ and suppose to the contrary that there is a structure M so that $M \vDash \Gamma \cup \{\neg A\}$. Since $M \vDash \Gamma$ and $\Gamma \vDash A$, $M \vDash A$. Also, since $M \vDash \Gamma \cup \{\neg A\}$, $M \vDash \neg A$, so we have both $M \vDash A$ and $M \nvDash A$, a contradiction. Hence, there can be no such structure M, so $\Gamma \cup \{\neg A\}$ is unsatisfiable.

For the reverse direction, suppose $\Gamma \cup \{\neg A\}$ is unsatisfiable. So for every structure M, either $M \nvDash \Gamma$ or $M \vDash A$. Hence, for every structure M with $M \vDash \Gamma$, $M \vDash A$, so $\Gamma \vDash A$. □

Proposition 7.28. *If* $\Gamma \subseteq \Gamma'$ *and* $\Gamma \vDash A$, *then* $\Gamma' \vDash A$.

Proof. Suppose that $\Gamma \subseteq \Gamma'$ and $\Gamma \vDash A$. Let M be a structure such that $M \vDash \Gamma'$; then $M \vDash \Gamma$, and since $\Gamma \vDash A$, we get that $M \vDash A$. Hence, whenever $M \vDash \Gamma'$, $M \vDash A$, so $\Gamma' \vDash A$. □

Theorem 7.29 (Semantic Deduction Theorem). $\Gamma \cup \{A\} \vDash B$ *iff* $\Gamma \vDash A \to B$.

Proof. For the forward direction, let $\Gamma \cup \{A\} \vDash B$ and let M be a structure so that $M \vDash \Gamma$. If $M \vDash A$, then $M \vDash \Gamma \cup \{A\}$, so since $\Gamma \cup \{A\}$ entails B, we get $M \vDash B$. Therefore, $M \vDash A \to B$, so $\Gamma \vDash A \to B$.

For the reverse direction, let $\Gamma \vDash A \to B$ and M be a structure so that $M \vDash \Gamma \cup \{A\}$. Then $M \vDash \Gamma$, so $M \vDash A \to B$, and since $M \vDash A$, $M \vDash B$. Hence, whenever $M \vDash \Gamma \cup \{A\}$, $M \vDash B$, so $\Gamma \cup \{A\} \vDash B$. □

Proposition 7.30. *Let M be a structure, and $A(x)$ a formula with one free variable x, and t a closed term. Then:*

1. $A(t) \vDash \exists x\, A(x)$

2. $\forall x\, A(x) \vDash A(t)$

Proof. 1. Suppose $M \vDash A(t)$. Let s be a variable assignment with $s(x) = \mathrm{Val}^M(t)$. Then $M, s \vDash A(t)$ since $A(t)$ is a sentence. By Proposition 7.22, $M, s \vDash A(x)$. By Proposition 7.18, $M \vDash \exists x\, A(x)$.

2. Exercise. □

Summary

The **semantics** for a first-order language is given by a **structure** for that language. It consists of a **domain** and elements of that domain are assigned to each constant symbol. Function symbols are interpreted by functions and relation symbols by relation on the domain. A function from the set of variables to the domain is a **variable assignment**. The relation of **satisfaction** relates structures, variable assignments and formulas; $M, s \vDash A$ is defined by induction on the structure of A. $M, s \vDash A$ only depends on the interpretation of the symbols actually occurring in A, and in particular does not depend on s if A contains no free variables. So if A is a sentence, $M \vDash A$ if $M, s \vDash A$ for any (or all) s.

The satisfaction relation is the basis for all semantic notions. A sentence is **valid**, $\vDash A$, if it is satisfied in every structure. A sentence A is **entailed** by set of sentences Γ, $\Gamma \vDash A$, iff $M \vDash A$ for all M which satisfy every sentence in Γ. A set Γ is **satisfiable** iff there is some structure that satisfies every sentence in Γ, otherwise **unsatisfiable**. These notions are interrelated, e.g., $\Gamma \vDash A$ iff $\Gamma \cup \{\neg A\}$ is unsatisfiable.

Problems

Problem 7.1. Is N, the standard model of arithmetic, covered? Explain.

Problem 7.2. Let $\mathscr{L} = \{c, f, A\}$ with one constant symbol, one one-place function symbol and one two-place predicate symbol, and let the structure M be given by

1. $|M| = \{1, 2, 3\}$
2. $c^M = 3$
3. $f^M(1) = 2, f^M(2) = 3, f^M(3) = 2$
4. $A^M = \{\langle 1, 2 \rangle, \langle 2, 3 \rangle, \langle 3, 3 \rangle\}$

(a) Let $s(v) = 1$ for all variables v. Find out whether

$$M, s \models \exists x \, (A(f(z), c) \to \forall y \, (A(y, x) \lor A(f(y), x)))$$

Explain why or why not.

(b) Give a different structure and variable assignment in which the formula is not satisfied.

Problem 7.3. Complete the proof of Proposition 7.14.

Problem 7.4. Prove Proposition 7.17

Problem 7.5. Prove Proposition 7.18.

Problem 7.6. Suppose $\mathscr{L}$ is a language without function symbols. Given a structure M, c a constant symbol and $a \in |M|$, define $M[a/c]$ to be the structure that is just like M, except that $c^{M[a/c]} = a$. Define $M \Vdash A$ for sentences A by:

1. $A \equiv \bot$: not $M \Vdash A$.
2. $A \equiv R(d_1, \ldots, d_n)$: $M \Vdash A$ iff $\langle d_1^M, \ldots, d_n^M \rangle \in R^M$.
3. $A \equiv d_1 = d_2$: $M \Vdash A$ iff $d_1^M = d_2^M$.
4. $A \equiv \neg B$: $M \Vdash A$ iff not $M \Vdash B$.
5. $A \equiv (B \land C)$: $M \Vdash A$ iff $M \Vdash B$ and $M \Vdash C$.

7.7. SEMANTIC NOTIONS

6. $A \equiv (B \lor C)$: $M \Vvdash A$ iff $M \Vvdash B$ or $M \Vvdash C$ (or both).

7. $A \equiv (B \to C)$: $M \Vvdash A$ iff not $M \Vvdash B$ or $M \Vvdash C$ (or both).

8. $A \equiv \forall x\, B$: $M \Vvdash A$ iff for all $a \in |M|$, $M[a/c] \Vvdash B[c/x]$, if c does not occur in B.

9. $A \equiv \exists x\, B$: $M \Vvdash A$ iff there is an $a \in |M|$ such that $M[a/c] \Vvdash B[c/x]$, if c does not occur in B.

Let $x_1, \ldots, x_n$ be all free variables in A, $c_1, \ldots, c_n$ constant symbols not in A, $a_1, \ldots, a_n \in |M|$, and $s(x_i) = a_i$.

Show that $M, s \vDash A$ iff $M[a_1/c_1, \ldots, a_n/c_n] \Vvdash A[c_1/x_1]\ldots[c_n/x_n]$.

(This problem shows that it is possible to give a semantics for first-order logic that makes do without variable assignments.)

Problem 7.7. Suppose that f is a function symbol not in $A(x, y)$. Show that there is a structure M such that $M \vDash \forall x\, \exists y\, A(x, y)$ iff there is an M' such that $M' \vDash \forall x\, A(x, f(x))$.

(This problem is a special case of what's known as Skolem's Theorem; $\forall x\, A(x, f(x))$ is called a *Skolem normal form* of $\forall x\, \exists y\, A(x, y)$.)

Problem 7.8. Carry out the proof of Proposition 7.19 in detail.

Problem 7.9. Prove Proposition 7.22

Problem 7.10.   1. Show that $\Gamma \vDash \bot$ iff Γ is unsatisfiable.

2. Show that $\Gamma \cup \{A\} \vDash \bot$ iff $\Gamma \vDash \neg A$.

3. Suppose c does not occur in A or Γ. Show that $\Gamma \vDash \forall x\, A$ iff $\Gamma \vDash A[c/x]$.

Problem 7.11. Complete the proof of Proposition 7.30.

CHAPTER 8
Theories and Their Models

8.1 Introduction

The development of the axiomatic method is a significant achievement in the history of science, and is of special importance in the history of mathematics. An axiomatic development of a field involves the clarification of many questions: What is the field about? What are the most fundamental concepts? How are they related? Can all the concepts of the field be defined in terms of these fundamental concepts? What laws do, and must, these concepts obey?

The axiomatic method and logic were made for each other. Formal logic provides the tools for formulating axiomatic theories, for proving theorems from the axioms of the theory in a precisely specified way, for studying the properties of all systems satisfying the axioms in a systematic way.

Definition 8.1. A set of sentences Γ is *closed* iff, whenever $\Gamma \vDash A$ then $A \in \Gamma$. The *closure* of a set of sentences Γ is $\{A : \Gamma \vDash A\}$.

We say that Γ is *axiomatized by* a set of sentences Δ if Γ is the closure of Δ.

8.1. INTRODUCTION

We can think of an axiomatic theory as the set of sentences that is axiomatized by its set of axioms Δ. In other words, when we have a first-order language which contains non-logical symbols for the primitives of the axiomatically developed science we wish to study, together with a set of sentences that express the fundamental laws of the science, we can think of the theory as represented by all the sentences in this language that are entailed by the axioms. This ranges from simple examples with only a single primitive and simple axioms, such as the theory of partial orders, to complex theories such as Newtonian mechanics.

The important logical facts that make this formal approach to the axiomatic method so important are the following. Suppose Γ is an axiom system for a theory, i.e., a set of sentences.

1. We can state precisely when an axiom system captures an intended class of structures. That is, if we are interested in a certain class of structures, we will successfully capture that class by an axiom system Γ iff the structures are exactly those M such that $M \vDash \Gamma$.

2. We may fail in this respect because there are M such that $M \vDash \Gamma$, but M is not one of the structures we intend. This may lead us to add axioms which are not true in M.

3. If we are successful at least in the respect that Γ is true in all the intended structures, then a sentence A is true in all intended structures whenever $\Gamma \vDash A$. Thus we can use logical tools (such as derivation methods) to show that sentences are true in all intended structures simply by showing that they are entailed by the axioms.

4. Sometimes we don't have intended structures in mind, but instead start from the axioms themselves: we begin with some primitives that we want to satisfy certain laws which we codify in an axiom system. One thing that we would like to verify right away is that the axioms do not contradict each other: if they do, there can be no concepts that obey

these laws, and we have tried to set up an incoherent theory. We can verify that this doesn't happen by finding a model of Γ. And if there are models of our theory, we can use logical methods to investigate them, and we can also use logical methods to construct models.

5. The independence of the axioms is likewise an important question. It may happen that one of the axioms is actually a consequence of the others, and so is redundant. We can prove that an axiom A in Γ is redundant by proving $\Gamma \setminus \{A\} \vDash A$. We can also prove that an axiom is not redundant by showing that $(\Gamma \setminus \{A\}) \cup \{\neg A\}$ is satisfiable. For instance, this is how it was shown that the parallel postulate is independent of the other axioms of geometry.

6. Another important question is that of definability of concepts in a theory: The choice of the language determines what the models of a theory consists of. But not every aspect of a theory must be represented separately in its models. For instance, every ordering $\leq$ determines a corresponding strict ordering $<$—given one, we can define the other. So it is not necessary that a model of a theory involving such an order must *also* contain the corresponding strict ordering. When is it the case, in general, that one relation can be defined in terms of others? When is it impossible to define a relation in terms of other (and hence must add it to the primitives of the language)?

8.2 Expressing Properties of Structures

It is often useful and important to express conditions on functions and relations, or more generally, that the functions and relations in a structure satisfy these conditions. For instance, we would like to have ways of distinguishing those structures for a language which "capture" what we want the predicate symbols to "mean" from those that do not. Of course we're completely

8.2. EXPRESSING PROPERTIES OF STRUCTURES

free to specify which structures we "intend," e.g., we can specify that the interpretation of the predicate symbol $\leq$ must be an ordering, or that we are only interested in interpretations of $\mathcal{L}$ in which the domain consists of sets and $\in$ is interpreted by the "is an element of" relation. But can we do this with sentences of the language? In other words, which conditions on a structure M can we express by a sentence (or perhaps a set of sentences) in the language of M? There are some conditions that we will not be able to express. For instance, there is no sentence of $\mathcal{L}_A$ which is only true in a structure M if $|M| = \mathbb{N}$. We cannot express "the domain contains only natural numbers." But there are "structural properties" of structures that we perhaps can express. Which properties of structures can we express by sentences? Or, to put it another way, which collections of structures can we describe as those making a sentence (or set of sentences) true?

Definition 8.2 (Model of a set). Let Γ be a set of sentences in a language $\mathcal{L}$. We say that a structure M *is a model of* Γ if $M \models A$ for all $A \in \Gamma$.

Example 8.3. The sentence $\forall x\, x \leq x$ is true in M iff $\leq^M$ is a reflexive relation. The sentence $\forall x \forall y\, ((x \leq y \wedge y \leq x) \rightarrow x = y)$ is true in M iff $\leq^M$ is anti-symmetric. The sentence $\forall x \forall y \forall z\, ((x \leq y \wedge y \leq z) \rightarrow x \leq z)$ is true in M iff $\leq^M$ is transitive. Thus, the models of

$$\{ \quad \forall x\, x \leq x,$$
$$\forall x \forall y\, ((x \leq y \wedge y \leq x) \rightarrow x = y),$$
$$\forall x \forall y \forall z\, ((x \leq y \wedge y \leq z) \rightarrow x \leq z) \quad \}$$

are exactly those structures in which $\leq^M$ is reflexive, anti-symmetric, and transitive, i.e., a partial order. Hence, we can take them as axioms for the *first-order theory of partial orders*.

8.3 Examples of First-Order Theories

Example 8.4. The theory of strict linear orders in the language $\mathscr{L}_<$ is axiomatized by the set

$$\forall x \, \neg x < x,$$
$$\forall x \, \forall y \, ((x < y \vee y < x) \vee x = y),$$
$$\forall x \, \forall y \, \forall z \, ((x < y \wedge y < z) \rightarrow x < z)$$

It completely captures the intended structures: every strict linear order is a model of this axiom system, and vice versa, if R is a linear order on a set X, then the structure M with $|M| = X$ and $<^M = R$ is a model of this theory.

Example 8.5. The theory of groups in the language 1 (constant symbol), $\cdot$ (two-place function symbol) is axiomatized by

$$\forall x \, (x \cdot 1) = x$$
$$\forall x \, \forall y \, \forall z \, (x \cdot (y \cdot z)) = ((x \cdot y) \cdot z)$$
$$\forall x \, \exists y \, (x \cdot y) = 1$$

Example 8.6. The theory of Peano arithmetic is axiomatized by the following sentences in the language of arithmetic $\mathscr{L}_A$.

$$\forall x \, \forall y \, (x' = y' \rightarrow x = y)$$
$$\forall x \, 0 \neq x'$$
$$\forall x \, (x + 0) = x$$
$$\forall x \, \forall y \, (x + y') = (x + y)'$$
$$\forall x \, (x \times 0) = 0$$
$$\forall x \, \forall y \, (x \times y') = ((x \times y) + x)$$
$$\forall x \, \forall y \, (x < y \leftrightarrow \exists z \, (z' + x) = y))$$

plus all sentences of the form

$$(A(0) \wedge \forall x \, (A(x) \rightarrow A(x'))) \rightarrow \forall x \, A(x)$$

8.3. EXAMPLES OF FIRST-ORDER THEORIES

Since there are infinitely many sentences of the latter form, this axiom system is infinite. The latter form is called the *induction schema*. (Actually, the induction schema is a bit more complicated than we let on here.)

The last axiom is an *explicit definition* of $<$.

Example 8.7. The theory of pure sets plays an important role in the foundations (and in the philosophy) of mathematics. A set is pure if all its elements are also pure sets. The empty set counts therefore as pure, but a set that has something as an element that is not a set would not be pure. So the pure sets are those that are formed just from the empty set and no "urelements," i.e., objects that are not themselves sets.

The following might be considered as an axiom system for a theory of pure sets:

$$\exists x \, \neg \exists y \, y \in x$$
$$\forall x \, \forall y \, (\forall z (z \in x \leftrightarrow z \in y) \to x = y)$$
$$\forall x \, \forall y \, \exists z \, \forall u \, (u \in z \leftrightarrow (u = x \lor u = y))$$
$$\forall x \, \exists y \, \forall z \, (z \in y \leftrightarrow \exists u \, (z \in u \land u \in x))$$

plus all sentences of the form

$$\exists x \, \forall y \, (y \in x \leftrightarrow A(y))$$

The first axiom says that there is a set with no elements (i.e., $\emptyset$ exists); the second says that sets are extensional; the third that for any sets X and Y, the set $\{X, Y\}$ exists; the fourth that for any set X, the set $\cup X$ exists, where $\cup X$ is the union of all the elements of X.

The sentences mentioned last are collectively called the *naive comprehension scheme*. It essentially says that for every $A(x)$, the set $\{x : A(x)\}$ exists—so at first glance a true, useful, and perhaps even necessary axiom. It is called "naive" because, as it turns out, it makes this theory unsatisfiable: if you take $A(y)$ to be $\neg y \in y$, you get the sentence

$$\exists x \, \forall y \, (y \in x \leftrightarrow \neg y \in y)$$

and this sentence is not satisfied in any structure.

Example 8.8. In the area of *mereology*, the relation of *parthood* is a fundamental relation. Just like theories of sets, there are theories of parthood that axiomatize various conceptions (sometimes conflicting) of this relation.

The language of mereology contains a single two-place predicate symbol P, and $P(x,y)$ "means" that x is a part of y. When we have this interpretation in mind, a structure for this language is called a *parthood structure*. Of course, not every structure for a single two-place predicate will really deserve this name. To have a chance of capturing "parthood," P^M must satisfy some conditions, which we can lay down as axioms for a theory of parthood. For instance, parthood is a partial order on objects: every object is a part (albeit an *improper* part) of itself; no two different objects can be parts of each other; a part of a part of an object is itself part of that object. Note that in this sense "is a part of" resembles "is a subset of," but does not resemble "is an element of" which is neither reflexive nor transitive.

$$\forall x\, P(x,x),$$
$$\forall x\, \forall y\, ((P(x,y) \land P(y,x)) \to x = y),$$
$$\forall x\, \forall y\, \forall z\, ((P(x,y) \land P(y,z)) \to P(x,z)),$$

Moreover, any two objects have a mereological sum (an object that has these two objects as parts, and is minimal in this respect).

$$\forall x\, \forall y\, \exists z\, \forall u\, (P(z,u) \leftrightarrow (P(x,u) \land P(y,u)))$$

These are only some of the basic principles of parthood considered by metaphysicians. Further principles, however, quickly become hard to formulate or write down without first introducing some defined relations. For instance, most metaphysicians interested in mereology also view the following as a valid principle: whenever an object x has a proper part y, it also has a part z that has no parts in common with y, and so that the fusion of y and z is x.

8.4 Expressing Relations in a Structure

One main use formulas can be put to is to express properties and relations in a structure M in terms of the primitives of the language $\mathscr{L}$ of M. By this we mean the following: the domain of M is a set of objects. The constant symbols, function symbols, and predicate symbols are interpreted in M by some objects in $|M|$, functions on $|M|$, and relations on $|M|$. For instance, if A_0^2 is in $\mathscr{L}$, then M assigns to it a relation $R = A_0^{2M}$. Then the formula $A_0^2(v_1, v_2)$ *expresses* that very relation, in the following sense: if a variable assignment s maps v_1 to $a \in |M|$ and v_2 to $b \in |M|$, then

$$Rab \quad \text{iff} \quad M, s \vDash A_0^2(v_1, v_2).$$

Note that we have to involve variable assignments here: we can't just say "Rab iff $M \vDash A_0^2(a, b)$" because a and b are not symbols of our language: they are elements of $|M|$.

Since we don't just have atomic formulas, but can combine them using the logical connectives and the quantifiers, more complex formulas can define other relations which aren't directly built into M. We're interested in how to do that, and specifically, which relations we can define in a structure.

Definition 8.9. Let $A(v_1, \ldots, v_n)$ be a formula of $\mathscr{L}$ in which only $v_1, \ldots, v_n$ occur free, and let M be a structure for $\mathscr{L}$. $A(v_1, \ldots, v_n)$ *expresses the relation* $R \subseteq |M|^n$ iff

$$Ra_1 \ldots a_n \quad \text{iff} \quad M, s \vDash A(v_1, \ldots, v_n)$$

for any variable assignment s with $s(v_i) = a_i$ $(i = 1, \ldots, n)$.

Example 8.10. In the standard model of arithmetic N, the formula $v_1 < v_2 \lor v_1 = v_2$ expresses the $\leq$ relation on $\mathbb{N}$. The formula $v_2 = v_1'$ expresses the successor relation, i.e., the relation $R \subseteq \mathbb{N}^2$ where Rnm holds if m is the successor of n. The formula $v_1 = v_2'$ expresses the predecessor relation. The formulas $\exists v_3 \, (v_3 \neq 0 \land v_2 = (v_1 + v_3))$ and $\exists v_3 \, (v_1 + v_3') = v_2$ both express the < relation. This means that the predicate symbol < is actually superfluous in the language of arithmetic; it can be defined.

This idea is not just interesting in specific structures, but generally whenever we use a language to describe an intended model or models, i.e., when we consider theories. These theories often only contain a few predicate symbols as basic symbols, but in the domain they are used to describe often many other relations play an important role. If these other relations can be systematically expressed by the relations that interpret the basic predicate symbols of the language, we say we can *define* them in the language.

8.5 The Theory of Sets

Almost all of mathematics can be developed in the theory of sets. Developing mathematics in this theory involves a number of things. First, it requires a set of axioms for the relation $\in$. A number of different axiom systems have been developed, sometimes with conflicting properties of $\in$. The axiom system known as **ZFC**, Zermelo-Fraenkel set theory with the axiom of choice stands out: it is by far the most widely used and studied, because it turns out that its axioms suffice to prove almost all the things mathematicians expect to be able to prove. But before that can be established, it first is necessary to make clear how we can even *express* all the things mathematicians would like to express. For starters, the language contains no constant symbols or function symbols, so it seems at first glance unclear that we can talk about particular sets (such as $\emptyset$ or $\mathbb{N}$), can talk about operations on sets (such as $X \cup Y$ and $\wp(X)$), let alone other constructions which involve things other than sets, such as relations and functions.

To begin with, "is an element of" is not the only relation we are interested in: "is a subset of" seems almost as important. But we can *define* "is a subset of" in terms of "is an element of." To do this, we have to find a formula $A(x,y)$ in the language of set theory which is satisfied by a pair of sets $\langle X, Y \rangle$ iff $X \subseteq Y$. But X

8.5. THE THEORY OF SETS

is a subset of Y just in case all elements of X are also elements of Y. So we can define $\subseteq$ by the formula

$$\forall z\, (z \in x \to z \in y)$$

Now, whenever we want to use the relation $\subseteq$ in a formula, we could instead use that formula (with x and y suitably replaced, and the bound variable z renamed if necessary). For instance, extensionality of sets means that if any sets x and y are contained in each other, then x and y must be the same set. This can be expressed by $\forall x\, \forall y\, ((x \subseteq y \land y \subseteq x) \to x = y)$, or, if we replace $\subseteq$ by the above definition, by

$$\forall x\, \forall y\, ((\forall z\, (z \in x \to z \in y) \land \forall z\, (z \in y \to z \in x)) \to x = y).$$

This is in fact one of the axioms of **ZFC**, the "axiom of extensionality."

There is no constant symbol for $\emptyset$, but we can express "x is empty" by $\neg \exists y\, y \in x$. Then "$\emptyset$ exists" becomes the sentence $\exists x\, \neg \exists y\, y \in x$. This is another axiom of **ZFC**. (Note that the axiom of extensionality implies that there is only one empty set.) Whenever we want to talk about $\emptyset$ in the language of set theory, we would write this as "there is a set that's empty and ..." As an example, to express the fact that $\emptyset$ is a subset of every set, we could write

$$\exists x\, (\neg \exists y\, y \in x \land \forall z\, x \subseteq z)$$

where, of course, $x \subseteq z$ would in turn have to be replaced by its definition.

To talk about operations on sets, such has $X \cup Y$ and $\wp(X)$, we have to use a similar trick. There are no function symbols in the language of set theory, but we can express the functional relations $X \cup Y = Z$ and $\wp(X) = Y$ by

$$\forall u\, ((u \in x \lor u \in y) \leftrightarrow u \in z)$$
$$\forall u\, (u \subseteq x \leftrightarrow u \in y)$$

since the elements of $X \cup Y$ are exactly the sets that are either elements of X or elements of Y, and the elements of $\wp(X)$ are exactly the subsets of X. However, this doesn't allow us to use $x \cup y$ or $\wp(x)$ as if they were terms: we can only use the entire formulas that define the relations $X \cup Y = Z$ and $\wp(X) = Y$. In fact, we do not know that these relations are ever satisfied, i.e., we do not know that unions and power sets always exist. For instance, the sentence $\forall x\, \exists y\, \wp(x) = y$ is another axiom of **ZFC** (the power set axiom).

Now what about talk of ordered pairs or functions? Here we have to explain how we can think of ordered pairs and functions as special kinds of sets. One way to define the ordered pair $\langle x, y \rangle$ is as the set $\{\{x\}, \{x, y\}\}$. But like before, we cannot introduce a function symbol that names this set; we can only define the relation $\langle x, y \rangle = z$, i.e., $\{\{x\}, \{x, y\}\} = z$:

$$\forall u\, (u \in z \leftrightarrow (\forall v\, (v \in u \leftrightarrow v = x) \vee \forall v\, (v \in u \leftrightarrow (v = x \vee v = y))))$$

This says that the elements u of z are exactly those sets which either have x as its only element or have x and y as its only elements (in other words, those sets that are either identical to $\{x\}$ or identical to $\{x, y\}$). Once we have this, we can say further things, e.g., that $X \times Y = Z$:

$$\forall z\, (z \in Z \leftrightarrow \exists x\, \exists y\, (x \in X \wedge y \in Y \wedge \langle x, y \rangle = z))$$

A function $f \colon X \to Y$ can be thought of as the relation $f(x) = y$, i.e., as the set of pairs $\{\langle x, y \rangle : f(x) = y\}$. We can then say that a set f is a function from X to Y if (a) it is a relation $\subseteq X \times Y$, (b) it is total, i.e., for all $x \in X$ there is some $y \in Y$ such that $\langle x, y \rangle \in f$ and (c) it is functional, i.e., whenever $\langle x, y \rangle, \langle x, y' \rangle \in f$, $y = y'$ (because values of functions must be unique). So "f is a function from X to Y" can be written as:

$$\forall u\, (u \in f \to \exists x\, \exists y\, (x \in X \wedge y \in Y \wedge \langle x, y \rangle = u)) \wedge$$
$$\forall x\, (x \in X \to (\exists y\, (y \in Y \wedge \mathrm{maps}(f, x, y)) \wedge$$
$$(\forall y\, \forall y'\, ((\mathrm{maps}(f, x, y) \wedge \mathrm{maps}(f, x, y')) \to y = y')))$$

8.5. THE THEORY OF SETS

where maps(f,x,y) abbreviates $\exists v\, (v \in f \wedge \langle x,y \rangle = v)$ (this formula expresses "$f(x) = y$").

It is now also not hard to express that $f: X \to Y$ is injective, for instance:

$$f: X \to Y \wedge \forall x\, \forall x'\, ((x \in X \wedge x' \in X \wedge$$
$$\exists y\, (\text{maps}(f,x,y) \wedge \text{maps}(f,x',y))) \to x = x')$$

A function $f: X \to Y$ is injective iff, whenever f maps $x, x' \in X$ to a single y, $x = x'$. If we abbreviate this formula as $\text{inj}(f, X, Y)$, we're already in a position to state in the language of set theory something as non-trivial as Cantor's theorem: there is no injective function from $\wp(X)$ to X:

$$\forall X\, \forall Y\, (\wp(X) = Y \to \neg \exists f\, \text{inj}(f, Y, X))$$

One might think that set theory requires another axiom that guarantees the existence of a set for every defining property. If $A(x)$ is a formula of set theory with the variable x free, we can consider the sentence

$$\exists y\, \forall x\, (x \in y \leftrightarrow A(x)).$$

This sentence states that there is a set y whose elements are all and only those x that satisfy $A(x)$. This schema is called the "comprehension principle." It looks very useful; unfortunately it is inconsistent. Take $A(x) \equiv \neg x \in x$, then the comprehension principle states

$$\exists y\, \forall x\, (x \in y \leftrightarrow x \notin x),$$

i.e., it states the existence of a set of all sets that are not elements of themselves. No such set can exist—this is Russell's Paradox. **ZFC**, in fact, contains a restricted—and consistent—version of this principle, the separation principle:

$$\forall z\, \exists y\, \forall x\, (x \in y \leftrightarrow (x \in z \wedge A(x))).$$

8.6 Expressing the Size of Structures

There are some properties of structures we can express even without using the non-logical symbols of a language. For instance, there are sentences which are true in a structure iff the domain of the structure has at least, at most, or exactly a certain number n of elements.

Proposition 8.11. *The sentence*

$$A_{\geq n} \equiv \exists x_1 \exists x_2 \ldots \exists x_n$$
$$(x_1 \neq x_2 \wedge x_1 \neq x_3 \wedge x_1 \neq x_4 \wedge \cdots \wedge x_1 \neq x_n \wedge$$
$$x_2 \neq x_3 \wedge x_2 \neq x_4 \wedge \cdots \wedge x_2 \neq x_n \wedge$$
$$\vdots$$
$$x_{n-1} \neq x_n)$$

is true in a structure M iff $|M|$ contains at least n elements. Consequently, $M \vDash \neg A_{\geq n+1}$ iff $|M|$ contains at most n elements.

Proposition 8.12. *The sentence*

$$A_{=n} \equiv \exists x_1 \exists x_2 \ldots \exists x_n$$
$$(x_1 \neq x_2 \wedge x_1 \neq x_3 \wedge x_1 \neq x_4 \wedge \cdots \wedge x_1 \neq x_n \wedge$$
$$x_2 \neq x_3 \wedge x_2 \neq x_4 \wedge \cdots \wedge x_2 \neq x_n \wedge$$
$$\vdots$$
$$x_{n-1} \neq x_n \wedge$$
$$\forall y\, (y = x_1 \vee \cdots \vee y = x_n))$$

is true in a structure M iff $|M|$ contains exactly n elements.

8.6. EXPRESSING THE SIZE OF STRUCTURES

Proposition 8.13. *A structure is infinite iff it is a model of*

$$\{A_{\geq 1}, A_{\geq 2}, A_{\geq 3}, \dots\}.$$

There is no single purely logical sentence which is true in M iff $|M|$ is infinite. However, one can give sentences with non-logical predicate symbols which only have infinite models (although not every infinite structure is a model of them). The property of being a finite structure, and the property of being a uncountable structure cannot even be expressed with an infinite set of sentences. These facts follow from the compactness and Löwenheim-Skolem theorems.

Summary

Sets of sentences in a sense describe the structures in which they are jointly true; these structures are their **models**. Conversely, if we start with a structure or set of structures, we might be interested in the set of sentences they are models of, this is the **theory** of the structure or set of structures. Any such set of sentences has the property that every sentence entailed by them is already in the set; they are **closed**. More generally, we call a set Γ a theory if it is closed under entailment, and say Γ is **axiomatized** by Δ is Γ consists of all sentences entailed by Δ.

Mathematics yields many examples of theories, e.g., the theories of linear orders, of groups, or theories of arithmetic, e.g., the theory axiomatized by Peano's axioms. But there are many examples of important theories in other disciplines as well, e.g., relational databases may be thought of as theories, and metaphysics concerns itself with theories of parthood which can be axiomatized.

One significant question when setting up a theory for study is whether its language is expressive enough to allow us to formulate everything we want the theory to talk about, and another is whether it is strong enough to prove what we want it to prove. To **express** a relation we need a formula with the requisite number of free variables. In **set theory**, we only have $\in$ as a relation symbol, but it allows us to express $x \subseteq y$ using $\forall u\,(u \in x \rightarrow u \in y)$. **Zermelo-Fraenkel set theory ZFC**, in fact, is strong enough to both express (almost) every mathematical claim and to (almost) prove every mathematical theorem using a handful of axioms and a chain of increasingly complicated definitions such as that of $\subseteq$.

Problems

Problem 8.1. Find formulas in $\mathscr{L}_A$ which define the following relations:

1. n is between i and j;

2. n evenly divides m (i.e., m is a multiple of n);

3. n is a prime number (i.e., no number other than 1 and n evenly divides n).

Problem 8.2. Suppose the formula $A(v_1, v_2)$ expresses the relation $R \subseteq |M|^2$ in a structure M. Find formulas that express the following relations:

1. the inverse R^{-1} of R;

2. the relative product $R \mid R$;

Can you find a way to express R^+, the transitive closure of R?

Problem 8.3. Let $\mathscr{L}$ be the language containing a 2-place predicate symbol $<$ only (no other constant symbols, function symbols or predicate symbols— except of course $=$). Let N be the structure such that $|N| = \mathbb{N}$, and $<^N = \{\langle n, m \rangle : n < m\}$. Prove the following:

1. $\{0\}$ is definable in N;

2. $\{1\}$ is definable in N;

8.6. EXPRESSING THE SIZE OF STRUCTURES

3. $\{2\}$ is definable in N;

4. for each $n \in \mathbb{N}$, the set $\{n\}$ is definable in N;

5. every finite subset of $|N|$ is definable in N;

6. every co-finite subset of $|N|$ is definable in N (where $X \subseteq \mathbb{N}$ is co-finite iff $\mathbb{N} \setminus X$ is finite).

Problem 8.4. Show that the comprehension principle is inconsistent by giving a derivation that shows

$$\exists y\, \forall x\, (x \in y \leftrightarrow x \notin x) \vdash \bot.$$

It may help to first show $(A \to \neg A) \land (\neg A \to A) \vdash \bot$.

CHAPTER 9
Derivation Systems

9.1 Introduction

Logics commonly have both a semantics and a derivation system. The semantics concerns concepts such as truth, satisfiability, validity, and entailment. The purpose of derivation systems is to provide a purely syntactic method of establishing entailment and validity. They are purely syntactic in the sense that a derivation in such a system is a finite syntactic object, usually a sequence (or other finite arrangement) of sentences or formulas. Good derivation systems have the property that any given sequence or arrangement of sentences or formulas can be verified mechanically to be "correct."

The simplest (and historically first) derivation systems for first-order logic were *axiomatic*. A sequence of formulas counts as a derivation in such a system if each individual formula in it is either among a fixed set of "axioms" or follows from formulas coming before it in the sequence by one of a fixed number of "inference rules"—and it can be mechanically verified if a formula is an axiom and whether it follows correctly from other formulas by one of the inference rules. Axiomatic derivation systems are easy to describe—and also easy to handle meta-theoretically—

but derivations in them are hard to read and understand, and are also hard to produce.

Other derivation systems have been developed with the aim of making it easier to construct derivations or easier to understand derivations once they are complete. Examples are natural deduction, truth trees, also known as tableaux proofs, and the sequent calculus. Some derivation systems are designed especially with mechanization in mind, e.g., the resolution method is easy to implement in software (but its derivations are essentially impossible to understand). Most of these other derivation systems represent derivations as trees of formulas rather than sequences. This makes it easier to see which parts of a derivation depend on which other parts.

So for a given logic, such as first-order logic, the different derivation systems will give different explications of what it is for a sentence to be a *theorem* and what it means for a sentence to be derivable from some others. However that is done (via axiomatic derivations, natural deductions, sequent derivations, truth trees, resolution refutations), we want these relations to match the semantic notions of validity and entailment. Let's write $\vdash A$ for "A is a theorem" and "$\Gamma \vdash A$" for "A is derivable from Γ." However $\vdash$ is defined, we want it to match up with $\vDash$, that is:

1. $\vdash A$ if and only if $\vDash A$

2. $\Gamma \vdash A$ if and only if $\Gamma \vDash A$

The "only if" direction of the above is called *soundness*. A derivation system is sound if derivability guarantees entailment (or validity). Every decent derivation system has to be sound; unsound derivation systems are not useful at all. After all, the entire purpose of a derivation is to provide a syntactic guarantee of validity or entailment. We'll prove soundness for the derivation systems we present.

The converse "if" direction is also important: it is called *completeness*. A complete derivation system is strong enough to show that A is a theorem whenever A is valid, and that $\Gamma \vdash A$ whenever $\Gamma \vDash A$. Completeness is harder to establish, and some logics have no complete derivation systems. First-order logic does. Kurt Gödel was the first one to prove completeness for a derivation system of first-order logic in his 1929 dissertation.

Another concept that is connected to derivation systems is that of *consistency*. A set of sentences is called inconsistent if anything whatsoever can be derived from it, and consistent otherwise. Inconsistency is the syntactic counterpart to unsatisfiablity: like unsatisfiable sets, inconsistent sets of sentences do not make good theories, they are defective in a fundamental way. Consistent sets of sentences may not be true or useful, but at least they pass that minimal threshold of logical usefulness. For different derivation systems the specific definition of consistency of sets of sentences might differ, but like $\vdash$, we want consistency to coincide with its semantic counterpart, satisfiability. We want it to always be the case that Γ is consistent if and only if it is satisfiable. Here, the "if" direction amounts to completeness (consistency guarantees satisfiability), and the "only if" direction amounts to soundness (satisfiability guarantees consistency). In fact, for classical first-order logic, the two versions of soundness and completeness are equivalent.

9.2 The Sequent Calculus

While many derivation systems operate with arrangements of sentences, the sequent calculus operates with *sequents*. A sequent is an expression of the form

$$A_1, \ldots, A_m \Rightarrow B_1, \ldots, B_m,$$

that is a pair of sequences of sentences, separated by the sequent symbol $\Rightarrow$. Either sequence may be empty. A derivation in the sequent calculus is a tree of sequents, where the topmost sequents are of a special form (they are called "initial sequents" or "axioms") and every other sequent follows from the sequents immediately above it by one of the rules of inference. The rules of inference either manipulate the sentences in the sequents (adding, removing, or rearranging them on either the left or the right), or they introduce a complex formula in the conclusion of the rule. For instance, the ∧L rule allows the inference from $A, \Gamma \Rightarrow \Delta$ to $A \wedge B, \Gamma \Rightarrow \Delta$, and the →R allows the inference from $A, \Gamma \Rightarrow \Delta, B$ to $\Gamma \Rightarrow \Delta, A \rightarrow B$, for any Γ, Δ, A, and B. (In particular, Γ and Δ may be empty.)

The ⊢ relation based on the sequent calculus is defined as follows: $\Gamma \vdash A$ iff there is some sequence Γ_0 such that every A in Γ_0 is in Γ and there is a derivation with the sequent $\Gamma_0 \Rightarrow A$ at its root. A is a theorem in the sequent calculus if the sequent $\Rightarrow A$ has a derivation. For instance, here is a derivation that shows that $\vdash (A \wedge B) \rightarrow A$:

$$\dfrac{\dfrac{\dfrac{}{A \Rightarrow A}}{A \wedge B \Rightarrow A} \wedge\text{L}}{\Rightarrow (A \wedge B) \rightarrow A} \rightarrow\text{R}$$

A set Γ is inconsistent in the sequent calculus if there is a derivation of $\Gamma_0 \Rightarrow$ (where every $A \in \Gamma_0$ is in Γ and the right side of the sequent is empty). Using the rule WR, any sentence can be derived from an inconsistent set.

The sequent calculus was invented in the 1930s by Gerhard Gentzen. Because of its systematic and symmetric design, it is a very useful formalism for developing a theory of derivations. It is relatively easy to find derivations in the sequent calculus, but these derivations are often hard to read and their connection to proofs are sometimes not easy to see. It has proved to be a very elegant approach to derivation systems, however, and many logics have sequent calculus systems.

9.3 Natural Deduction

Natural deduction is a derivation system intended to mirror actual reasoning (especially the kind of regimented reasoning employed by mathematicians). Actual reasoning proceeds by a number of "natural" patterns. For instance, proof by cases allows us to establish a conclusion on the basis of a disjunctive premise, by establishing that the conclusion follows from either of the disjuncts. Indirect proof allows us to establish a conclusion by showing that its negation leads to a contradiction. Conditional proof establishes a conditional claim "if ... then ... " by showing that the consequent follows from the antecedent. Natural deduction is a formalization of some of these natural inferences. Each of the logical connectives and quantifiers comes with two rules, an introduction and an elimination rule, and they each correspond to one such natural inference pattern. For instance, →Intro corresponds to conditional proof, and ∨Elim to proof by cases. A particularly simple rule is ∧Elim which allows the inference from $A \wedge B$ to A (or B).

One feature that distinguishes natural deduction from other derivation systems is its use of assumptions. A derivation in natural deduction is a tree of formulas. A single formula stands at the root of the tree of formulas, and the "leaves" of the tree are formulas from which the conclusion is derived. In natural deduction, some leaf formulas play a role inside the derivation but are "used up" by the time the derivation reaches the conclusion. This corresponds to the practice, in actual reasoning, of introducing hypotheses which only remain in effect for a short while. For instance, in a proof by cases, we assume the truth of each of the disjuncts; in conditional proof, we assume the truth of the antecedent; in indirect proof, we assume the truth of the negation of the conclusion. This way of introducing hypothetical assumptions and then doing away with them in the service of establishing an intermediate step is a hallmark of natural deduction. The formulas at the leaves of a natural deduction derivation are called assumptions, and some of the rules of inference may "discharge" them. For instance, if we have a derivation of B from some assumptions which include A, then the →Intro rule allows us to infer $A \rightarrow B$ and discharge any assumption of the form A. (To keep track of which assumptions are discharged at which inferences, we label the inference and the assumptions it discharges with a number.) The assumptions that remain undischarged at the end of the derivation are together sufficient for the truth of the conclusion, and so a derivation establishes that its undischarged assumptions entail its conclusion.

The relation $\Gamma \vdash A$ based on natural deduction holds iff there is a derivation in which A is the last sentence in the tree, and every leaf which is undischarged is in Γ. A is a theorem in natural deduction iff there is a derivation in which A is the last sentence and all assumptions are discharged. For instance, here is a derivation that shows that $\vdash (A \wedge B) \to A$:

$$1\dfrac{\dfrac{[A \wedge B]^1}{A}\wedge\text{Elim}}{(A \wedge B) \to A}\to\text{Intro}$$

The label 1 indicates that the assumption $A \wedge B$ is discharged at the $\to$Intro inference.

A set Γ is inconsistent iff $\Gamma \vdash \bot$ in natural deduction. The rule $\bot_I$ makes it so that from an inconsistent set, any sentence can be derived.

Natural deduction systems were developed by Gerhard Gentzen and Stanisław Jaśkowski in the 1930s, and later developed by Dag Prawitz and Frederic Fitch. Because its inferences mirror natural methods of proof, it is favored by philosophers. The versions developed by Fitch are often used in introductory logic textbooks. In the philosophy of logic, the rules of natural deduction have sometimes been taken to give the meanings of the logical operators ("proof-theoretic semantics").

9.4 Tableaux

While many derivation systems operate with arrangements of sentences, tableaux operate with signed formulas. A signed formula is a pair consisting of a truth value sign ($\mathbb{T}$ or $\mathbb{F}$) and a sentence

$$\mathbb{T}\, A \text{ or } \mathbb{F}\, A.$$

A tableau consists of signed formulas arranged in a downward-branching tree. It begins with a number of *assumptions* and continues with signed formulas which result from one of the signed formulas above it by applying one of the rules of inference. Each rule allows us to add one or more signed formulas to the end of a branch, or two signed formulas side by side—in this case a branch splits into two, with the two added signed formulas forming the ends of the two branches.

A rule applied to a complex signed formula results in the addition of signed formulas which are immediate sub-formulas. They come in pairs, one rule for each of the two signs. For instance, the $\wedge\mathbb{T}$ rule applies to $\mathbb{T}\,A \wedge B$, and allows the addition of both the two signed formulas $\mathbb{T}\,A$ and $\mathbb{T}\,B$ to the end of any branch containing $\mathbb{T}\,A \wedge B$, and the rule $A \wedge B\mathbb{F}$ allows a branch to be split by adding $\mathbb{F}\,A$ and $\mathbb{F}\,B$ side-by-side. A tableau is closed if every one of its branches contains a matching pair of signed formulas $\mathbb{T}\,A$ and $\mathbb{F}\,A$.

The $\vdash$ relation based on tableaux is defined as follows: $\Gamma \vdash A$ iff there is some finite set $\Gamma_0 = \{B_1, \ldots, B_n\} \subseteq \Gamma$ such that there is a closed tableau for the assumptions

$$\{\mathbb{F}\,A, \mathbb{T}\,B_1, \ldots, \mathbb{T}\,B_n\}$$

For instance, here is a closed tableau that shows that $\vdash (A \wedge B) \to A$:

1.	$\mathbb{F}\,(A \wedge B) \to A$	Assumption
2.	$\mathbb{T}\,A \wedge B$	$\to \mathbb{F}\,1$
3.	$\mathbb{F}\,A$	$\to \mathbb{F}\,1$
4.	$\mathbb{T}\,A$	$\to \mathbb{T}\,2$
5.	$\mathbb{T}\,B$	$\to \mathbb{T}\,2$
	$\otimes$	

A set Γ is inconsistent in the tableau calculus if there is a closed tableau for assumptions

$$\{\mathbb{T}\,B_1, \ldots, \mathbb{T}\,B_n\}$$

for some $B_i \in \Gamma$.

Tableaux were invented in the 1950s independently by Evert Beth and Jaakko Hintikka, and simplified and popularized by Raymond Smullyan. They are very easy to use, since constructing a tableau is a very systematic procedure. Because of the systematic nature of tableaux, they also lend themselves to implementation by computer. However, a tableau is often hard to read and their connection to proofs are sometimes not easy to see. The approach is also quite general, and many different logics have tableau systems. Tableaux also help us to find structures that satisfy given (sets of) sentences: if the set is satisfiable, it won't have a closed tableau, i.e., any tableau will have an open branch. The satisfying structure can be "read off" an open branch, provided every rule it is possible to apply has been applied on that branch. There is also a very close connection to the sequent calculus: essentially, a closed tableau is a condensed derivation in the sequent calculus, written upside-down.

9.5 Axiomatic Derivations

Axiomatic derivations are the oldest and simplest logical derivation systems. Its derivations are simply sequences of sentences. A sequence of sentences counts as a correct derivation if every sentence A in it satisfies one of the following conditions:

1. A is an axiom, or

2. A is an element of a given set Γ of sentences, or

3. A is justified by a rule of inference.

To be an axiom, A has to have the form of one of a number of fixed sentence schemas. There are many sets of axiom schemas that provide a satisfactory (sound and complete) derivation system for first-order logic. Some are organized according to the connectives they govern, e.g., the schemas

$$A \to (B \to A) \qquad B \to (B \lor C) \qquad (B \land C) \to B$$

are common axioms that govern $\to$, $\vee$ and $\wedge$. Some axiom systems aim at a minimal number of axioms. Depending on the connectives that are taken as primitives, it is even possible to find axiom systems that consist of a single axiom.

A rule of inference is a conditional statement that gives a sufficient condition for a sentence in a derivation to be justified. Modus ponens is one very common such rule: it says that if A and $A \to B$ are already justified, then B is justified. This means that a line in a derivation containing the sentence B is justified, provided that both A and $A \to B$ (for some sentence A) appear in the derivation before B.

The $\vdash$ relation based on axiomatic derivations is defined as follows: $\Gamma \vdash A$ iff there is a derivation with the sentence A as its last formula (and Γ is taken as the set of sentences in that derivation which are justified by (2) above). A is a theorem if A has a derivation where Γ is empty, i.e., every sentence in the derivation is justfied either by (1) or (3). For instance, here is a derivation that shows that $\vdash A \to (B \to (B \vee A))$:

1. $B \to (B \vee A)$
2. $(B \to (B \vee A)) \to (A \to (B \to (B \vee A)))$
3. $A \to (B \to (B \vee A))$

The sentence on line 1 is of the form of the axiom $A \to (A \vee B)$ (with the roles of A and B reversed). The sentence on line 2 is of the form of the axiom $A \to (B \to A)$. Thus, both lines are justified. Line 3 is justified by modus ponens: if we abbreviate it as D, then line 2 has the form $C \to D$, where C is $B \to (B \vee A)$, i.e., line 1.

A set Γ is inconsistent if $\Gamma \vdash \bot$. A complete axiom system will also prove that $\bot \to A$ for any A, and so if Γ is inconsistent, then $\Gamma \vdash A$ for any A.

Systems of axiomatic derivations for logic were first given by Gottlob Frege in his 1879 *Begriffsschrift*, which for this reason is often considered the first work of modern logic. They were perfected in Alfred North Whitehead and Bertrand Russell's *Principia Mathematica* and by David Hilbert and his students in the

9.5. AXIOMATIC DERIVATIONS

1920s. They are thus often called "Frege systems" or "Hilbert systems." They are very versatile in that it is often easy to find an axiomatic system for a logic. Because derivations have a very simple structure and only one or two inference rules, it is also relatively easy to prove things *about* them. However, they are very hard to use in practice, i.e., it is difficult to find and write proofs.

CHAPTER 10

The Sequent Calculus

10.1 Rules and Derivations

For the following, let $\Gamma, \Delta, \Pi, \Lambda$ represent finite sequences of sentences.

> **Definition 10.1 (Sequent).** A *sequent* is an expression of the form
> $$\Gamma \Rightarrow \Delta$$
> where Γ and Δ are finite (possibly empty) sequences of sentences of the language $\mathscr{L}$. Γ is called the *antecedent*, while Δ is the *succedent*.

The intuitive idea behind a sequent is: if all of the sentences in the antecedent hold, then at least one of the sentences in the succedent holds. That is, if $\Gamma = \langle A_1, \ldots, A_m \rangle$ and $\Delta = \langle B_1, \ldots, B_n \rangle$, then $\Gamma \Rightarrow \Delta$ holds iff

$$(A_1 \wedge \cdots \wedge A_m) \rightarrow (B_1 \vee \cdots \vee B_n)$$

holds. There are two special cases: where Γ is empty and when Δ is empty. When Γ is empty, i.e., $m = 0$, $\Rightarrow \Delta$ holds iff $B_1 \vee \cdots \vee$

10.2. PROPOSITIONAL RULES

B_n holds. When Δ is empty, i.e., $n = 0$, $\Gamma \Rightarrow$ holds iff $\neg(A_1 \wedge \cdots \wedge A_m)$ does. We say a sequent is valid iff the corresponding sentence is valid.

If Γ is a sequence of sentences, we write Γ, A for the result of appending A to the right end of Γ (and A, Γ for the result of appending A to the left end of Γ). If Δ is a sequence of sentences also, then Γ, Δ is the concatenation of the two sequences.

Definition 10.2 (Initial Sequent). An *initial sequent* is a sequent of one of the following forms:

1. $A \Rightarrow A$

2. $\bot \Rightarrow$

for any sentence A in the language.

Derivations in the sequent calculus are certain trees of sequents, where the topmost sequents are initial sequents, and if a sequent stands below one or two other sequents, it must follow correctly by a rule of inference. The rules for **LK** are divided into two main types: *logical* rules and *structural* rules. The logical rules are named for the main operator of the sentence containing A and/or B in the lower sequent. Each one comes in two versions, one for inferring a sequent with the sentence containing the logical operator on the left, and one with the sentence on the right.

10.2 Propositional Rules

Rules for $\neg$

$$\frac{\Gamma \Rightarrow \Delta, A}{\neg A, \Gamma \Rightarrow \Delta} \neg\text{L} \qquad \frac{A, \Gamma \Rightarrow \Delta}{\Gamma \Rightarrow \Delta, \neg A} \neg\text{R}$$

Rules for $\wedge$

$$\frac{A,\Gamma \Rightarrow \Delta}{A \wedge B,\Gamma \Rightarrow \Delta} \wedge\text{L}$$

$$\frac{B,\Gamma \Rightarrow \Delta}{A \wedge B,\Gamma \Rightarrow \Delta} \wedge\text{L}$$

$$\frac{\Gamma \Rightarrow \Delta,A \quad \Gamma \Rightarrow \Delta,B}{\Gamma \Rightarrow \Delta,A \wedge B} \wedge\text{R}$$

Rules for $\vee$

$$\frac{A,\Gamma \Rightarrow \Delta \quad B,\Gamma \Rightarrow \Delta}{A \vee B,\Gamma \Rightarrow \Delta} \vee\text{L}$$

$$\frac{\Gamma \Rightarrow \Delta,A}{\Gamma \Rightarrow \Delta,A \vee B} \vee\text{R}$$

$$\frac{\Gamma \Rightarrow \Delta,B}{\Gamma \Rightarrow \Delta,A \vee B} \vee\text{R}$$

Rules for $\to$

$$\frac{\Gamma \Rightarrow \Delta,A \quad B,\Pi \Rightarrow \Lambda}{A \to B,\Gamma,\Pi \Rightarrow \Delta,\Lambda} \to\text{L}$$

$$\frac{A,\Gamma \Rightarrow \Delta,B}{\Gamma \Rightarrow \Delta,A \to B} \to\text{R}$$

10.3 Quantifier Rules

Rules for $\forall$

$$\frac{A(t),\Gamma \Rightarrow \Delta}{\forall x\, A(x),\Gamma \Rightarrow \Delta} \forall\text{L}$$

$$\frac{\Gamma \Rightarrow \Delta,A(a)}{\Gamma \Rightarrow \Delta,\forall x\, A(x)} \forall\text{R}$$

In $\forall$L, t is a closed term (i.e., one without variables). In $\forall$R, a is a constant symbol which must not occur anywhere in the lower sequent of the $\forall$R rule. We call a the *eigenvariable* of the $\forall$R inference.

Rules for $\exists$

$$\frac{A(a),\Gamma \Rightarrow \Delta}{\exists x\, A(x),\Gamma \Rightarrow \Delta} \exists\text{L}$$

$$\frac{\Gamma \Rightarrow \Delta,A(t)}{\Gamma \Rightarrow \Delta,\exists x\, A(x)} \exists\text{R}$$

10.4. STRUCTURAL RULES

Again, t is a closed term, and a is a constant symbol which does not occur in the lower sequent of the $\exists L$ rule. We call a the *eigenvariable* of the $\exists L$ inference.

The condition that an eigenvariable not occur in the lower sequent of the $\forall R$ or $\exists L$ inference is called the *eigenvariable condition*.

We use the term "eigenvariable" even though a in the above rules is a constant symbol. This has historical reasons.

In $\exists R$ and $\forall L$ there are no restrictions on the term t. On the other hand, in the $\exists L$ and $\forall R$ rules, the eigenvariable condition requires that the constant symbol a does not occur anywhere outside of $A(a)$ in the upper sequent. It is necessary to ensure that the system is sound, i.e., only derives sequents that are valid. Without this condition, the following would be allowed:

$$\cfrac{\cfrac{A(a) \Rightarrow A(a)}{\exists x\, A(x) \Rightarrow A(a)}{}^{*\exists L}}{\exists x\, A(x) \Rightarrow \forall x\, A(x)}\forall R \qquad \cfrac{\cfrac{A(a) \Rightarrow A(a)}{A(a) \Rightarrow \forall x\, A(x)}{}^{*\forall R}}{\exists x\, A(x) \Rightarrow \forall x\, A(x)}\exists L$$

However, $\exists x\, A(x) \Rightarrow \forall x\, A(x)$ is not valid.

10.4 Structural Rules

We also need a few rules that allow us to rearrange sentences in the left and right side of a sequent. Since the logical rules require that the sentences in the premise which the rule acts upon stand either to the far left or to the far right, we need an "exchange" rule that allows us to move sentences to the right position. It's also important sometimes to be able to combine two identical sentences into one, and to add a sentence on either side.

Weakening

$$\cfrac{\Gamma \Rightarrow \Delta}{A, \Gamma \Rightarrow \Delta}\,WL \qquad\qquad \cfrac{\Gamma \Rightarrow \Delta}{\Gamma \Rightarrow \Delta, A}\,WR$$

Contraction

$$\dfrac{A,A,\Gamma \Rightarrow \Delta}{A,\Gamma \Rightarrow \Delta}\ \text{CL} \qquad \dfrac{\Gamma \Rightarrow \Delta,A,A}{\Gamma \Rightarrow \Delta,A}\ \text{CR}$$

Exchange

$$\dfrac{\Gamma,A,B,\Pi \Rightarrow \Delta}{\Gamma,B,A,\Pi \Rightarrow \Delta}\ \text{XL} \qquad \dfrac{\Gamma \Rightarrow \Delta,A,B,\Lambda}{\Gamma \Rightarrow \Delta,B,A,\Lambda}\ \text{XR}$$

A series of weakening, contraction, and exchange inferences will often be indicated by double inference lines.

The following rule, called "cut," is not strictly speaking necessary, but makes it a lot easier to reuse and combine derivations.

$$\dfrac{\Gamma \Rightarrow \Delta,A \quad A,\Pi \Rightarrow \Lambda}{\Gamma,\Pi \Rightarrow \Delta,\Lambda}\ \text{Cut}$$

10.5 Derivations

We've said what an initial sequent looks like, and we've given the rules of inference. Derivations in the sequent calculus are inductively generated from these: each derivation either is an initial sequent on its own, or consists of one or two derivations followed by an inference.

Definition 10.3 (LK derivation). An **LK**-*derivation* of a sequent S is a tree of sequents satisfying the following conditions:

1. The topmost sequents of the tree are initial sequents.

2. The bottommost sequent of the tree is S.

3. Every sequent in the tree except S is a premise of a correct

10.5. DERIVATIONS

application of an inference rule whose conclusion stands directly below that sequent in the tree.

We then say that S is the *end-sequent* of the derivation and that S is *derivable in* **LK** (or **LK**-derivable).

Example 10.4. Every initial sequent, e.g., $C \Rightarrow C$ is a derivation. We can obtain a new derivation from this by applying, say, the WL rule,

$$\frac{\Gamma \Rightarrow \Delta}{A, \Gamma \Rightarrow \Delta} \text{ WL}$$

The rule, however, is meant to be general: we can replace the A in the rule with any sentence, e.g., also with D. If the premise matches our initial sequent $C \Rightarrow C$, that means that both Γ and Δ are just C, and the conclusion would then be $D, C \Rightarrow C$. So, the following is a derivation:

$$\frac{C \Rightarrow C}{D, C \Rightarrow C} \text{ WL}$$

We can now apply another rule, say XL, which allows us to switch two sentences on the left. So, the following is also a correct derivation:

$$\frac{\dfrac{C \Rightarrow C}{D, C \Rightarrow C} \text{ WL}}{C, D \Rightarrow C} \text{ XL}$$

In this application of the rule, which was given as

$$\frac{\Gamma, A, B, \Pi \Rightarrow \Delta}{\Gamma, B, A, \Pi \Rightarrow \Delta,} \text{ XL}$$

both Γ and Π were empty, Δ is C, and the roles of A and B are played by D and C, respectively. In much the same way, we also see that

$$\frac{D \Rightarrow D}{C, D \Rightarrow D} \text{ WL}$$

is a derivation. Now we can take these two derivations, and combine them using ∧R. That rule was

$$\frac{\Gamma \Rightarrow \Delta, A \quad \Gamma \Rightarrow \Delta, B}{\Gamma \Rightarrow \Delta, A \wedge B} \wedge \text{R}$$

In our case, the premises must match the last sequents of the derivations ending in the premises. That means that Γ is C, D, Δ is empty, A is C and B is D. So the conclusion, if the inference should be correct, is $C, D \Rightarrow C \wedge D$.

$$\cfrac{\cfrac{\cfrac{C \Rightarrow C}{D, C \Rightarrow C}\text{WL}}{C, D \Rightarrow C}\text{XL} \quad \cfrac{D \Rightarrow D}{C, D \Rightarrow D}\text{WL}}{C, D \Rightarrow C \wedge D} \wedge \text{R}$$

Of course, we can also reverse the premises, then A would be D and B would be C.

$$\cfrac{\cfrac{D \Rightarrow D}{C, D \Rightarrow D}\text{WL} \quad \cfrac{\cfrac{C \Rightarrow C}{D, C \Rightarrow C}\text{WL}}{C, D \Rightarrow C}\text{XL}}{C, D \Rightarrow D \wedge C} \wedge \text{R}$$

10.6 Examples of Derivations

Example 10.5. Give an **LK**-derivation for the sequent $A \wedge B \Rightarrow A$.

We begin by writing the desired end-sequent at the bottom of the derivation.

$$\overline{A \wedge B \Rightarrow A}$$

Next, we need to figure out what kind of inference could have a lower sequent of this form. This could be a structural rule, but it is a good idea to start by looking for a logical rule. The only logical connective occurring in the lower sequent is ∧, so we're looking for an ∧ rule, and since the ∧ symbol occurs in the antecedent, we're looking at the ∧L rule.

10.6. EXAMPLES OF DERIVATIONS

$$\frac{}{A \wedge B \Rightarrow A} \wedge \text{L}$$

There are two options for what could have been the upper sequent of the ∧L inference: we could have an upper sequent of $A \Rightarrow A$, or of $B \Rightarrow A$. Clearly, $A \Rightarrow A$ is an initial sequent (which is a good thing), while $B \Rightarrow A$ is not derivable in general. We fill in the upper sequent:

$$\frac{A \Rightarrow A}{A \wedge B \Rightarrow A} \wedge \text{L}$$

We now have a correct **LK**-derivation of the sequent $A \wedge B \Rightarrow A$.

Example 10.6. Give an **LK**-derivation for the sequent $\neg A \vee B \Rightarrow A \rightarrow B$.

Begin by writing the desired end-sequent at the bottom of the derivation.

$$\overline{\neg A \vee B \Rightarrow A \rightarrow B}$$

To find a logical rule that could give us this end-sequent, we look at the logical connectives in the end-sequent: ¬, ∨, and →. We only care at the moment about ∨ and → because they are main operators of sentences in the end-sequent, while ¬ is inside the scope of another connective, so we will take care of it later. Our options for logical rules for the final inference are therefore the ∨L rule and the →R rule. We could pick either rule, really, but let's pick the →R rule (if for no reason other than it allows us to put off splitting into two branches). According to the form of →R inferences which can yield the lower sequent, this must look like:

$$\frac{\overline{A, \neg A \vee B \Rightarrow B}}{\neg A \vee B \Rightarrow A \rightarrow B} \rightarrow \text{R}$$

If we move $\neg A \vee B$ to the outside of the antecedent, we can apply the ∨L rule. According to the schema, this must split into two upper sequents as follows:

$$\dfrac{\dfrac{\neg A, A \Rightarrow B \quad B, A \Rightarrow B}{\neg A \vee B, A \Rightarrow B} \vee L}{\dfrac{A, \neg A \vee B \Rightarrow B}{\neg A \vee B \Rightarrow A \to B} \to R} XR$$

Remember that we are trying to wind our way up to initial sequents; we seem to be pretty close! The right branch is just one weakening and one exchange away from an initial sequent and then it is done:

$$\dfrac{\neg A, A \Rightarrow B \quad \dfrac{\dfrac{\dfrac{B \Rightarrow B}{A, B \Rightarrow B} WL}{B, A \Rightarrow B} XL}{\neg A \vee B, A \Rightarrow B} \vee L}{\dfrac{A, \neg A \vee B \Rightarrow B}{\neg A \vee B \Rightarrow A \to B} \to R} XR$$

Now looking at the left branch, the only logical connective in any sentence is the ¬ symbol in the antecedent sentences, so we're looking at an instance of the ¬L rule.

$$\dfrac{\dfrac{\dfrac{A \Rightarrow B, A}{\neg A, A \Rightarrow B} \neg L \quad \dfrac{\dfrac{B \Rightarrow B}{A, B \Rightarrow B} WL}{B, A \Rightarrow B} XL}{\neg A \vee B, A \Rightarrow B} \vee L}{\dfrac{A, \neg A \vee B \Rightarrow B}{\neg A \vee B \Rightarrow A \to B} \to R} XR$$

Similarly to how we finished off the right branch, we are just one weakening and one exchange away from finishing off this left branch as well.

$$\dfrac{\dfrac{\dfrac{\dfrac{\dfrac{A \Rightarrow A}{A \Rightarrow A, B} WR}{A \Rightarrow B, A} XR}{\neg A, A \Rightarrow B} \neg L \quad \dfrac{\dfrac{B \Rightarrow B}{A, B \Rightarrow B} WL}{B, A \Rightarrow B} XL}{\neg A \vee B, A \Rightarrow B} \vee L}{\dfrac{A, \neg A \vee B \Rightarrow B}{\neg A \vee B \Rightarrow A \to B} \to R} XR$$

10.6. EXAMPLES OF DERIVATIONS

Example 10.7. Give an **LK**-derivation of the sequent $\neg A \lor \neg B \Rightarrow \neg(A \land B)$

Using the techniques from above, we start by writing the desired end-sequent at the bottom.

$$\overline{\neg A \lor \neg B \;\Rightarrow\; \neg(A \land B)}$$

The available main connectives of sentences in the end-sequent are the $\lor$ symbol and the $\neg$ symbol. It would work to apply either the $\lor$L or the $\neg$R rule here, but we start with the $\neg$R rule because it avoids splitting up into two branches for a moment:

$$\cfrac{\overline{A \land B, \neg A \lor \neg B \;\Rightarrow\;}}{\neg A \lor \neg B \;\Rightarrow\; \neg(A \land B)}\;\neg\text{R}$$

Now we have a choice of whether to look at the $\land$L or the $\lor$L rule. Let's see what happens when we apply the $\land$L rule: we have a choice to start with either the sequent $A, \neg A \lor B \Rightarrow$ or the sequent $B, \neg A \lor B \Rightarrow$. Since the derivation is symmetric with regards to A and B, let's go with the former:

$$\cfrac{\cfrac{\overline{A, \neg A \lor \neg B \;\Rightarrow\;}}{A \land B, \neg A \lor \neg B \;\Rightarrow\;}\;\land\text{L}}{\neg A \lor \neg B \;\Rightarrow\; \neg(A \land B)}\;\neg\text{R}$$

Continuing to fill in the derivation, we see that we run into a problem:

$$\cfrac{\cfrac{\cfrac{\cfrac{\cfrac{\cfrac{A \Rightarrow A}{\neg A, A \;\Rightarrow\;}\;\neg\text{L} \quad \cfrac{\overline{A \Rightarrow B}^{\;?}}{\neg B, A \;\Rightarrow\;}\;\neg\text{L}}{\neg A \lor \neg B, A \;\Rightarrow\;}\;\lor\text{L}}{A, \neg A \lor \neg B \;\Rightarrow\;}\;\text{XL}}{A \land B, \neg A \lor \neg B \;\Rightarrow\;}\;\land\text{L}}{\neg A \lor \neg B \;\Rightarrow\; \neg(A \land B)}\;\neg\text{R}$$

The top of the right branch cannot be reduced any further, and it cannot be brought by way of structural inferences to an initial sequent, so this is not the right path to take. So clearly, it was a mistake to apply the ∧L rule above. Going back to what we had before and carrying out the ∨L rule instead, we get

$$\dfrac{\dfrac{\dfrac{\dfrac{\neg A, A \wedge B \Rightarrow \qquad \neg B, A \wedge B \Rightarrow}{\neg A \vee \neg B, A \wedge B \Rightarrow} \vee L}{A \wedge B, \neg A \vee \neg B \Rightarrow} XL}{\neg A \vee \neg B \Rightarrow \neg (A \wedge B)} \neg R$$

Completing each branch as we've done before, we get

$$\dfrac{\dfrac{\dfrac{\dfrac{\dfrac{A \Rightarrow A}{A \wedge B \Rightarrow A} \wedge L}{\neg A, A \wedge B \Rightarrow} \neg L \qquad \dfrac{\dfrac{B \Rightarrow B}{A \wedge B \Rightarrow B} \wedge L}{\neg B, A \wedge B \Rightarrow} \neg L}{\neg A \vee \neg B, A \wedge B \Rightarrow} \vee L}{A \wedge B, \neg A \vee \neg B \Rightarrow} XL}{\neg A \vee \neg B \Rightarrow \neg (A \wedge B)} \neg R$$

(We could have carried out the ∧ rules lower than the ¬ rules in these steps and still obtained a correct derivation).

Example 10.8. So far we haven't used the contraction rule, but it is sometimes required. Here's an example where that happens. Suppose we want to prove $\Rightarrow A \vee \neg A$. Applying ∨R backwards would give us one of these two derivations:

$$\dfrac{\Rightarrow A}{\Rightarrow A \vee \neg A} \vee R \qquad \dfrac{\dfrac{A \Rightarrow}{\Rightarrow \neg A} \neg R}{\Rightarrow A \vee \neg A} \vee R$$

Neither of these of course ends in an initial sequent. The trick is to realize that the contraction rule allows us to combine two copies of a sentence into one—and when we're searching for a proof, i.e., going from bottom to top, we can keep a copy of $A \vee \neg A$ in the premise, e.g.,

$$\dfrac{\dfrac{\Rightarrow A \vee \neg A, A}{\Rightarrow A \vee \neg A, A \vee \neg A} \vee R}{\Rightarrow A \vee \neg A} CR$$

Now we can apply ∨R a second time, and also get ¬A, which leads to a complete derivation.

$$
\frac{\frac{\frac{\frac{\frac{A \Rightarrow A}{\Rightarrow A, \neg A} \neg R}{\Rightarrow A, A \vee \neg A} \vee R}{\Rightarrow A \vee \neg A, A} XR}{\Rightarrow A \vee \neg A, A \vee \neg A} \vee R}{\Rightarrow A \vee \neg A} CR
$$

10.7 Derivations with Quantifiers

Example 10.9. Give an **LK**-derivation of the sequent $\exists x \, \neg A(x) \Rightarrow \neg \forall x \, A(x)$.

When dealing with quantifiers, we have to make sure not to violate the eigenvariable condition, and sometimes this requires us to play around with the order of carrying out certain inferences. In general, it helps to try and take care of rules subject to the eigenvariable condition first (they will be lower down in the finished proof). Also, it is a good idea to try and look ahead and try to guess what the initial sequent might look like. In our case, it will have to be something like $A(a) \Rightarrow A(a)$. That means that when we are "reversing" the quantifier rules, we will have to pick the same term—what we will call a—for both the ∀ and the ∃ rule. If we picked different terms for each rule, we would end up with something like $A(a) \Rightarrow A(b)$, which, of course, is not derivable.

Starting as usual, we write

$$\overline{\exists x \, \neg A(x) \Rightarrow \neg \forall x \, A(x)}$$

We could either carry out the ∃L rule or the ¬R rule. Since the ∃L rule is subject to the eigenvariable condition, it's a good idea to take care of it sooner rather than later, so we'll do that one first.

$$\frac{\overline{\neg A(a) \Rightarrow \neg \forall x\, A(x)}}{\exists x\, \neg A(x) \Rightarrow \neg \forall x\, A(x)} \exists L$$

Applying the ¬L and ¬R rules backwards, we get

$$\frac{\dfrac{\overline{\forall x\, A(x) \Rightarrow A(a)}}{\dfrac{\neg A(a), \forall x\, A(x) \Rightarrow}{\dfrac{\forall x\, A(x), \neg A(a) \Rightarrow}{\dfrac{\neg A(a) \Rightarrow \neg \forall x A(x)}{\exists x \neg A(x) \Rightarrow \neg \forall x A(x)} \exists L} \neg R} XL} \neg L}$$

At this point, our only option is to carry out the ∀L rule. Since this rule is not subject to the eigenvariable restriction, we're in the clear. Remember, we want to try and obtain an initial sequent (of the form $A(a) \Rightarrow A(a)$), so we should choose a as our argument for A when we apply the rule.

$$\frac{\dfrac{\dfrac{\overline{A(a) \Rightarrow A(a)}}{\forall x\, A(x) \Rightarrow A(a)} \forall L}{\dfrac{\neg A(a), \forall x\, A(x) \Rightarrow}{\dfrac{\forall x\, A(x), \neg A(a) \Rightarrow}{\dfrac{\neg A(a) \Rightarrow \neg \forall x\, A(x)}{\exists x\, \neg A(x) \Rightarrow \neg \forall x\, A(x)} \exists L} \neg R} XL} \neg L}$$

It is important, especially when dealing with quantifiers, to double check at this point that the eigenvariable condition has not been violated. Since the only rule we applied that is subject to the eigenvariable condition was ∃L, and the eigenvariable a does not occur in its lower sequent (the end-sequent), this is a correct derivation.

10.8 Proof-Theoretic Notions

Just as we've defined a number of important semantic notions (validity, entailment, satisfiabilty), we now define corresponding *proof-theoretic notions*. These are not defined by appeal to satisfaction of sentences in structures, but by appeal to the derivability

10.8. PROOF-THEORETIC NOTIONS

or non-derivability of certain sequents. It was an important discovery that these notions coincide. That they do is the content of the *soundness* and *completeness theorem*.

Definition 10.10 (Theorems). A sentence A is a *theorem* if there is a derivation in **LK** of the sequent $\Rightarrow A$. We write $\vdash A$ if A is a theorem and $\nvdash A$ if it is not.

Definition 10.11 (Derivability). A sentence A is *derivable from* a set of sentences Γ, $\Gamma \vdash A$, iff there is a finite subset $\Gamma_0 \subseteq \Gamma$ and a sequence Γ_0' of the sentences in Γ_0 such that **LK** derives $\Gamma_0' \Rightarrow A$. If A is not derivable from Γ we write $\Gamma \nvdash A$.

Because of the contraction, weakening, and exchange rules, the order and number of sentences in Γ_0' does not matter: if a sequent $\Gamma_0' \Rightarrow A$ is derivable, then so is $\Gamma_0'' \Rightarrow A$ for any Γ_0'' that contains the same sentences as Γ_0'. For instance, if $\Gamma_0 = \{B, C\}$ then both $\Gamma_0' = \langle B, B, C \rangle$ and $\Gamma_0'' = \langle C, C, B \rangle$ are sequences containing just the sentences in Γ_0. If a sequent containing one is derivable, so is the other, e.g.:

$$\frac{\vdots}{\dfrac{\dfrac{\dfrac{B,B,C \Rightarrow A}{B,C \Rightarrow A}\text{CL}}{C,B \Rightarrow A}\text{XL}}{C,C,B \Rightarrow A}\text{WL}}$$

From now on we'll say that if Γ_0 is a finite set of sentences then $\Gamma_0 \Rightarrow A$ is any sequent where the antecedent is a sequence of sentences in Γ_0 and tacitly include contractions, exchanges, and weakenings if necessary.

Definition 10.12 (Consistency). A set of sentences Γ is *inconsistent* iff there is a finite subset $\Gamma_0 \subseteq \Gamma$ such that **LK** derives $\Gamma_0 \Rightarrow$. If Γ is not inconsistent, i.e., if for every finite $\Gamma_0 \subseteq \Gamma$, **LK** does not derive $\Gamma_0 \Rightarrow$, we say it is *consistent*.

Proposition 10.13 (Reflexivity). *If $A \in \Gamma$, then $\Gamma \vdash A$.*

Proof. The initial sequent $A \Rightarrow A$ is derivable, and $\{A\} \subseteq \Gamma$. □

Proposition 10.14 (Monotony). *If $\Gamma \subseteq \Delta$ and $\Gamma \vdash A$, then $\Delta \vdash A$.*

Proof. Suppose $\Gamma \vdash A$, i.e., there is a finite $\Gamma_0 \subseteq \Gamma$ such that $\Gamma_0 \Rightarrow A$ is derivable. Since $\Gamma \subseteq \Delta$, then Γ_0 is also a finite subset of Δ. The derivation of $\Gamma_0 \Rightarrow A$ thus also shows $\Delta \vdash A$. □

Proposition 10.15 (Transitivity). *If $\Gamma \vdash A$ and $\{A\} \cup \Delta \vdash B$, then $\Gamma \cup \Delta \vdash B$.*

Proof. If $\Gamma \vdash A$, there is a finite $\Gamma_0 \subseteq \Gamma$ and a derivation π_0 of $\Gamma_0 \Rightarrow A$. If $\{A\} \cup \Delta \vdash B$, then for some finite subset $\Delta_0 \subseteq \Delta$, there is a derivation π_1 of $A, \Delta_0 \Rightarrow B$. Consider the following derivation:

$$\cfrac{\vdots \pi_0 \qquad \vdots \pi_1}{\cfrac{\Gamma_0 \Rightarrow A \qquad A, \Delta_0 \Rightarrow B}{\Gamma_0, \Delta_0 \Rightarrow B}} \text{Cut}$$

Since $\Gamma_0 \cup \Delta_0 \subseteq \Gamma \cup \Delta$, this shows $\Gamma \cup \Delta \vdash B$. □

Note that this means that in particular if $\Gamma \vdash A$ and $A \vdash B$, then $\Gamma \vdash B$. It follows also that if $A_1, \ldots, A_n \vdash B$ and $\Gamma \vdash A_i$ for each i, then $\Gamma \vdash B$.

Proposition 10.16. Γ *is inconsistent iff* $\Gamma \vdash A$ *for every sentence A.*

Proof. Exercise. □

Proposition 10.17 (Compactness). *1. If $\Gamma \vdash A$ then there is a finite subset $\Gamma_0 \subseteq \Gamma$ such that $\Gamma_0 \vdash A$.*

  2. If every finite subset of Γ is consistent, then Γ is consistent.

Proof. 1. If $\Gamma \vdash A$, then there is a finite subset $\Gamma_0 \subseteq \Gamma$ such that the sequent $\Gamma_0 \Rightarrow A$ has a derivation. Consequently, $\Gamma_0 \vdash A$.

2. If Γ is inconsistent, there is a finite subset $\Gamma_0 \subseteq \Gamma$ such that **LK** derives $\Gamma_0 \Rightarrow$. But then Γ_0 is a finite subset of Γ that is inconsistent. □

10.9 Derivability and Consistency

We will now establish a number of properties of the derivability relation. They are independently interesting, but each will play a role in the proof of the completeness theorem.

Proposition 10.18. *If $\Gamma \vdash A$ and $\Gamma \cup \{A\}$ is inconsistent, then Γ is inconsistent.*

Proof. There are finite Γ_0 and $\Gamma_1 \subseteq \Gamma$ such that **LK** derives $\Gamma_0 \Rightarrow A$ and $A, \Gamma_1 \Rightarrow$. Let the **LK**-derivation of $\Gamma_0 \Rightarrow A$ be π_0 and the **LK**-derivation of $\Gamma_1, A \Rightarrow$ be π_1. We can then derive

$$\cfrac{\vdots\pi_0 \qquad \vdots\pi_1}{\cfrac{\Gamma_0 \Rightarrow A \qquad A, \Gamma_1 \Rightarrow}{\Gamma_0, \Gamma_1 \Rightarrow}} \text{Cut}$$

Since $\Gamma_0 \subseteq \Gamma$ and $\Gamma_1 \subseteq \Gamma$, $\Gamma_0 \cup \Gamma_1 \subseteq \Gamma$, hence Γ is inconsistent. □

Proposition 10.19. *$\Gamma \vdash A$ iff $\Gamma \cup \{\neg A\}$ is inconsistent.*

Proof. First suppose $\Gamma \vdash A$, i.e., there is a derivation π_0 of $\Gamma \Rightarrow A$. By adding a $\neg$L rule, we obtain a derivation of $\neg A, \Gamma \Rightarrow$, i.e., $\Gamma \cup \{\neg A\}$ is inconsistent.

If $\Gamma \cup \{\neg A\}$ is inconsistent, there is a derivation π_1 of $\neg A, \Gamma \Rightarrow$. The following is a derivation of $\Gamma \Rightarrow A$:

$$\cfrac{\cfrac{\cfrac{A \Rightarrow A}{\Rightarrow A, \neg A} \neg\text{R}}{} \quad \cfrac{\vdots \pi_1}{\neg A, \Gamma \Rightarrow}}{\Gamma \Rightarrow A} \text{Cut}$$

□

Proposition 10.20. *If $\Gamma \vdash A$ and $\neg A \in \Gamma$, then Γ is inconsistent.*

Proof. Suppose $\Gamma \vdash A$ and $\neg A \in \Gamma$. Then there is a derivation π of a sequent $\Gamma_0 \Rightarrow A$. The sequent $\neg A, \Gamma_0 \Rightarrow$ is also derivable:

$$\cfrac{\cfrac{\vdots \pi}{\Gamma_0 \Rightarrow A} \quad \cfrac{\cfrac{\cfrac{A \Rightarrow A}{\neg A, A \Rightarrow} \neg\text{L}}{A, \neg A \Rightarrow} \text{XL}}{}}{\Gamma, \neg A \Rightarrow} \text{Cut}$$

Since $\neg A \in \Gamma$ and $\Gamma_0 \subseteq \Gamma$, this shows that Γ is inconsistent. □

Proposition 10.21. *If $\Gamma \cup \{A\}$ and $\Gamma \cup \{\neg A\}$ are both inconsistent, then Γ is inconsistent.*

Proof. There are finite sets $\Gamma_0 \subseteq \Gamma$ and $\Gamma_1 \subseteq \Gamma$ and **LK**-derivations π_0 and π_1 of $A, \Gamma_0 \Rightarrow$ and $\neg A, \Gamma_1 \Rightarrow$, respectively. We can then derive

10.10 Derivability and the Propositional Connectives

$$\begin{array}{c} \vdots\, \pi_0 \\ A, \Gamma_0 \Rightarrow \\ \hline \Gamma_0 \Rightarrow \neg A \end{array} \neg R \qquad \begin{array}{c} \vdots\, \pi_1 \\ \neg A, \Gamma_1 \Rightarrow \end{array}$$
$$\overline{\Gamma_0, \Gamma_1 \Rightarrow} \text{ Cut}$$

Since $\Gamma_0 \subseteq \Gamma$ and $\Gamma_1 \subseteq \Gamma$, $\Gamma_0 \cup \Gamma_1 \subseteq \Gamma$. Hence Γ is inconsistent. □

10.10 Derivability and the Propositional Connectives

We establish that the derivability relation ⊢ of the sequent calculus is strong enough to establish some basic facts involving the propositional connectives, such as that $A \wedge B \vdash A$ and $A, A \rightarrow B \vdash B$ (modus ponens). These facts are needed for the proof of the completeness theorem.

Proposition 10.22. *1. Both $A \wedge B \vdash A$ and $A \wedge B \vdash B$.*

2. $A, B \vdash A \wedge B$.

Proof. 1. Both sequents $A \wedge B \Rightarrow A$ and $A \wedge B \Rightarrow B$ are derivable:

$$\frac{A \Rightarrow A}{A \wedge B \Rightarrow A} \wedge L \qquad \frac{B \Rightarrow B}{A \wedge B \Rightarrow B} \wedge L$$

2. Here is a derivation of the sequent $A, B \Rightarrow A \wedge B$:

$$\frac{A \Rightarrow A \quad B \Rightarrow B}{A, B \Rightarrow A \wedge B} \wedge R \qquad \square$$

Proposition 10.23. *1. $A \vee B, \neg A, \neg B$ is inconsistent.*

2. Both $A \vdash A \vee B$ and $B \vdash A \vee B$.

Proof. 1. We give a derivation of the sequent $A \vee B, \neg A, \neg B \Rightarrow$:

$$\dfrac{\dfrac{A \Rightarrow A}{\neg A, A \Rightarrow} \neg L}{\dfrac{A, \neg A, \neg B \Rightarrow}{A \vee B, \neg A, \neg B \Rightarrow}} \quad \dfrac{\dfrac{B \Rightarrow B}{\neg B, B \Rightarrow} \neg L}{B, \neg A, \neg B \Rightarrow} \quad \vee L$$

(Recall that double inference lines indicate several weakening, contraction, and exchange inferences.)

2. Both sequents $A \Rightarrow A \vee B$ and $B \Rightarrow A \vee B$ have derivations:

$$\dfrac{A \Rightarrow A}{A \Rightarrow A \vee B} \vee R \qquad \dfrac{B \Rightarrow B}{B \Rightarrow A \vee B} \vee R \qquad \square$$

Proposition 10.24. *1. $A, A \to B \vdash B$.*

2. Both $\neg A \vdash A \to B$ and $B \vdash A \to B$.

Proof. 1. The sequent $A \to B, A \Rightarrow B$ is derivable:

$$\dfrac{A \Rightarrow A \quad B \Rightarrow B}{A \to B, A \Rightarrow B} \to L$$

2. Both sequents $\neg A \Rightarrow A \to B$ and $B \Rightarrow A \to B$ are derivable:

$$\dfrac{\dfrac{\dfrac{\dfrac{A \Rightarrow A}{\neg A, A \Rightarrow} \neg L}{A, \neg A \Rightarrow} XL}{A, \neg A \Rightarrow B} WR}{\neg A \Rightarrow A \to B} \to R \qquad \dfrac{\dfrac{\dfrac{B \Rightarrow B}{A, B \Rightarrow B} WL}{B \Rightarrow A \to B}}{B \Rightarrow A \to B} \to R \qquad \square$$

10.11 Derivability and the Quantifiers

The completeness theorem also requires that the sequent calculus rules rules yield the facts about ⊢ established in this section.

10.12. SOUNDNESS

Theorem 10.25. *If c is a constant not occurring in Γ or $A(x)$ and $\Gamma \vdash A(c)$, then $\Gamma \vdash \forall x\, A(x)$.*

Proof. Let π_0 be an **LK**-derivation of $\Gamma_0 \Rightarrow A(c)$ for some finite $\Gamma_0 \subseteq \Gamma$. By adding a $\forall$R inference, we obtain a derivation of $\Gamma_0 \Rightarrow \forall x\, A(x)$, since c does not occur in Γ or $A(x)$ and thus the eigenvariable condition is satisfied. □

Proposition 10.26. 1. $A(t) \vdash \exists x\, A(x)$.

2. $\forall x\, A(x) \vdash A(t)$.

Proof.   1. The sequent $A(t) \Rightarrow \exists x\, A(x)$ is derivable:

$$\dfrac{A(t) \Rightarrow A(t)}{A(t) \Rightarrow \exists x\, A(x)}\ \exists\mathrm{R}$$

2. The sequent $\forall x\, A(x) \Rightarrow A(t)$ is derivable:

$$\dfrac{A(t) \Rightarrow A(t)}{\forall x\, A(x) \Rightarrow A(t)}\ \forall\mathrm{L}$$

□

10.12 Soundness

A derivation system, such as the sequent calculus, is *sound* if it cannot derive things that do not actually hold. Soundness is thus a kind of guaranteed safety property for derivation systems. Depending on which proof theoretic property is in question, we would like to know for instance, that

1. every derivable A is valid;

2. if a sentence is derivable from some others, it is also a consequence of them;

3. if a set of sentences is inconsistent, it is unsatisfiable.

These are important properties of a derivation system. If any of them do not hold, the derivation system is deficient—it would derive too much. Consequently, establishing the soundness of a derivation system is of the utmost importance.

Because all these proof-theoretic properties are defined via derivability in the sequent calculus of certain sequents, proving (1)–(3) above requires proving something about the semantic properties of derivable sequents. We will first define what it means for a sequent to be *valid*, and then show that every derivable sequent is valid. (1)–(3) then follow as corollaries from this result.

Definition 10.27. A structure M *satisfies* a sequent $\Gamma \Rightarrow \Delta$ iff either $M \nvDash A$ for some $A \in \Gamma$ or $M \vDash A$ for some $A \in \Delta$.

A sequent is *valid* iff every structure M satisfies it.

Theorem 10.28 (Soundness). *If* LK *derives* $\Theta \Rightarrow \Xi$*, then* $\Theta \Rightarrow \Xi$ *is valid.*

Proof. Let π be a derivation of $\Theta \Rightarrow \Xi$. We proceed by induction on the number of inferences n in π.

If the number of inferences is 0, then π consists only of an initial sequent. Every initial sequent $A \Rightarrow A$ is obviously valid, since for every M, either $M \nvDash A$ or $M \vDash A$.

If the number of inferences is greater than 0, we distinguish cases according to the type of the lowermost inference. By induction hypothesis, we can assume that the premises of that inference are valid, since the number of inferences in the derivation of any premise is smaller than n.

First, we consider the possible inferences with only one premise.

1. The last inference is a weakening. Then $\Theta \Rightarrow \Xi$ is either $A, \Gamma \Rightarrow \Delta$ (if the last inference is WL) or $\Gamma \Rightarrow \Delta, A$ (if it's WR), and the derivation ends in one of

10.12. SOUNDNESS

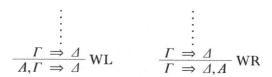

By induction hypothesis, $\Gamma \Rightarrow \Delta$ is valid, i.e., for every structure M, either there is some $C \in \Gamma$ such that $M \nvDash C$ or there is some $C \in \Delta$ such that $M \vDash C$.

If $M \nvDash C$ for some $C \in \Gamma$, then $C \in \Theta$ as well since $\Theta = A, \Gamma$, and so $M \nvDash C$ for some $C \in \Theta$. Similarly, if $M \vDash C$ for some $C \in \Delta$, as $C \in \Xi$, $M \vDash C$ for some $C \in \Xi$. Consequently, $\Theta \Rightarrow \Xi$ is valid.

2. The last inference is $\neg$L: Then the premise of the last inference is $\Gamma \Rightarrow \Delta, A$ and the conclusion is $\neg A, \Gamma \Rightarrow \Delta$, i.e., the derivation ends in

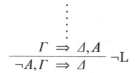

and $\Theta = \neg A, \Gamma$ while $\Xi = \Delta$.

The induction hypothesis tells us that $\Gamma \Rightarrow \Delta, A$ is valid, i.e., for every M, either (a) for some $C \in \Gamma$, $M \nvDash C$, or (b) for some $C \in \Delta$, $M \vDash C$, or (c) $M \vDash A$. We want to show that $\Theta \Rightarrow \Xi$ is also valid. Let M be a structure. If (a) holds, then there is $C \in \Gamma$ so that $M \nvDash C$, but $C \in \Theta$ as well. If (b) holds, there is $C \in \Delta$ such that $M \vDash C$, but $C \in \Xi$ as well. Finally, if $M \vDash A$, then $M \nvDash \neg A$. Since $\neg A \in \Theta$, there is $C \in \Theta$ such that $M \nvDash C$. Consequently, $\Theta \Rightarrow \Xi$ is valid.

3. The last inference is $\neg$R: Exercise.

4. The last inference is $\wedge$L: There are two variants: $A \wedge B$ may be inferred on the left from A or from B on the left side of the premise. In the first case, the π ends in

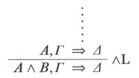

and $\Theta = A \wedge B, \Gamma$ while $\Xi = \Delta$. Consider a structure M. Since by induction hypothesis, $A, \Gamma \Rightarrow \Delta$ is valid, (a) $M \nvDash A$, (b) $M \nvDash C$ for some $C \in \Gamma$, or (c) $M \vDash C$ for some $C \in \Delta$. In case (a), $M \nvDash A \wedge B$, so there is $C \in \Theta$ (namely, $A \wedge B$) such that $M \nvDash C$. In case (b), there is $C \in \Gamma$ such that $M \nvDash C$, and $C \in \Theta$ as well. In case (c), there is $C \in \Delta$ such that $M \vDash C$, and $C \in \Xi$ as well since $\Xi = \Delta$. So in each case, M satisfies $A \wedge B, \Gamma \Rightarrow \Delta$. Since M was arbitrary, $\Gamma \Rightarrow \Delta$ is valid. The case where $A \wedge B$ is inferred from B is handled the same, changing A to B.

5. The last inference is $\vee$R: There are two variants: $A \vee B$ may be inferred on the right from A or from B on the right side of the premise. In the first case, π ends in

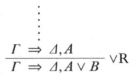

Now $\Theta = \Gamma$ and $\Xi = \Delta, A \vee B$. Consider a structure M. Since $\Gamma \Rightarrow \Delta, A$ is valid, (a) $M \vDash A$, (b) $M \nvDash C$ for some $C \in \Gamma$, or (c) $M \vDash C$ for some $C \in \Delta$. In case (a), $M \vDash A \vee B$. In case (b), there is $C \in \Gamma$ such that $M \nvDash C$. In case (c), there is $C \in \Delta$ such that $M \vDash C$. So in each case, M satisfies $\Gamma \Rightarrow \Delta, A \vee B$, i.e., $\Theta \Rightarrow \Xi$. Since M was arbitrary, $\Theta \Rightarrow \Xi$ is valid. The case where $A \vee B$ is inferred from B is handled the same, changing A to B.

6. The last inference is $\rightarrow$R: Then π ends in

10.12. SOUNDNESS

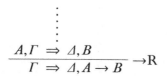

$$\frac{A, \Gamma \Rightarrow \Delta, B}{\Gamma \Rightarrow \Delta, A \to B} \to \text{R}$$

Again, the induction hypothesis says that the premise is valid; we want to show that the conclusion is valid as well. Let M be arbitrary. Since $A, \Gamma \Rightarrow \Delta, B$ is valid, at least one of the following cases obtains: (a) $M \not\models A$, (b) $M \models B$, (c) $M \not\models C$ for some $C \in \Gamma$, or (d) $M \models C$ for some $C \in \Delta$. In cases (a) and (b), $M \models A \to B$ and so there is a $C \in \Delta, A \to B$ such that $M \models C$. In case (c), for some $C \in \Gamma$, $M \not\models C$. In case (d), for some $C \in \Delta$, $M \models C$. In each case, M satisfies $\Gamma \Rightarrow \Delta, A \to B$. Since M was arbitrary, $\Gamma \Rightarrow \Delta, A \to B$ is valid.

7. The last inference is $\forall$L: Then there is a formula $A(x)$ and a closed term t such that π ends in

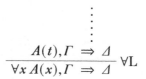

$$\frac{A(t), \Gamma \Rightarrow \Delta}{\forall x\, A(x), \Gamma \Rightarrow \Delta} \forall \text{L}$$

We want to show that the conclusion $\forall x\, A(x), \Gamma \Rightarrow \Delta$ is valid. Consider a structure M. Since the premise $A(t), \Gamma \Rightarrow \Delta$ is valid, (a) $M \not\models A(t)$, (b) $M \not\models C$ for some $C \in \Gamma$, or (c) $M \models C$ for some $C \in \Delta$. In case (a), by Proposition 7.30, if $M \models \forall x\, A(x)$, then $M \models A(t)$. Since $M \not\models A(t)$, $M \not\models \forall x\, A(x)$. In case (b) and (c), M also satisfies $\forall x\, A(x), \Gamma \Rightarrow \Delta$. Since M was arbitrary, $\forall x\, A(x), \Gamma \Rightarrow \Delta$ is valid.

8. The last inference is $\exists$R: Exercise.

9. The last inference is $\forall$R: Then there is a formula $A(x)$ and a constant symbol a such that π ends in

$$\frac{\vdots}{\Gamma \Rightarrow \Delta, A(a)} \forall R$$
$$\overline{\Gamma \Rightarrow \Delta, \forall x\, A(x)}$$

where the eigenvariable condition is satisfied, i.e., a does not occur in $A(x)$, Γ, or Δ. By induction hypothesis, the premise of the last inference is valid. We have to show that the conclusion is valid as well, i.e., that for any structure M, (a) $M \vDash \forall x\, A(x)$, (b) $M \nvDash C$ for some $C \in \Gamma$, or (c) $M \vDash C$ for some $C \in \Delta$.

Suppose M is an arbitrary structure. If (b) or (c) holds, we are done, so suppose neither holds: for all $C \in \Gamma$, $M \vDash C$, and for all $C \in \Delta$, $M \nvDash C$. We have to show that (a) holds, i.e., $M \vDash \forall x\, A(x)$. By Proposition 7.18, if suffices to show that $M, s \vDash A(x)$ for all variable assignments s. So let s be an arbitrary variable assignment. Consider the structure M' which is just like M except $a^{M'} = s(x)$. By Corollary 7.20, for any $C \in \Gamma$, $M' \vDash C$ since a does not occur in Γ, and for any $C \in \Delta$, $M' \nvDash C$. But the premise is valid, so $M' \vDash A(a)$. By Proposition 7.17, $M', s \vDash A(a)$, since $A(a)$ is a sentence. Now $s \sim_x s$ with $s(x) = \mathrm{Val}_s^{M'}(a)$, since we've defined M' in just this way. So Proposition 7.22 applies, and we get $M', s \vDash A(x)$. Since a does not occur in $A(x)$, by Proposition 7.19, $M, s \vDash A(x)$. Since s was arbitrary, we've completed the proof that $M, s \vDash A(x)$ for all variable assignments.

10. The last inference is $\exists L$: Exercise.

Now let's consider the possible inferences with two premises.

1. The last inference is a cut: then π ends in

10.12. SOUNDNESS

$$\frac{\Gamma \Rightarrow \Delta, A \quad A, \Pi \Rightarrow \Lambda}{\Gamma, \Pi \Rightarrow \Delta, \Lambda} \text{ Cut}$$

Let M be a structure. By induction hypothesis, the premises are valid, so M satisfies both premises. We distinguish two cases: (a) $M \not\models A$ and (b) $M \models A$. In case (a), in order for M to satisfy the left premise, it must satisfy $\Gamma \Rightarrow \Delta$. But then it also satisfies the conclusion. In case (b), in order for M to satisfy the right premise, it must satisfy $\Pi \setminus \Lambda$. Again, M satisfies the conclusion.

2. The last inference is $\wedge$R. Then π ends in

$$\frac{\Gamma \Rightarrow \Delta, A \quad \Gamma \Rightarrow \Delta, B}{\Gamma \Rightarrow \Delta, A \wedge B} \wedge\text{R}$$

Consider a structure M. If M satisfies $\Gamma \Rightarrow \Delta$, we are done. So suppose it doesn't. Since $\Gamma \Rightarrow \Delta, A$ is valid by induction hypothesis, $M \models A$. Similarly, since $\Gamma \Rightarrow \Delta, B$ is valid, $M \models B$. But then $M \models A \wedge B$.

3. The last inference is $\vee$L: Exercise.

4. The last inference is $\to$L. Then π ends in

$$\frac{\Gamma \Rightarrow \Delta, A \quad B, \Pi \Rightarrow \Lambda}{A \to B, \Gamma, \Pi \Rightarrow \Delta, \Lambda} \to\text{L}$$

Again, consider a structure M and suppose M doesn't satisfy $\Gamma, \Pi \Rightarrow \Delta, \Lambda$. We have to show that $M \not\models A \to B$. If M

doesn't satisfy $\Gamma, \Pi \Rightarrow \Delta, \Lambda$, it satisfies neither $\Gamma \Rightarrow \Delta$ nor $\Pi \Rightarrow \Lambda$. Since, $\Gamma \Rightarrow \Delta, A$ is valid, we have $M \vDash A$. Since $B, \Pi \Rightarrow \Lambda$ is valid, we have $M \nvDash B$. But then $M \nvDash A \to B$, which is what we wanted to show. □

Corollary 10.29. *If $\vdash A$ then A is valid.*

Corollary 10.30. *If $\Gamma \vdash A$ then $\Gamma \vDash A$.*

Proof. If $\Gamma \vdash A$ then for some finite subset $\Gamma_0 \subseteq \Gamma$, there is a derivation of $\Gamma_0 \Rightarrow A$. By Theorem 10.28, every structure M either makes some $B \in \Gamma_0$ false or makes A true. Hence, if $M \vDash \Gamma$ then also $M \vDash A$. □

Corollary 10.31. *If Γ is satisfiable, then it is consistent.*

Proof. We prove the contrapositive. Suppose that Γ is not consistent. Then there is a finite $\Gamma_0 \subseteq \Gamma$ and a derivation of $\Gamma_0 \Rightarrow$. By Theorem 10.28, $\Gamma_0 \Rightarrow$ is valid. In other words, for every structure M, there is $C \in \Gamma_0$ so that $M \nvDash C$, and since $\Gamma_0 \subseteq \Gamma$, that C is also in Γ. Thus, no M satisfies Γ, and Γ is not satisfiable. □

10.13 Derivations with Identity predicate

Derivations with identity predicate require additional initial sequents and inference rules.

Definition 10.32 (Initial sequents for =). If t is a closed term, then $\Rightarrow t = t$ is an initial sequent.

The rules for = are (t_1 and t_2 are closed terms):

$$\frac{t_1 = t_2, \Gamma \Rightarrow \Delta, A(t_1)}{t_1 = t_2, \Gamma \Rightarrow \Delta, A(t_2)} = \qquad \frac{t_1 = t_2, \Gamma \Rightarrow \Delta, A(t_2)}{t_1 = t_2, \Gamma \Rightarrow \Delta, A(t_1)} =$$

Example 10.33. If s and t are closed terms, then $s = t, A(s) \vdash A(t)$:

$$\frac{\dfrac{A(s) \Rightarrow A(s)}{s = t, A(s) \Rightarrow A(s)} \text{WL}}{s = t, A(s) \Rightarrow A(t)} =$$

This may be familiar as the principle of substitutability of identicals, or Leibniz' Law.

LK proves that = is symmetric and transitive:

$$\frac{\dfrac{\Rightarrow t_1 = t_1}{t_1 = t_2 \Rightarrow t_1 = t_1} \text{WL}}{t_1 = t_2 \Rightarrow t_2 = t_1} = \qquad \frac{\dfrac{\dfrac{t_1 = t_2 \Rightarrow t_1 = t_2}{t_2 = t_3, t_1 = t_2 \Rightarrow t_1 = t_2} \text{WL}}{t_2 = t_3, t_1 = t_2 \Rightarrow t_1 = t_3} =}{t_1 = t_2, t_2 = t_3 \Rightarrow t_1 = t_3} \text{XL}$$

In the derivation on the left, the formula $x = t_1$ is our $A(x)$. On the right, we take $A(x)$ to be $t_1 = x$.

10.14 Soundness with Identity predicate

Proposition 10.34. **LK** *with initial sequents and rules for identity is sound.*

Proof. Initial sequents of the form $\Rightarrow t = t$ are valid, since for every structure M, $M \vDash t = t$. (Note that we assume the term t to be closed, i.e., it contains no variables, so variable assignments are irrelevant).

Suppose the last inference in a derivation is =. Then the premise is $t_1 = t_2, \Gamma \Rightarrow \Delta, A(t_1)$ and the conclusion is $t_1 = t_2, \Gamma \Rightarrow \Delta, A(t_2)$. Consider a structure M. We need to show that the conclusion is valid, i.e., if $M \vDash t_1 = t_2$ and $M \vDash \Gamma$, then either $M \vDash C$ for some $C \in \Delta$ or $M \vDash A(t_2)$.

By induction hypothesis, the premise is valid. This means that if $M \vDash t_1 = t_2$ and $M \vDash \Gamma$ either (a) for some $C \in \Delta$, $M \vDash C$ or (b) $M \vDash A(t_1)$. In case (a) we are done. Consider case (b). Let s be a variable assignment with $s(x) = \mathrm{Val}^M(t_1)$. By Proposition 7.17, $M, s \vDash A(t_1)$. Since $s \sim_x s$, by Proposition 7.22, $M, s \vDash A(x)$. since $M \vDash t_1 = t_2$, we have $\mathrm{Val}^M(t_1) = \mathrm{Val}^M(t_2)$, and hence $s(x) = \mathrm{Val}^M(t_2)$. By applying Proposition 7.22 again, we also have $M, s \vDash A(t_2)$. By Proposition 7.17, $M \vDash A(t_2)$. □

Summary

Proof systems provide purely syntactic methods for characterizing consequence and compatibility between sentences. The **sequent calculus** is one such proof system. A **derivation** in it consists of a tree of sequents (a sequent $\Gamma \Rightarrow \Delta$ consists of two sequences of formulas separated by $\Rightarrow$). The topmost sequents in a derivation are **initial sequents** of the form $A \Rightarrow A$. All other sequents, for the derivation to be correct, must be correctly justified by one of a number of **inference rules**. These come in pairs; a rule for operating on the left and on the right side of a sequent for each connective and quantifier. For instance, if a sequent $\Gamma \Rightarrow \Delta, A \rightarrow B$ is justified by the $\rightarrow$R rule, the preceding sequent (the **premise**) must be $A, \Gamma \Rightarrow \Delta, B$. Some rules also allow the order or number of sentences in a sequent to be manipulated, e.g., the XR rule allows two formulas on the right side of a sequent to be switched.

If there is a derivation of the sequent $\Rightarrow A$, we say A is a **theorem** and write $\vdash A$. If there is a derivation of $\Gamma_0 \Rightarrow A$ where every B in Γ_0 is in Γ, we say A is **derivable from** Γ and write $\Gamma \vdash A$. If there is a derivation of $\Gamma_0 \Rightarrow$ where every B in Γ_0 is in Γ, we say Γ is **inconsistent**, otherwise **consistent**. These notions are interrelated, e.g., $\Gamma \vdash A$ iff $\Gamma \cup \{\neg A\}$ is inconsistent. They are also related to the corresponding semantic notions, e.g., if $\Gamma \vdash A$ then $\Gamma \vDash A$. This property of proof systems—what can be derived from Γ is guaranteed to be entailed by Γ—is called

10.14. SOUNDNESS WITH IDENTITY PREDICATE

soundness. The **soundness theorem** is proved by induction on the length of derivations, showing that each individual inference preserves validity of the conclusion sequent provided the premise sequents are valid.

Problems

Problem 10.1. Give derivations of the following sequents:

1. $\Rightarrow \neg(A \to B) \to (A \land \neg B)$
2. $(A \land B) \to C \Rightarrow (A \to C) \lor (B \to C)$

Problem 10.2. Give derivations of the following sequents:

1. $\forall x\, (A(x) \to B) \Rightarrow (\exists y\, A(y) \to B)$
2. $\exists x\, (A(x) \to \forall y\, A(y))$

Problem 10.3. Prove Proposition 10.16

Problem 10.4. Prove that $\Gamma \vdash \neg A$ iff $\Gamma \cup \{A\}$ is inconsistent.

Problem 10.5. Complete the proof of Theorem 10.28.

Problem 10.6. Give derivations of the following sequents:

1. $\Rightarrow \forall x\, \forall y\, ((x = y \land A(x)) \to A(y))$
2. $\exists x\, A(x) \land \forall y\, \forall z\, ((A(y) \land A(z)) \to y = z) \Rightarrow \exists x\, (A(x) \land \forall y\, (A(y) \to y = x))$

CHAPTER 11

Natural Deduction

11.1 Rules and Derivations

Natural deduction systems are meant to closely parallel the informal reasoning used in mathematical proof (hence it is somewhat "natural"). Natural deduction proofs begin with assumptions. Inference rules are then applied. Assumptions are "discharged" by the ¬Intro, →Intro, ∨Elim and ∃Elim inference rules, and the label of the discharged assumption is placed beside the inference for clarity.

> **Definition 11.1 (Assumption).** An *assumption* is any sentence in the topmost position of any branch.

Derivations in natural deduction are certain trees of sentences, where the topmost sentences are assumptions, and if a sentence stands below one, two, or three other sequents, it must follow correctly by a rule of inference. The sentences at the top of the inference are called the *premises* and the sentence below the *conclusion* of the inference. The rules come in pairs, an introduction and an elimination rule for each logical operator. They introduce a logical operator in the conclusion or remove

11.2. PROPOSITIONAL RULES

a logical operator from a premise of the rule. Some of the rules allow an assumption of a certain type to be *discharged*. To indicate which assumption is discharged by which inference, we also assign labels to both the assumption and the inference. This is indicated by writing the assumption as "$[A]^n$."

It is customary to consider rules for all the logical operators $\wedge, \vee, \rightarrow, \neg$, and $\bot$, even if some of those are defined.

11.2 Propositional Rules

Rules for $\wedge$

$$\frac{A \quad B}{A \wedge B} \wedge \text{Intro}$$

$$\frac{A \wedge B}{A} \wedge \text{Elim}$$

$$\frac{A \wedge B}{B} \wedge \text{Elim}$$

Rules for $\vee$

$$\frac{A}{A \vee B} \vee \text{Intro}$$

$$\frac{B}{A \vee B} \vee \text{Intro}$$

$$n \frac{A \vee B \quad \begin{array}{c}[A]^n \\ \vdots \\ C\end{array} \quad \begin{array}{c}[B]^n \\ \vdots \\ C\end{array}}{C} \vee \text{Elim}$$

Rules for $\rightarrow$

$$n \frac{\begin{array}{c}[A]^n \\ \vdots \\ B\end{array}}{A \rightarrow B} \rightarrow \text{Intro}$$

$$\frac{A \rightarrow B \quad A}{B} \rightarrow \text{Elim}$$

Rules for $\neg$

$$n \frac{\begin{array}{c}[A]^n \\ \vdots \\ \bot\end{array}}{\neg A} \neg\text{Intro} \qquad \frac{\neg A \quad A}{\bot} \neg\text{Elim}$$

Rules for $\bot$

$$\frac{\bot}{A} \bot_I \qquad n \frac{\begin{array}{c}[\neg A]^n \\ \vdots \\ \bot\end{array}}{A} \bot_C$$

Note that ¬Intro and $\bot_C$ are very similar: The difference is that ¬Intro derives a negated sentence $\neg A$ but $\bot_C$ a positive sentence A.

Whenever a rule indicates that some assumption may be discharged, we take this to be a permission, but not a requirement. E.g., in the →Intro rule, we may discharge any number of assumptions of the form A in the derivation of the premise B, including zero.

11.3 Quantifier Rules

Rules for ∀

$$\frac{A(a)}{\forall x\, A(x)} \forall\text{Intro} \qquad \frac{\forall x\, A(x)}{A(t)} \forall\text{Elim}$$

In the rules for ∀, t is a ground term (a term that does not contain any variables), and a is a constant symbol which does not occur in the conclusion $\forall x\, A(x)$, or in any assumption which is undischarged in the derivation ending with the premise $A(a)$. We call a the *eigenvariable* of the ∀Intro inference.

11.3. QUANTIFIER RULES

Rules for $\exists$

$$\frac{A(t)}{\exists x\, A(x)}\ \exists\text{Intro} \qquad\qquad n\,\frac{\exists x\, A(x) \quad \begin{array}{c}[A(a)]^n\\ \vdots\\ C\end{array}}{C}\ \exists\text{Elim}$$

Again, t is a ground term, and a is a constant which does not occur in the premise $\exists x\, A(x)$, in the conclusion C, or any assumption which is undischarged in the derivations ending with the two premises (other than the assumptions $A(a)$). We call a the *eigenvariable* of the $\exists$Elim inference.

The condition that an eigenvariable neither occur in the premises nor in any assumption that is undischarged in the derivations leading to the premises for the $\forall$Intro or $\exists$Elim inference is called the *eigenvariable condition*.

We use the term "eigenvariable" even though a in the above rules is a constant. This has historical reasons.

In $\exists$Intro and $\forall$Elim there are no restrictions, and the term t can be anything, so we do not have to worry about any conditions. On the other hand, in the $\exists$Elim and $\forall$Intro rules, the eigenvariable condition requires that the constant symbol a does not occur anywhere in the conclusion or in an undischarged assumption. The condition is necessary to ensure that the system is sound, i.e., only derives sentences from undischarged assumptions from which they follow. Without this condition, the following would be allowed:

$$\frac{\exists x\, A(x) \quad \dfrac{[A(a)]^1}{\forall x\, A(x)}\ {}^*\forall\text{Intro}}{\forall x\, A(x)}\ \exists\text{Elim}$$

However, $\exists x\, A(x) \nvDash \forall x\, A(x)$.

11.4 Derivations

We've said what an assumption is, and we've given the rules of inference. Derivations in natural deduction are inductively generated from these: each derivation either is an assumption on its own, or consists of one, two, or three derivations followed by a correct inference.

Definition 11.2 (Derivation). A *derivation* of a sentence A from assumptions Γ is a tree of sentences satisfying the following conditions:

1. The topmost sentences of the tree are either in Γ or are discharged by an inference in the tree.

2. The bottommost sentence of the tree is A.

3. Every sentence in the tree except the sentence A at the bottom is a premise of a correct application of an inference rule whose conclusion stands directly below that sentence in the tree.

We then say that A is the *conclusion* of the derivation and that A is *derivable* from Γ.

Example 11.3. Every assumption on its own is a derivation. So, e.g., C by itself is a derivation, and so is D by itself. We can obtain a new derivation from these by applying, say, the ∧Intro rule,

$$\frac{A \quad B}{A \wedge B} \wedge\text{Intro}$$

These rules are meant to be general: we can replace the A and B in it with any sentences, e.g., by C and D. Then the conclusion would be $C \wedge D$, and so

$$\frac{C \quad D}{C \wedge D} \wedge\text{Intro}$$

is a correct derivation. Of course, we can also switch the assumptions, so that D plays the role of A and C that of B. Thus,

$$\frac{D \quad C}{D \wedge C} \wedge \text{Intro}$$

is also a correct derivation.

We can now apply another rule, say, $\rightarrow$Intro, which allows us to conclude a conditional and allows us to discharge any assumption that is identical to the antecedent of that conditional. So both of the following would be correct derivations:

$$1\frac{\dfrac{[C]^1 \quad D}{C \wedge D}\wedge\text{Intro}}{C \rightarrow (C \wedge D)}\rightarrow\text{Intro} \qquad 1\frac{\dfrac{C \quad [D]^1}{C \wedge D}\wedge\text{Intro}}{D \rightarrow (C \wedge D)}\rightarrow\text{Intro}$$

Remember that discharging of assumptions is a permission, not a requirement: we don't have to discharge the assumptions. In particular, we can apply a rule even if the assumptions are not present in the derivation. For instance, the following is legal, even though there is no assumption A to be discharged:

$$1\frac{B}{A \rightarrow B}\rightarrow\text{Intro}$$

11.5 Examples of Derivations

Example 11.4. Let's give a derivation of the sentence $(A \wedge B) \rightarrow A$.

We begin by writing the desired conclusion at the bottom of the derivation.

$$\overline{(A \wedge B) \rightarrow A}$$

Next, we need to figure out what kind of inference could result in a sentence of this form. The main operator of the conclusion is $\rightarrow$, so we'll try to arrive at the conclusion using the $\rightarrow$Intro rule. It is best to write down the assumptions involved and label the inference rules as you progress, so it is easy to see whether all assumptions have been discharged at the end of the proof.

$$
1 \frac{\begin{array}{c} [A \wedge B]^1 \\ \vdots \\ A \end{array}}{(A \wedge B) \to A} \to \text{Intro}
$$

We now need to fill in the steps from the assumption $A \wedge B$ to A. Since we only have one connective to deal with, $\wedge$, we must use the $\wedge$ elim rule. This gives us the following proof:

$$
1 \frac{\dfrac{[A \wedge B]^1}{A} \wedge \text{Elim}}{(A \wedge B) \to A} \to \text{Intro}
$$

We now have a correct derivation of $(A \wedge B) \to A$.

Example 11.5. Now let's give a derivation of $(\neg A \vee B) \to (A \to B)$.

We begin by writing the desired conclusion at the bottom of the derivation.

$$
\overline{(\neg A \vee B) \to (A \to B)}
$$

To find a logical rule that could give us this conclusion, we look at the logical connectives in the conclusion: $\neg$, $\vee$, and $\to$. We only care at the moment about the first occurence of $\to$ because it is the main operator of the sentence in the end-sequent, while $\neg$, $\vee$ and the second occurence of $\to$ are inside the scope of another connective, so we will take care of those later. We therefore start with the $\to$Intro rule. A correct application must look like this:

$$
1 \frac{\begin{array}{c} [\neg A \vee B]^1 \\ \vdots \\ A \to B \end{array}}{(\neg A \vee B) \to (A \to B)} \to \text{Intro}
$$

This leaves us with two possibilities to continue. Either we can keep working from the bottom up and look for another application of the →Intro rule, or we can work from the top down and apply a ∨Elim rule. Let us apply the latter. We will use the assumption $\neg A \lor B$ as the leftmost premise of ∨Elim. For a valid application of ∨Elim, the other two premises must be identical to the conclusion $A \rightarrow B$, but each may be derived in turn from another assumption, namely the two disjuncts of $\neg A \lor B$. So our derivation will look like this:

$$
{\scriptstyle 1}\,\dfrac{{\scriptstyle 2}\,\dfrac{[\neg A \lor B]^1 \qquad \begin{matrix}[\neg A]^2 \\ \vdots \\ A \rightarrow B\end{matrix} \qquad \begin{matrix}[B]^2 \\ \vdots \\ A \rightarrow B\end{matrix}}{A \rightarrow B}\;\lor\text{Elim}}{(\neg A \lor B) \rightarrow (A \rightarrow B)}\;\rightarrow\text{Intro}
$$

In each of the two branches on the right, we want to derive $A \rightarrow B$, which is best done using →Intro.

$$
{\scriptstyle 1}\,\dfrac{{\scriptstyle 2}\,\dfrac{[\neg A \lor B]^1 \qquad {\scriptstyle 3}\,\dfrac{\begin{matrix}[\neg A]^2, [A]^3 \\ \vdots \\ B\end{matrix}}{A \rightarrow B}\;\rightarrow\text{Intro} \qquad {\scriptstyle 4}\,\dfrac{\begin{matrix}[B]^2, [A]^4 \\ \vdots \\ B\end{matrix}}{A \rightarrow B}\;\rightarrow\text{Intro}}{A \rightarrow B}\;\lor\text{Elim}}{(\neg A \lor B) \rightarrow (A \rightarrow B)}\;\rightarrow\text{Intro}
$$

For the two missing parts of the derivation, we need derivations of B from $\neg A$ and A in the middle, and from A and B on the left. Let's take the former first. $\neg A$ and A are the two premises of ¬Elim:

$$
\begin{matrix}\dfrac{[\neg A]^2 \qquad [A]^3}{\bot}\;\neg\text{Elim} \\ \vdots \\ B\end{matrix}
$$

By using $\bot_I$, we can obtain B as a conclusion and complete the branch.

$$2\dfrac{[\neg A \vee B]^1 \qquad \dfrac{\dfrac{[\neg A]^2 \quad [A]^3}{\bot}\bot\text{Intro}}{3\dfrac{B}{A \to B}\bot_I}\to\text{Intro} \qquad \dfrac{\overset{[B]^2, [A]^4}{\vdots}}{4\dfrac{B}{A \to B}}\to\text{Intro}}{1\dfrac{A \to B}{(\neg A \vee B) \to (A \to B)}\to\text{Intro}}\vee\text{Elim}$$

Let's now look at the rightmost branch. Here it's important to realize that the definition of derivation *allows assumptions to be discharged* but *does not require* them to be. In other words, if we can derive B from one of the assumptions A and B without using the other, that's ok. And to derive B from B is trivial: B by itself is such a derivation, and no inferences are needed. So we can simply delete the assumption A.

$$2\dfrac{[\neg A \vee B]^1 \qquad \dfrac{\dfrac{[\neg A]^2 \quad [A]^3}{\bot}\neg\text{Elim}}{3\dfrac{B}{A \to B}\bot_I}\to\text{Intro} \qquad \dfrac{[B]^2}{A \to B}\to\text{Intro}}{1\dfrac{A \to B}{(\neg A \vee B) \to (A \to B)}\to\text{Intro}}\vee\text{Elim}$$

Note that in the finished derivation, the rightmost $\to$Intro inference does not actually discharge any assumptions.

Example 11.6. So far we have not needed the $\bot_C$ rule. It is special in that it allows us to discharge an assumption that isn't a sub-formula of the conclusion of the rule. It is closely related to the $\bot_I$ rule. In fact, the $\bot_I$ rule is a special case of the $\bot_C$ rule— there is a logic called "intuitionistic logic" in which only $\bot_I$ is allowed. The $\bot_C$ rule is a last resort when nothing else works. For instance, suppose we want to derive $A \vee \neg A$. Our usual strategy would be to attempt to derive $A \vee \neg A$ using $\vee$Intro. But this would require us to derive either A or $\neg A$ from no assumptions, and this can't be done. $\bot_C$ to the rescue!

11.5. EXAMPLES OF DERIVATIONS

$$
\begin{array}{c}
[\neg(A \vee \neg A)]^1 \\
\vdots \\
1 \dfrac{\bot}{A \vee \neg A} \, \bot_C
\end{array}
$$

Now we're looking for a derivation of $\bot$ from $\neg(A \vee \neg A)$. Since $\bot$ is the conclusion of $\neg$Elim we might try that:

$$
\cfrac{\begin{array}{cc} [\neg(A \vee \neg A)]^1 & [\neg(A \vee \neg A)]^1 \\ \vdots & \vdots \\ \neg A & A \end{array} \; \neg\text{Elim}}{1 \; \dfrac{\bot}{A \vee \neg A} \, \bot_C}
$$

Our strategy for finding a derivation of $\neg A$ calls for an application of $\neg$Intro:

$$
\cfrac{\begin{array}{cc} [\neg(A \vee \neg A)]^1, [A]^2 & [\neg(A \vee \neg A)]^1 \\ \vdots & \vdots \\ 2 \, \dfrac{\bot}{\neg A} \, \neg\text{Intro} & A \end{array} \; \neg\text{Elim}}{1 \, \dfrac{\bot}{A \vee \neg A} \, \bot_C}
$$

Here, we can get $\bot$ easily by applying $\neg$Elim to the assumption $\neg(A \vee \neg A)$ and $A \vee \neg A$ which follows from our new assumption A by $\vee$Intro:

$$
\cfrac{\cfrac{[\neg(A \vee \neg A)]^1 \quad \cfrac{[A]^2}{A \vee \neg A} \, \vee\text{Intro}}{2 \, \dfrac{\bot}{\neg A} \, \neg\text{Intro}} \, \neg\text{Elim} \qquad \begin{array}{c}[\neg(A \vee \neg A)]^1 \\ \vdots \\ A\end{array}}{1 \, \dfrac{\bot}{A \vee \neg A} \, \bot_C} \; \neg\text{Elim}
$$

On the right side we use the same strategy, except we get A by $\bot_C$:

$$\cfrac{[\neg(A \vee \neg A)]^1 \quad \cfrac{\cfrac{[A]^2}{A \vee \neg A}\vee\text{Intro}}{\cfrac{\bot}{\neg A}\ \neg\text{Intro}}\ \neg\text{Elim}}{2\ \cfrac{\bot}{\neg A}\ \neg\text{Intro}} \qquad \cfrac{[\neg(A \vee \neg A)]^1 \quad \cfrac{\cfrac{[\neg A]^3}{A \vee \neg A}\vee\text{Intro}}{\cfrac{\bot}{A}\ \bot_C}\ \neg\text{Elim}}{3\ \cfrac{\bot}{A}\ \bot_C}\ \neg\text{Elim}$$
$$1\ \cfrac{\bot}{A \vee \neg A}\ \bot_C$$

11.6 Derivations with Quantifiers

Example 11.7. When dealing with quantifiers, we have to make sure not to violate the eigenvariable condition, and sometimes this requires us to play around with the order of carrying out certain inferences. In general, it helps to try and take care of rules subject to the eigenvariable condition first (they will be lower down in the finished proof).

Let's see how we'd give a derivation of the formula $\exists x\, \neg A(x) \to \neg \forall x\, A(x)$. Starting as usual, we write

$$\overline{\exists x\, \neg A(x) \to \neg \forall x\, A(x)}$$

We start by writing down what it would take to justify that last step using the →Intro rule.

$$\begin{array}{c}[\exists x\, \neg A(x)]^1 \\ \vdots \\ \neg \forall x\, A(x) \\ \hline \exists x\, \neg A(x) \to \neg \forall x\, A(x)\end{array} \to\text{Intro}$$

Since there is no obvious rule to apply to $\neg \forall x\, A(x)$, we will proceed by setting up the derivation so we can use the ∃Elim rule. Here we must pay attention to the eigenvariable condition, and choose a constant that does not appear in $\exists x\, A(x)$ or any assumptions that it depends on. (Since no constant symbols appear, however, any choice will do fine.)

11.6. DERIVATIONS WITH QUANTIFIERS

$$\cfrac{[\exists x \neg A(x)]^1 \quad \cfrac{[\neg A(a)]^2 \\ \vdots \\ \neg \forall x\, A(x)}{}}{\cfrac{\neg \forall x\, A(x)}{\exists x \neg A(x) \to \neg \forall x\, A(x)} \to\text{Intro} \; 1} \; \exists\text{Elim} \; 2$$

In order to derive $\neg \forall x\, A(x)$, we will attempt to use the $\neg$Intro rule: this requires that we derive a contradiction, possibly using $\forall x\, A(x)$ as an additional assumption. Of course, this contradiction may involve the assumption $\neg A(a)$ which will be discharged by the $\exists$Elim inference. We can set it up as follows:

$$\cfrac{[\exists x \neg A(x)]^1 \quad \cfrac{\cfrac{[\neg A(a)]^2, [\forall x\, A(x)]^3 \\ \vdots \\ \bot}{\neg \forall x\, A(x)} \, \neg\text{Intro} \; 3}{\neg \forall x\, A(x)} \, \exists\text{Elim} \; 2}{\exists x \neg A(x) \to \neg \forall x\, A(x)} \to\text{Intro} \; 1$$

It looks like we are close to getting a contradiction. The easiest rule to apply is the $\forall$Elim, which has no eigenvariable conditions. Since we can use any term we want to replace the universally quantified x, it makes the most sense to continue using a so we can reach a contradiction.

$$\cfrac{[\exists x \neg A(x)]^1 \quad \cfrac{\cfrac{[\neg A(a)]^2 \quad \cfrac{[\forall x\, A(x)]^3}{A(a)} \, \forall\text{Elim}}{\bot} \, \neg\text{Elim}}{\cfrac{\neg \forall x\, A(x)}{} \, \neg\text{Intro} \; 3}}{\cfrac{\neg \forall x\, A(x)}{\exists x \neg A(x) \to \neg \forall x\, A(x)} \to\text{Intro} \; 1} \, \exists\text{Elim} \; 2$$

It is important, especially when dealing with quantifiers, to double check at this point that the eigenvariable condition has not been violated. Since the only rule we applied that is subject to the eigenvariable condition was $\exists$Elim, and the eigenvariable a

does not occur in any assumptions it depends on, this is a correct derivation.

Example 11.8. Sometimes we may derive a formula from other formulas. In these cases, we may have undischarged assumptions. It is important to keep track of our assumptions as well as the end goal.

Let's see how we'd give a derivation of the formula $\exists x\, C(x, b)$ from the assumptions $\exists x\, (A(x) \land B(x))$ and $\forall x\, (B(x) \to C(x, b))$. Starting as usual, we write the conclusion at the bottom.

$$\overline{\exists x\, C(x, b)}$$

We have two premises to work with. To use the first, i.e., try to find a derivation of $\exists x\, C(x, b)$ from $\exists x\, (A(x) \land B(x))$ we would use the $\exists$Elim rule. Since it has an eigenvariable condition, we will apply that rule first. We get the following:

$$\cfrac{\exists x\, (A(x) \land B(x)) \qquad \begin{array}{c}[A(a) \land B(a)]^1 \\ \vdots \\ \exists x\, C(x, b)\end{array}}{\exists x\, C(x, b)}\; \exists\text{Elim}\quad 1$$

The two assumptions we are working with share B. It may be useful at this point to apply $\land$Elim to separate out $B(a)$.

$$\cfrac{\exists x\, (A(x) \land B(x)) \qquad \begin{array}{c}\cfrac{[A(a) \land B(a)]^1}{B(a)}\; \land\text{Elim} \\ \vdots \\ \exists x\, C(x, b)\end{array}}{\exists x\, C(x, b)}\; \exists\text{Elim}\quad 1$$

The second assumption we have to work with is $\forall x\, (B(x) \to C(x, b))$. Since there is no eigenvariable condition we can instantiate x with the constant symbol a using $\forall$Elim to get $B(a) \to$

11.6. DERIVATIONS WITH QUANTIFIERS

$C(a, b)$. We now have both $B(a) \to C(a, b)$ and $B(a)$. Our next move should be a straightforward application of the $\to$Elim rule.

$$\cfrac{\cfrac{\forall x \, (B(x) \to C(x,b))}{B(a) \to C(a,b)} \, \forall\text{Elim} \quad \cfrac{[A(a) \wedge B(a)]^1}{B(a)} \, \wedge\text{Elim}}{C(a,b)} \, \to\text{Elim}$$

$$\vdots$$

$$1 \cfrac{\exists x \, (A(x) \wedge B(x)) \quad \exists x \, C(x,b)}{\exists x \, C(x,b)} \, \exists\text{Elim}$$

We are so close! One application of $\exists$Intro and we have reached our goal.

$$1 \cfrac{\exists x \, (A(x) \wedge B(x)) \quad \cfrac{\cfrac{\forall x \, (B(x) \to C(x,b))}{B(a) \to C(a,b)} \, \forall\text{Elim} \quad \cfrac{[A(a) \wedge B(a)]^1}{B(a)} \, \wedge\text{Elim}}{\cfrac{C(a,b)}{\exists x \, C(x,b)} \, \exists\text{Intro}} \, \to\text{Elim}}{\exists x \, C(x,b)} \, \exists\text{Elim}$$

Since we ensured at each step that the eigenvariable conditions were not violated, we can be confident that this is a correct derivation.

Example 11.9. Give a derivation of the formula $\neg \forall x \, A(x)$ from the assumptions $\forall x \, A(x) \to \exists y \, B(y)$ and $\neg \exists y \, B(y)$. Starting as usual, we write the target formula at the bottom.

$$\overline{\neg \forall x \, A(x)}$$

The last line of the derivation is a negation, so let's try using $\neg$Intro. This will require that we figure out how to derive a contradiction.

$$[\forall x \, A(x)]^1$$
$$\vdots$$
$$1 \cfrac{\bot}{\neg \forall x \, A(x)} \, \neg\text{Intro}$$

So far so good. We can use ∀Elim but it's not obvious if that will help us get to our goal. Instead, let's use one of our assumptions. $\forall x\, A(x) \to \exists y\, B(y)$ together with $\forall x\, A(x)$ will allow us to use the →Elim rule.

$$\dfrac{\dfrac{\forall x\, A(x) \to \exists y\, B(y) \qquad [\forall x\, A(x)]^1}{\exists y\, B(y)} \to\text{Elim}}{}$$

$$1\,\dfrac{\bot}{\neg \forall x\, A(x)}\,\neg\text{Intro}$$

We now have one final assumption to work with, and it looks like this will help us reach a contradiction by using ¬Elim.

$$\dfrac{\neg \exists y\, B(y) \qquad \dfrac{\forall x\, A(x) \to \exists y\, B(y) \qquad [\forall x\, A(x)]^1}{\exists y\, B(y)}\to\text{Elim}}{1\,\dfrac{\bot}{\neg \forall x\, A(x)}\,\neg\text{Intro}}\,\neg\text{Elim}$$

11.7 Proof-Theoretic Notions

Just as we've defined a number of important semantic notions (validity, entailment, satisfiabilty), we now define corresponding *proof-theoretic notions*. These are not defined by appeal to satisfaction of sentences in structures, but by appeal to the derivability or non-derivability of certain sentences from others. It was an important discovery that these notions coincide. That they do is the content of the *soundness* and *completeness theorems*.

Definition 11.10 (Theorems). A sentence A is a *theorem* if there is a derivation of A in natural deduction in which all assumptions are discharged. We write $\vdash A$ if A is a theorem and $\nvdash A$ if it is not.

11.7. PROOF-THEORETIC NOTIONS

Definition 11.11 (Derivability). A sentence A is *derivable from* a set of sentences Γ, $\Gamma \vdash A$, if there is a derivation with conclusion A and in which every assumption is either discharged or is in Γ. If A is not derivable from Γ we write $\Gamma \nvdash A$.

Definition 11.12 (Consistency). A set of sentences Γ is *inconsistent* iff $\Gamma \vdash \bot$. If Γ is not inconsistent, i.e., if $\Gamma \nvdash \bot$, we say it is *consistent*.

Proposition 11.13 (Reflexivity). *If $A \in \Gamma$, then $\Gamma \vdash A$.*

Proof. The assumption A by itself is a derivation of A where every undischarged assumption (i.e., A) is in Γ. □

Proposition 11.14 (Monotony). *If $\Gamma \subseteq \Delta$ and $\Gamma \vdash A$, then $\Delta \vdash A$.*

Proof. Any derivation of A from Γ is also a derivation of A from Δ. □

Proposition 11.15 (Transitivity). *If $\Gamma \vdash A$ and $\{A\} \cup \Delta \vdash B$, then $\Gamma \cup \Delta \vdash B$.*

Proof. If $\Gamma \vdash A$, there is a derivation δ_0 of A with all undischarged assumptions in Γ. If $\{A\} \cup \Delta \vdash B$, then there is a derivation δ_1 of B with all undischarged assumptions in $\{A\} \cup \Delta$. Now consider:

$$\cfrac{\cfrac{\begin{array}{c}\Delta, [A]^1 \\ \vdots\, \delta_1 \\ B\end{array}}{A \to B}\,{\to}\text{Intro}^1 \quad \cfrac{\begin{array}{c}\Gamma \\ \vdots\, \delta_0 \\ A\end{array}}{}}{B}\,{\to}\text{Elim}$$

The undischarged assumptions are now all among $\Gamma \cup \Delta$, so this shows $\Gamma \cup \Delta \vdash B$. □

When $\Gamma = \{A_1, A_2, \ldots, A_k\}$ is a finite set we may use the simplified notation $A_1, A_2, \ldots, A_k \vdash B$ for $\Gamma \vdash B$, in particular $A \vdash B$ means that $\{A\} \vdash B$.

Note that if $\Gamma \vdash A$ and $A \vdash B$, then $\Gamma \vdash B$. It follows also that if $A_1, \ldots, A_n \vdash B$ and $\Gamma \vdash A_i$ for each i, then $\Gamma \vdash B$.

Proposition 11.16. *The following are equivalent.*

1. *Γ is inconsistent.*

2. *$\Gamma \vdash A$ for every sentence A.*

3. *$\Gamma \vdash A$ and $\Gamma \vdash \neg A$ for some sentence A.*

Proof. Exercise. □

Proposition 11.17 (Compactness).   1. *If $\Gamma \vdash A$ then there is a finite subset $\Gamma_0 \subseteq \Gamma$ such that $\Gamma_0 \vdash A$.*

2. *If every finite subset of Γ is consistent, then Γ is consistent.*

Proof.   1. If $\Gamma \vdash A$, then there is a derivation δ of A from Γ. Let Γ_0 be the set of undischarged assumptions of δ. Since any derivation is finite, Γ_0 can only contain finitely many sentences. So, δ is a derivation of A from a finite $\Gamma_0 \subseteq \Gamma$.

2. This is the contrapositive of (1) for the special case $A \equiv \bot$. □

11.8 Derivability and Consistency

We will now establish a number of properties of the derivability relation. They are independently interesting, but each will play a role in the proof of the completeness theorem.

11.8. DERIVABILITY AND CONSISTENCY

Proposition 11.18. *If $\Gamma \vdash A$ and $\Gamma \cup \{A\}$ is inconsistent, then Γ is inconsistent.*

Proof. Let the derivation of A from Γ be δ_1 and the derivation of $\bot$ from $\Gamma \cup \{A\}$ be δ_2. We can then derive:

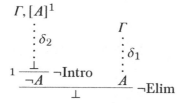

In the new derivation, the assumption A is discharged, so it is a derivation from Γ. □

Proposition 11.19. *$\Gamma \vdash A$ iff $\Gamma \cup \{\neg A\}$ is inconsistent.*

Proof. First suppose $\Gamma \vdash A$, i.e., there is a derivation δ_0 of A from undischarged assumptions Γ. We obtain a derivation of $\bot$ from $\Gamma \cup \{\neg A\}$ as follows:

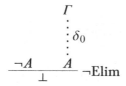

Now assume $\Gamma \cup \{\neg A\}$ is inconsistent, and let δ_1 be the corresponding derivation of $\bot$ from undischarged assumptions in $\Gamma \cup \{\neg A\}$. We obtain a derivation of A from Γ alone by using $\bot_C$:

$$\begin{array}{c} \Gamma, [\neg A]^1 \\ \vdots \, \delta_1 \\ \dfrac{\bot}{A} \, \bot_C \end{array}$$

□

Proposition 11.20. *If $\Gamma \vdash A$ and $\neg A \in \Gamma$, then Γ is inconsistent.*

Proof. Suppose $\Gamma \vdash A$ and $\neg A \in \Gamma$. Then there is a derivation δ of A from Γ. Consider this simple application of the $\neg$Elim rule:

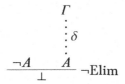

Since $\neg A \in \Gamma$, all undischarged assumptions are in Γ, this shows that $\Gamma \vdash \bot$. □

Proposition 11.21. *If $\Gamma \cup \{A\}$ and $\Gamma \cup \{\neg A\}$ are both inconsistent, then Γ is inconsistent.*

Proof. There are derivations δ_1 and δ_2 of $\bot$ from $\Gamma \cup \{A\}$ and $\bot$ from $\Gamma \cup \{\neg A\}$, respectively. We can then derive

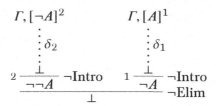

Since the assumptions A and $\neg A$ are discharged, this is a derivation of $\bot$ from Γ alone. Hence Γ is inconsistent. □

11.9 Derivability and the Propositional Connectives

We establish that the derivability relation $\vdash$ of natural deduction is strong enough to establish some basic facts involving the propositional connectives, such as that $A \wedge B \vdash A$ and $A, A \rightarrow B \vdash B$ (modus ponens). These facts are needed for the proof of the completeness theorem.

11.9. DERIVABILITY AND THE PROPOSITIONAL CONNECTIVES

Proposition 11.22.   *1. Both $A \wedge B \vdash A$ and $A \wedge B \vdash B$*

2. $A, B \vdash A \wedge B$.

Proof.   1. We can derive both

$$\frac{A \wedge B}{A} \wedge\text{Elim} \qquad \frac{A \wedge B}{B} \wedge\text{Elim}$$

2. We can derive:

$$\frac{A \quad B}{A \wedge B} \wedge\text{Intro} \qquad \square$$

Proposition 11.23.   *1. $A \vee B, \neg A, \neg B$ is inconsistent.*

2. Both $A \vdash A \vee B$ and $B \vdash A \vee B$.

Proof.   1. Consider the following derivation:

$$\cfrac{A \vee B \quad \cfrac{\neg A \quad [A]^1}{\bot}\neg\text{Elim} \quad \cfrac{\neg B \quad [B]^1}{\bot}\neg\text{Elim}}{\bot}{}_1 \vee\text{Elim}$$

This is a derivation of $\bot$ from undischarged assumptions $A \vee B$, $\neg A$, and $\neg B$.

2. We can derive both

$$\frac{A}{A \vee B} \vee\text{Intro} \qquad \frac{B}{A \vee B} \vee\text{Intro} \qquad \square$$

Proposition 11.24.   *1. $A, A \to B \vdash B$.*

2. Both $\neg A \vdash A \to B$ and $B \vdash A \to B$.

Proof.   1. We can derive:

$$\frac{A \to B \quad A}{B} \to\text{Elim}$$

2. This is shown by the following two derivations:

$$
1\frac{\dfrac{\neg A \quad [A]^1}{\bot}\neg\text{Elim}}{\dfrac{B}{A \to B}\bot_I}\to\text{Intro} \qquad \dfrac{B}{A \to B}\to\text{Intro}
$$

Note that →Intro may, but does not have to, discharge the assumption A. □

11.10 Derivability and the Quantifiers

The completeness theorem also requires that the natural deduction rules yield the facts about ⊢ established in this section.

Theorem 11.25. *If c is a constant not occurring in Γ or $A(x)$ and $\Gamma \vdash A(c)$, then $\Gamma \vdash \forall x\, A(x)$.*

Proof. Let δ be a derivation of $A(c)$ from Γ. By adding a ∀Intro inference, we obtain a derivation of $\forall x\, A(x)$. Since c does not occur in Γ or $A(x)$, the eigenvariable condition is satisfied. □

Proposition 11.26. *1. $A(t) \vdash \exists x\, A(x)$.*

2. $\forall x\, A(x) \vdash A(t)$.

Proof. 1. The following is a derivation of $\exists x\, A(x)$ from $A(t)$:

$$\dfrac{A(t)}{\exists x\, A(x)}\exists\text{Intro}$$

2. The following is a derivation of $A(t)$ from $\forall x\, A(x)$:

$$\dfrac{\forall x\, A(x)}{A(t)}\forall\text{Elim}$$

□

11.11 Soundness

A derivation system, such as natural deduction, is *sound* if it cannot derive things that do not actually follow. Soundness is thus a kind of guaranteed safety property for derivation systems. Depending on which proof theoretic property is in question, we would like to know for instance, that

1. every derivable sentence is valid;

2. if a sentence is derivable from some others, it is also a consequence of them;

3. if a set of sentences is inconsistent, it is unsatisfiable.

These are important properties of a derivation system. If any of them do not hold, the derivation system is deficient—it would derive too much. Consequently, establishing the soundness of a derivation system is of the utmost importance.

Theorem 11.27 (Soundness). *If A is derivable from the undischarged assumptions Γ, then $\Gamma \vDash A$.*

Proof. Let δ be a derivation of A. We proceed by induction on the number of inferences in δ.

For the induction basis we show the claim if the number of inferences is 0. In this case, δ consists only of a single sentence A, i.e., an assumption. That assumption is undischarged, since assumptions can only be discharged by inferences, and there are no inferences. So, any structure M that satisfies all of the undischarged assumptions of the proof also satisfies A.

Now for the inductive step. Suppose that δ contains n inferences. The premise(s) of the lowermost inference are derived using sub-derivations, each of which contains fewer than n inferences. We assume the induction hypothesis: The premises of the lowermost inference follow from the undischarged assumptions of the sub-derivations ending in those premises. We have to show that the conclusion A follows from the undischarged assumptions of the entire proof.

We distinguish cases according to the type of the lowermost inference. First, we consider the possible inferences with only one premise.

1. Suppose that the last inference is $\neg$Intro: The derivation has the form

$$
\begin{array}{c}
\Gamma, [A]^n \\
\vdots\; \delta_1 \\
n\; \dfrac{\bot}{\neg A}\; \neg\text{Intro}
\end{array}
$$

By inductive hypothesis, $\bot$ follows from the undischarged assumptions $\Gamma \cup \{A\}$ of δ_1. Consider a structure M. We need to show that, if $M \vDash \Gamma$, then $M \vDash \neg A$. Suppose for reductio that $M \vDash \Gamma$, but $M \nvDash \neg A$, i.e., $M \vDash A$. This would mean that $M \vDash \Gamma \cup \{A\}$. This is contrary to our inductive hypothesis. So, $M \vDash \neg A$.

2. The last inference is $\wedge$Elim: There are two variants: A or B may be inferred from the premise $A \wedge B$. Consider the first case. The derivation δ looks like this:

$$
\begin{array}{c}
\Gamma \\
\vdots\; \delta_1 \\
\dfrac{A \wedge B}{A}\; \wedge\text{Elim}
\end{array}
$$

By inductive hypothesis, $A \wedge B$ follows from the undischarged assumptions Γ of δ_1. Consider a structure M. We need to show that, if $M \vDash \Gamma$, then $M \vDash A$. Suppose $M \vDash \Gamma$. By our inductive hypothesis ($\Gamma \vDash A \wedge B$), we know that $M \vDash A \wedge B$. By definition, $M \vDash A \wedge B$ iff $M \vDash A$ and $M \vDash B$.

11.11. SOUNDNESS

(The case where B is inferred from $A \wedge B$ is handled similarly.)

3. The last inference is $\vee$Intro: There are two variants: $A \vee B$ may be inferred from the premise A or the premise B. Consider the first case. The derivation has the form

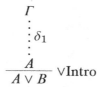

By inductive hypothesis, A follows from the undischarged assumptions Γ of δ_1. Consider a structure M. We need to show that, if $M \vDash \Gamma$, then $M \vDash A \vee B$. Suppose $M \vDash \Gamma$; then $M \vDash A$ since $\Gamma \vDash A$ (the inductive hypothesis). So it must also be the case that $M \vDash A \vee B$. (The case where $A \vee B$ is inferred from B is handled similarly.)

4. The last inference is $\rightarrow$Intro: $A \rightarrow B$ is inferred from a subproof with assumption A and conclusion B, i.e.,

$$\begin{array}{c} \Gamma, [A]^n \\ \vdots \, \delta_1 \\ \vdots \\ n \, \dfrac{B}{A \rightarrow B} \, \rightarrow\text{Intro} \end{array}$$

By inductive hypothesis, B follows from the undischarged assumptions of δ_1, i.e., $\Gamma \cup \{A\} \vDash B$. Consider a structure M. The undischarged assumptions of δ are just Γ, since A is discharged at the last inference. So we need to show that $\Gamma \vDash A \rightarrow B$. For reductio, suppose that for some structure M, $M \vDash \Gamma$ but $M \nvDash A \rightarrow B$. So, $M \vDash A$ and $M \nvDash B$. But by hypothesis, B is a consequence of $\Gamma \cup \{A\}$, i.e., $M \vDash B$, which is a contradiction. So, $\Gamma \vDash A \rightarrow B$.

5. The last inference is $\bot_I$: Here, δ ends in

By induction hypothesis, $\Gamma \vDash \bot$. We have to show that $\Gamma \vDash A$. Suppose not; then for some M we have $M \vDash \Gamma$ and $M \nvDash A$. But we always have $M \nvDash \bot$, so this would mean that $\Gamma \nvDash \bot$, contrary to the induction hypothesis.

6. The last inference is $\bot_C$: Exercise.

7. The last inference is ∀Intro: Then δ has the form

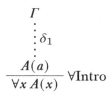

The premise $A(a)$ is a consequence of the undischarged assumptions Γ by induction hypothesis. Consider some structure, M, such that $M \vDash \Gamma$. We need to show that $M \vDash \forall x\, A(x)$. Since $\forall x\, A(x)$ is a sentence, this means we have to show that for every variable assignment s, $M, s \vDash A(x)$ (Proposition 7.18). Since Γ consists entirely of sentences, $M, s \vDash B$ for all $B \in \Gamma$ by Definition 7.11. Let M' be like M except that $a^{M'} = s(x)$. Since a does not occur in Γ, $M' \vDash \Gamma$ by Corollary 7.20. Since $\Gamma \vDash A(a)$, $M' \vDash A(a)$. Since $A(a)$ is a sentence, $M', s \vDash A(a)$ by Proposition 7.17. $M', s \vDash A(x)$ iff $M' \vDash A(a)$ by Proposition 7.22 (recall that $A(a)$ is just $A(x)[a/x]$). So, $M', s \vDash A(x)$. Since a does not occur in $A(x)$, by Proposition 7.19, $M, s \vDash A(x)$. But s was an arbitrary variable assignment, so $M \vDash \forall x\, A(x)$.

11.11. SOUNDNESS

8. The last inference is ∃Intro: Exercise.

9. The last inference is ∀Elim: Exercise.

Now let's consider the possible inferences with several premises: ∨Elim, ∧Intro, →Elim, and ∃Elim.

1. The last inference is ∧Intro. $A \wedge B$ is inferred from the premises A and B and δ has the form

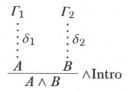

By induction hypothesis, A follows from the undischarged assumptions Γ_1 of δ_1 and B follows from the undischarged assumptions Γ_2 of δ_2. The undischarged assumptions of δ are $\Gamma_1 \cup \Gamma_2$, so we have to show that $\Gamma_1 \cup \Gamma_2 \vDash A \wedge B$. Consider a structure M with $M \vDash \Gamma_1 \cup \Gamma_2$. Since $M \vDash \Gamma_1$, it must be the case that $M \vDash A$ as $\Gamma_1 \vDash A$, and since $M \vDash \Gamma_2$, $M \vDash B$ since $\Gamma_2 \vDash B$. Together, $M \vDash A \wedge B$.

2. The last inference is ∨Elim: Exercise.

3. The last inference is →Elim. B is inferred from the premises $A \to B$ and A. The derivation δ looks like this:

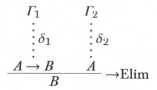

By induction hypothesis, $A \to B$ follows from the undischarged assumptions Γ_1 of δ_1 and A follows from the undischarged assumptions Γ_2 of δ_2. Consider a structure M. We need to show that, if $M \vDash \Gamma_1 \cup \Gamma_2$, then $M \vDash B$. Suppose $M \vDash \Gamma_1 \cup \Gamma_2$. Since $\Gamma_1 \vDash A \rightarrow B$, $M \vDash A \rightarrow B$. Since $\Gamma_2 \vDash A$, we have $M \vDash A$. This means that $M \vDash B$ (For if $M \nvDash B$, since $M \vDash A$, we'd have $M \nvDash A \rightarrow B$, contradicting $M \vDash A \rightarrow B$).

4. The last inference is ¬Elim: Exercise.

5. The last inference is ∃Elim: Exercise. □

Corollary 11.28. *If $\vdash A$, then A is valid.*

Corollary 11.29. *If Γ is satisfiable, then it is consistent.*

Proof. We prove the contrapositive. Suppose that Γ is not consistent. Then $\Gamma \vdash \bot$, i.e., there is a derivation of $\bot$ from undischarged assumptions in Γ. By Theorem 11.27, any structure M that satisfies Γ must satisfy $\bot$. Since $M \nvDash \bot$ for every structure M, no M can satisfy Γ, i.e., Γ is not satisfiable. □

11.12 Derivations with Identity predicate

Derivations with identity predicate require additional inference rules.

$$\overline{t = t} \text{ =Intro} \qquad \frac{t_1 = t_2 \quad A(t_1)}{A(t_2)} \text{=Elim}$$

$$\frac{t_1 = t_2 \quad A(t_2)}{A(t_1)} \text{=Elim}$$

In the above rules, t, t_1, and t_2 are closed terms. The =Intro rule allows us to derive any identity statement of the form $t = t$ outright, from no assumptions.

Example 11.30. If s and t are closed terms, then $A(s), s = t \vdash A(t)$:

11.12. DERIVATIONS WITH IDENTITY PREDICATE

$$\frac{s = t \quad A(s)}{A(t)} = \text{Elim}$$

This may be familiar as the "principle of substitutability of identicals," or Leibniz' Law.

Example 11.31. We derive the sentence

$$\forall x \, \forall y \, ((A(x) \wedge A(y)) \to x = y)$$

from the sentence

$$\exists x \, \forall y \, (A(y) \to y = x)$$

We develop the derivation backwards:

$$\exists x \, \forall y \, (A(y) \to y = x) \quad [A(a) \wedge A(b)]^1$$
$$\vdots$$
$$1 \frac{\dfrac{a = b}{((A(a) \wedge A(b)) \to a = b)} \to \text{Intro}}{\dfrac{\forall y \, ((A(a) \wedge A(y)) \to a = y)}{\forall x \, \forall y \, ((A(x) \wedge A(y)) \to x = y)} \, \forall \text{Intro}} \, \forall \text{Intro}$$

We'll now have to use the main assumption: since it is an existential formula, we use ∃Elim to derive the intermediary conclusion $a = b$.

$$[\forall y \, (A(y) \to y = c)]^2$$
$$[A(a) \wedge A(b)]^1$$
$$\vdots$$

$$2 \frac{\exists x \, \forall y \, (A(y) \to y = x) \quad a = b}{\dfrac{1 \dfrac{a = b}{((A(a) \wedge A(b)) \to a = b)} \to \text{Intro}}{\dfrac{\forall y \, ((A(a) \wedge A(y)) \to a = y)}{\forall x \, \forall y \, ((A(x) \wedge A(y)) \to x = y)} \, \forall \text{Intro}} \, \forall \text{Intro}} \, \exists \text{Elim}$$

The sub-derivation on the top right is completed by using its assumptions to show that $a = c$ and $b = c$. This requires two separate derivations. The derivation for $a = c$ is as follows:

$$\cfrac{\cfrac{[\forall y\,(A(y) \rightarrow y = c)]^2}{A(a) \rightarrow a = c}\,\forall\text{Elim} \quad \cfrac{[A(a) \wedge A(b)]^1}{A(a)}\,\wedge\text{Elim}}{a = c}\,\rightarrow\text{Elim}$$

From $a = c$ and $b = c$ we derive $a = b$ by =Elim.

11.13 Soundness with Identity predicate

Proposition 11.32. *Natural deduction with rules for = is sound.*

Proof. Any formula of the form $t = t$ is valid, since for every structure M, $M \vDash t = t$. (Note that we assume the term t to be ground, i.e., it contains no variables, so variable assignments are irrelevant).

Suppose the last inference in a derivation is =Elim, i.e., the derivation has the following form:

$$\cfrac{\begin{array}{cc} \Gamma_1 & \Gamma_2 \\ \vdots\,\delta_1 & \vdots\,\delta_2 \\ t_1 = t_2 & A(t_1) \end{array}}{A(t_2)}\,=\text{Elim}$$

The premises $t_1 = t_2$ and $A(t_1)$ are derived from undischarged assumptions Γ_1 and Γ_2, respectively. We want to show that $A(t_2)$ follows from $\Gamma_1 \cup \Gamma_2$. Consider a structure M with $M \vDash \Gamma_1 \cup \Gamma_2$. By induction hypothesis, $M \vDash A(t_1)$ and $M \vDash t_1 = t_2$. Therefore, $\text{Val}^M(t_1) = \text{Val}^M(t_2)$. Let s be any variable assignment, and $m = \text{Val}^M(t_1) = \text{Val}^M(t_2)$. By Proposition 7.22, $M, s \vDash A(t_1)$ iff $M, s[m/x] \vDash A(x)$ iff $M, s \vDash A(t_2)$. Since $M \vDash A(t_1)$, we have $M \vDash A(t_2)$. □

Summary

Proof systems provide purely syntactic methods for characterizing consequence and compatibility between sentences. **Natural deduction** is one such proof system. A **derivation** in it consists of a tree formulas. The topmost formulas in a derivation are **assumptions**. All other formulas, for the derivation to be correct, must be correctly justified by one of a number of **inference rules**. These come in pairs; an introduction and an elimination rule for each connective and quantifier. For instance, if a formula A is justified by a $\to$Elim rule, the preceding formulas (the **premises**) must be $B \to A$ and B (for some B). Some inference rules also allow assumptions to be **discharged**. For instance, if $A \to B$ is inferred from B using $\to$Intro, any occurrences of A as assumptions in the derivation leading to the premise B may be discharged, and is given a label that is also recorded at the inference.

If there is a derivation with end formula A and all assumptions are discharged, we say A is a **theorem** and write $\vdash A$. If all undischarged assumptions are in some set Γ, we say A is **derivable from** Γ and write $\Gamma \vdash A$. If $\Gamma \vdash \bot$ we say Γ is **inconsistent**, otherwise **consistent**. These notions are interrelated, e.g., $\Gamma \vdash A$ iff $\Gamma \cup \{\neg A\}$ is inconsistent. They are also related to the corresponding semantic notions, e.g., if $\Gamma \vdash A$ then $\Gamma \vDash A$. This property of proof systems—what can be derived from Γ is guaranteed to be entailed by Γ—is called **soundness**. The **soundness theorem** is proved by induction on the length of derivations, showing that each individual inference preserves entailment of its conclusion from open assumptions provided its premises are entailed by their undischarged assumptions.

Problems

Problem 11.1. Give derivations of the following:

1. $\neg(A \to B) \to (A \land \neg B)$

2. $(A \to C) \vee (B \to C)$ from the assumption $(A \wedge B) \to C$

3. $\neg\neg A \to A$

4. $\neg A \to \neg B$ from the assumption $B \to A$

5. $\neg A$ from the assumption $(A \to \neg A)$

6. A from the assumptions $B \to A$ and $\neg B \to A$

Problem 11.2. Give derivations of the following:

1. $\exists y\, A(y) \to B$ from the assumption $\forall x\, (A(x) \to B)$

2. $\exists x\, (A(x) \to \forall y\, A(y))$

Problem 11.3. Prove Proposition 11.16

Problem 11.4. Prove that $\Gamma \vdash \neg A$ iff $\Gamma \cup \{A\}$ is inconsistent.

Problem 11.5. Complete the proof of Theorem 11.27.

Problem 11.6. Prove that $=$ is both symmetric and transitive, i.e., give derivations of $\forall x\, \forall y\, (x = y \to y = x)$ and $\forall x\, \forall y\, \forall z((x = y \wedge y = z) \to x = z)$

Problem 11.7. Give derivations of the following formulas:

1. $\forall x\, \forall y\, ((x = y \wedge A(x)) \to A(y))$

2. $\exists x\, A(x) \wedge \forall y\, \forall z\, ((A(y) \wedge A(z)) \to y = z) \to \exists x\, (A(x) \wedge \forall y\, (A(y) \to y = x))$

CHAPTER 12

The Completeness Theorem

12.1 Introduction

The completeness theorem is one of the most fundamental results about logic. It comes in two formulations, the equivalence of which we'll prove. In its first formulation it says something fundamental about the relationship between semantic consequence and our derivation system: if a sentence A follows from some sentences Γ, then there is also a derivation that establishes $\Gamma \vdash A$. Thus, the derivation system is as strong as it can possibly be without proving things that don't actually follow.

In its second formulation, it can be stated as a model existence result: every consistent set of sentences is satisfiable. Consistency is a proof-theoretic notion: it says that our derivation system is unable to produce certain derivations. But who's to say that just because there are no derivations of a certain sort from Γ, it's guaranteed that there is a structure M? Before the completeness theorem was first proved—in fact before we had the derivation systems we now do—the great German mathematician David Hilbert held the view that consistency of mathematical theories guarantees the existence of the objects they are about. He put it as follows in a letter to Gottlob Frege:

> If the arbitrarily given axioms do not contradict one another with all their consequences, then they are true and the things defined by the axioms exist. This is for me the criterion of truth and existence.

Frege vehemently disagreed. The second formulation of the completeness theorem shows that Hilbert was right in at least the sense that if the axioms are consistent, then *some* structure exists that makes them all true.

These aren't the only reasons the completeness theorem—or rather, its proof—is important. It has a number of important consequences, some of which we'll discuss separately. For instance, since any derivation that shows $\Gamma \vdash A$ is finite and so can only use finitely many of the sentences in Γ, it follows by the completeness theorem that if A is a consequence of Γ, it is already a consequence of a finite subset of Γ. This is called *compactness*. Equivalently, if every finite subset of Γ is consistent, then Γ itself must be consistent.

Although the compactness theorem follows from the completeness theorem via the detour through derivations, it is also possible to use the *the proof of* the completeness theorem to establish it directly. For what the proof does is take a set of sentences with a certain property—consistency—and constructs a structure out of this set that has certain properties (in this case, that it satisfies the set). Almost the very same construction can be used to directly establish compactness, by starting from "finitely satisfiable" sets of sentences instead of consistent ones. The construction also yields other consequences, e.g., that any satisfiable set of sentences has a finite or countably infinite model. (This result is called the Löwenheim-Skolem theorem.) In general, the construction of structures from sets of sentences is used often in logic, and sometimes even in philosophy.

12.2 Outline of the Proof

The proof of the completeness theorem is a bit complex, and upon first reading it, it is easy to get lost. So let us outline the proof. The first step is a shift of perspective, that allows us to see a route to a proof. When completeness is thought of as "whenever $\Gamma \vDash A$ then $\Gamma \vdash A$," it may be hard to even come up with an idea: for to show that $\Gamma \vdash A$ we have to find a derivation, and it does not look like the hypothesis that $\Gamma \vDash A$ helps us for this in any way. For some proof systems it is possible to directly construct a derivation, but we will take a slightly different approach. The shift in perspective required is this: completeness can also be formulated as: "if Γ is consistent, it is satisfiable." Perhaps we can use the information in Γ together with the hypothesis that it is consistent to construct a structure that satisfies every sentence in Γ. After all, we know what kind of structure we are looking for: one that is as Γ describes it!

If Γ contains only atomic sentences, it is easy to construct a model for it. Suppose the atomic sentences are all of the form $P(a_1, \ldots, a_n)$ where the a_i are constant symbols. All we have to do is come up with a domain $|M|$ and an assignment for P so that $M \vDash P(a_1, \ldots, a_n)$. But that's not very hard: put $|M| = \mathbb{N}$, $c_i^M = i$, and for every $P(a_1, \ldots, a_n) \in \Gamma$, put the tuple $\langle k_1, \ldots, k_n \rangle$ into P^M, where k_i is the index of the constant symbol a_i (i.e., $a_i \equiv c_{k_i}$).

Now suppose Γ contains some formula $\neg B$, with B atomic. We might worry that the construction of M interferes with the possibility of making $\neg B$ true. But here's where the consistency of Γ comes in: if $\neg B \in \Gamma$, then $B \notin \Gamma$, or else Γ would be inconsistent. And if $B \notin \Gamma$, then according to our construction of M, $M \nvDash B$, so $M \vDash \neg B$. So far so good.

What if Γ contains complex, non-atomic formulas? Say it contains $A \wedge B$. To make that true, we should proceed as if both A and B were in Γ. And if $A \vee B \in \Gamma$, then we will have to make at least one of them true, i.e., proceed as if one of them was in Γ.

This suggests the following idea: we add additional formulas to Γ so as to (a) keep the resulting set consistent and (b) make sure that for every possible atomic sentence A, either A is in the resulting set, or $\neg A$ is, and (c) such that, whenever $A \wedge B$ is in the set, so are both A and B, if $A \vee B$ is in the set, at least one of A or B is also, etc. We keep doing this (potentially forever). Call the set of all formulas so added Γ^*. Then our construction above would provide us with a structure M for which we could prove, by induction, that it satisfies all sentences in Γ^*, and hence also all sentence in Γ since $\Gamma \subseteq \Gamma^*$. It turns out that guaranteeing (a) and (b) is enough. A set of sentences for which (b) holds is called *complete*. So our task will be to extend the consistent set Γ to a consistent and complete set Γ^*.

There is one wrinkle in this plan: if $\exists x\, A(x) \in \Gamma$ we would hope to be able to pick some constant symbol c and add $A(c)$ in this process. But how do we know we can always do that? Perhaps we only have a few constant symbols in our language, and for each one of them we have $\neg A(c) \in \Gamma$. We can't also add $A(c)$, since this would make the set inconsistent, and we wouldn't know whether M has to make $A(c)$ or $\neg A(c)$ true. Moreover, it might happen that Γ contains only sentences in a language that has no constant symbols at all (e.g., the language of set theory).

The solution to this problem is to simply add infinitely many constants at the beginning, plus sentences that connect them with the quantifiers in the right way. (Of course, we have to verify that this cannot introduce an inconsistency.)

Our original construction works well if we only have constant symbols in the atomic sentences. But the language might also contain function symbols. In that case, it might be tricky to find the right functions on $\mathbb{N}$ to assign to these function symbols to make everything work. So here's another trick: instead of using i to interpret c_i, just take the set of constant symbols itself as the domain. Then M can assign every constant symbol to itself: $c_i^M = c_i$. But why not go all the way: let $|M|$ be all *terms* of the language! If we do this, there is an obvious assignment of functions (that take terms as arguments and have terms as values) to function symbols: we assign to the function symbol f_i^n the

12.2. OUTLINE OF THE PROOF

function which, given n terms t_1, ..., t_n as input, produces the term $f_i^n(t_1,\ldots,t_n)$ as value.

The last piece of the puzzle is what to do with $=$. The predicate symbol $=$ has a fixed interpretation: $M \models t = t'$ iff $\text{Val}^M(t) = \text{Val}^M(t')$. Now if we set things up so that the value of a term t is t itself, then this structure will make *no* sentence of the form $t = t'$ true unless t and t' are one and the same term. And of course this is a problem, since basically every interesting theory in a language with function symbols will have as theorems sentences $t = t'$ where t and t' are not the same term (e.g., in theories of arithmetic: $(0 + 0) = 0$). To solve this problem, we change the domain of M: instead of using terms as the objects in $|M|$, we use sets of terms, and each set is so that it contains all those terms which the sentences in Γ require to be equal. So, e.g., if Γ is a theory of arithmetic, one of these sets will contain: 0, $(0 + 0)$, (0×0), etc. This will be the set we assign to 0, and it will turn out that this set is also the value of all the terms in it, e.g., also of $(0 + 0)$. Therefore, the sentence $(0 + 0) = 0$ will be true in this revised structure.

So here's what we'll do. First we investigate the properties of complete consistent sets, in particular we prove that a complete consistent set contains $A \wedge B$ iff it contains both A and B, $A \vee B$ iff it contains at least one of them, etc. (Proposition 12.2). Then we define and investigate "saturated" sets of sentences. A saturated set is one which contains conditionals that link each quantified sentence to instances of it (Definition 12.5). We show that any consistent set Γ can always be extended to a saturated set Γ' (Lemma 12.6). If a set is consistent, saturated, and complete it also has the property that it contains $\exists x\, A(x)$ iff it contains $A(t)$ for some closed term t and $\forall x\, A(x)$ iff it contains $A(t)$ for all closed terms t (Proposition 12.7). We'll then take the saturated consistent set Γ' and show that it can be extended to a saturated, consistent, and complete set Γ^* (Lemma 12.8). This set Γ^* is what we'll use to define our term model $M(\Gamma^*)$. The term model has the set of closed terms as its domain, and the interpretation of its predicate symbols is given by the atomic sentences in Γ^* (Definition 12.9). We'll use the properties of saturated, complete consistent sets to show that indeed $M(\Gamma^*) \vDash A$ iff $A \in \Gamma^*$ (Lemma 12.11), and thus in particular, $M(\Gamma^*) \vDash \Gamma$. Finally, we'll consider how to define a term model if Γ contains = as well (Definition 12.15) and show that it satisfies Γ^* (Lemma 12.17).

12.3 Complete Consistent Sets of Sentences

Definition 12.1 (Complete set). A set Γ of sentences is *complete* iff for any sentence A, either $A \in \Gamma$ or $\neg A \in \Gamma$.

Complete sets of sentences leave no questions unanswered. For any sentence A, Γ "says" if A is true or false. The importance of complete sets extends beyond the proof of the completeness theorem. A theory which is complete and axiomatizable, for instance, is always decidable.

Complete consistent sets are important in the completeness proof since we can guarantee that every consistent set of sentences Γ is contained in a complete consistent set Γ^*. A complete consistent set contains, for each sentence A, either A or its negation $\neg A$, but not both. This is true in particular for atomic sentences, so from a complete consistent set in a language suitably expanded by constant symbols, we can construct a structure where the interpretation of predicate symbols is defined according to which atomic sentences are in Γ^*. This structure can then be shown to make all sentences in Γ^* (and hence also all those in Γ) true. The proof of this latter fact requires that $\neg A \in \Gamma^*$ iff $A \notin \Gamma^*$, $(A \lor B) \in \Gamma^*$ iff $A \in \Gamma^*$ or $B \in \Gamma^*$, etc.

In what follows, we will often tacitly use the properties of reflexivity, monotonicity, and transitivity of $\vdash$ (see sections 10.8 and 11.7).

Proposition 12.2. *Suppose Γ is complete and consistent. Then:*

1. If $\Gamma \vdash A$, then $A \in \Gamma$.

> 2. $A \wedge B \in \Gamma$ *iff both $A \in \Gamma$ and $B \in \Gamma$.*
>
> 3. $A \vee B \in \Gamma$ *iff either $A \in \Gamma$ or $B \in \Gamma$.*
>
> 4. $A \to B \in \Gamma$ *iff either $A \notin \Gamma$ or $B \in \Gamma$.*

Proof. Let us suppose for all of the following that Γ is complete and consistent.

1. If $\Gamma \vdash A$, then $A \in \Gamma$.

    Suppose that $\Gamma \vdash A$. Suppose to the contrary that $A \notin \Gamma$. Since Γ is complete, $\neg A \in \Gamma$. By Propositions 10.20 and 11.20, Γ is inconsistent. This contradicts the assumption that Γ is consistent. Hence, it cannot be the case that $A \notin \Gamma$, so $A \in \Gamma$.

2. Exercise.

3. First we show that if $A \vee B \in \Gamma$, then either $A \in \Gamma$ or $B \in \Gamma$. Suppose $A \vee B \in \Gamma$ but $A \notin \Gamma$ and $B \notin \Gamma$. Since Γ is complete, $\neg A \in \Gamma$ and $\neg B \in \Gamma$. By Propositions 10.23 and 11.23, item (1), Γ is inconsistent, a contradiction. Hence, either $A \in \Gamma$ or $B \in \Gamma$.

    For the reverse direction, suppose that $A \in \Gamma$ or $B \in \Gamma$. By Propositions 10.23 and 11.23, item (2), $\Gamma \vdash A \vee B$. By (1), $A \vee B \in \Gamma$, as required.

4. Exercise. □

12.4 Henkin Expansion

Part of the challenge in proving the completeness theorem is that the model we construct from a complete consistent set Γ must make all the quantified formulas in Γ true. In order to guarantee this, we use a trick due to Leon Henkin. In essence, the trick consists in expanding the language by infinitely many constant symbols and adding, for each formula with one free variable $A(x)$ a formula of the form $\exists x\, A(x) \to A(c)$, where c is one of the new constant symbols. When we construct the structure satisfying Γ, this will guarantee that each true existential sentence has a witness among the new constants.

Proposition 12.3. *If Γ is consistent in $\mathscr{L}$ and $\mathscr{L}'$ is obtained from $\mathscr{L}$ by adding a countably infinite set of new constant symbols d_0, d_1, ..., then Γ is consistent in $\mathscr{L}'$.*

Definition 12.4 (Saturated set). A set Γ of formulas of a language $\mathscr{L}$ is *saturated* iff for each formula $A(x) \in \mathrm{Frm}(\mathscr{L})$ with one free variable x there is a constant symbol $c \in \mathscr{L}$ such that $\exists x\, A(x) \to A(c) \in \Gamma$.

The following definition will be used in the proof of the next theorem.

Definition 12.5. Let $\mathscr{L}'$ be as in Proposition 12.3. Fix an enumeration $A_0(x_0), A_1(x_1), \ldots$ of all formulas $A_i(x_i)$ of $\mathscr{L}'$ in which one variable (x_i) occurs free. We define the sentences D_n by induction on n.

Let c_0 be the first constant symbol among the d_i we added to $\mathscr{L}$ which does not occur in $A_0(x_0)$. Assuming that $D_0, \ldots, D_{n-1}$ have already been defined, let c_n be the first among the new constant symbols d_i that occurs neither in $D_0, \ldots, D_{n-1}$ nor in $A_n(x_n)$.

Now let D_n be the formula $\exists x_n\, A_n(x_n) \to A_n(c_n)$.

Lemma 12.6. *Every consistent set Γ can be extended to a saturated consistent set Γ'.*

12.4. HENKIN EXPANSION

Proof. Given a consistent set of sentences Γ in a language $\mathscr{L}$, expand the language by adding a countably infinite set of new constant symbols to form $\mathscr{L}'$. By Proposition 12.3, Γ is still consistent in the richer language. Further, let D_i be as in Definition 12.5. Let

$$\Gamma_0 = \Gamma$$
$$\Gamma_{n+1} = \Gamma_n \cup \{D_n\}$$

i.e., $\Gamma_{n+1} = \Gamma \cup \{D_0, \ldots, D_n\}$, and let $\Gamma' = \bigcup_n \Gamma_n$. Γ' is clearly saturated.

If Γ' were inconsistent, then for some n, Γ_n would be inconsistent (Exercise: explain why). So to show that Γ' is consistent it suffices to show, by induction on n, that each set Γ_n is consistent.

The induction basis is simply the claim that $\Gamma_0 = \Gamma$ is consistent, which is the hypothesis of the theorem. For the induction step, suppose that Γ_n is consistent but $\Gamma_{n+1} = \Gamma_n \cup \{D_n\}$ is inconsistent. Recall that D_n is $\exists x_n A_n(x_n) \to A_n(c_n)$, where $A_n(x_n)$ is a formula of $\mathscr{L}'$ with only the variable x_n free. By the way we've chosen the c_n (see Definition 12.5), c_n does not occur in $A_n(x_n)$ nor in Γ_n.

If $\Gamma_n \cup \{D_n\}$ is inconsistent, then $\Gamma_n \vdash \neg D_n$, and hence both of the following hold:

$$\Gamma_n \vdash \exists x_n A_n(x_n) \qquad \Gamma_n \vdash \neg A_n(c_n)$$

Since c_n does not occur in Γ_n or in $A_n(x_n)$, Theorems 10.25 and 11.25 applies. From $\Gamma_n \vdash \neg A_n(c_n)$, we obtain $\Gamma_n \vdash \forall x_n \neg A_n(x_n)$. Thus we have that both $\Gamma_n \vdash \exists x_n A_n(x_n)$ and $\Gamma_n \vdash \forall x_n \neg A_n(x_n)$, so Γ_n itself is inconsistent. (Note that $\forall x_n \neg A_n(x_n) \vdash \neg \exists x_n A_n(x_n)$.) Contradiction: Γ_n was supposed to be consistent. Hence $\Gamma_n \cup \{D_n\}$ is consistent. $\square$

We'll now show that *complete*, consistent sets which are saturated have the property that it contains a universally quantified sentence iff it contains all its instances and it contains an existentially quantified sentence iff it contains at least one instance. We'll use this to show that the structure we'll generate from a complete, consistent, saturated set makes all its quantified sentences true.

Proposition 12.7. *Suppose Γ is complete, consistent, and saturated.*

1. *$\exists x\, A(x) \in \Gamma$ iff $A(t) \in \Gamma$ for at least one closed term t.*

2. *$\forall x\, A(x) \in \Gamma$ iff $A(t) \in \Gamma$ for all closed terms t.*

Proof. 1. First suppose that $\exists x\, A(x) \in \Gamma$. Because Γ is saturated, $(\exists x\, A(x) \to A(c)) \in \Gamma$ for some constant symbol c. By Propositions 10.24 and 11.24, item (1), and Proposition 12.2(1), $A(c) \in \Gamma$.

For the other direction, saturation is not necessary: Suppose $A(t) \in \Gamma$. Then $\Gamma \vdash \exists x\, A(x)$ by Propositions 10.26 and 11.26, item (1). By Proposition 12.2(1), $\exists x\, A(x) \in \Gamma$.

2. Exercise. □

12.5 Lindenbaum's Lemma

We now prove a lemma that shows that any consistent set of sentences is contained in some set of sentences which is not just consistent, but also complete. The proof works by adding one sentence at a time, guaranteeing at each step that the set remains consistent. We do this so that for every A, either A or $\neg A$ gets added at some stage. The union of all stages in that construction then contains either A or its negation $\neg A$ and is thus complete. It is also consistent, since we made sure at each stage not to introduce an inconsistency.

Lemma 12.8 (Lindenbaum's Lemma). *Every consistent set Γ in a language $\mathcal{L}$ can be extended to a complete and consistent set Γ^*.*

Proof. Let Γ be consistent. Let A_0, A_1, ... be an enumeration of all the sentences of $\mathcal{L}$. Define $\Gamma_0 = \Gamma$, and

$$\Gamma_{n+1} = \begin{cases} \Gamma_n \cup \{A_n\} & \text{if } \Gamma_n \cup \{A_n\} \text{ is consistent;} \\ \Gamma_n \cup \{\neg A_n\} & \text{otherwise.} \end{cases}$$

Let $\Gamma^* = \bigcup_{n \geq 0} \Gamma_n$.

Each Γ_n is consistent: Γ_0 is consistent by definition. If $\Gamma_{n+1} = \Gamma_n \cup \{A_n\}$, this is because the latter is consistent. If it isn't, $\Gamma_{n+1} = \Gamma_n \cup \{\neg A_n\}$. We have to verify that $\Gamma_n \cup \{\neg A_n\}$ is consistent. Suppose it's not. Then *both* $\Gamma_n \cup \{A_n\}$ and $\Gamma_n \cup \{\neg A_n\}$ are inconsistent. This means that Γ_n would be inconsistent by Propositions 10.20 and 11.20, contrary to the induction hypothesis.

For every n and every $i < n$, $\Gamma_i \subseteq \Gamma_n$. This follows by a simple induction on n. For $n = 0$, there are no $i < 0$, so the claim holds automatically. For the inductive step, suppose it is true for n. We have $\Gamma_{n+1} = \Gamma_n \cup \{A_n\}$ or $= \Gamma_n \cup \{\neg A_n\}$ by construction. So $\Gamma_n \subseteq \Gamma_{n+1}$. If $i < n$, then $\Gamma_i \subseteq \Gamma_n$ by inductive hypothesis, and so $\subseteq \Gamma_{n+1}$ by transitivity of $\subseteq$.

From this it follows that every finite subset of Γ^* is a subset of Γ_n for some n, since each $B \in \Gamma^*$ not already in Γ_0 is added at some stage i. If n is the last one of these, then all B in the finite subset are in Γ_n. So, every finite subset of Γ^* is consistent. By Propositions 10.17 and 11.17, Γ^* is consistent.

Every sentence of $\text{Frm}(\mathcal{L})$ appears on the list used to define Γ^*. If $A_n \notin \Gamma^*$, then that is because $\Gamma_n \cup \{A_n\}$ was inconsistent. But then $\neg A_n \in \Gamma^*$, so Γ^* is complete. □

12.6 Construction of a Model

Right now we are not concerned about =, i.e., we only want to show that a consistent set Γ of sentences not containing = is satisfiable. We first extend Γ to a consistent, complete, and saturated set Γ^*. In this case, the definition of a model $M(\Gamma^*)$ is simple: We take the set of closed terms of $\mathcal{L}'$ as the domain. We assign every constant symbol to itself, and make sure that more generally, for every closed term t, $\mathrm{Val}^{M(\Gamma^*)}(t) = t$. The predicate symbols are assigned extensions in such a way that an atomic sentence is true in $M(\Gamma^*)$ iff it is in Γ^*. This will obviously make all the atomic sentences in Γ^* true in $M(\Gamma^*)$. The rest are true provided the Γ^* we start with is consistent, complete, and saturated.

Definition 12.9 (Term model). Let Γ^* be a complete and consistent, saturated set of sentences in a language $\mathscr{L}$. The *term model* $M(\Gamma^*)$ of Γ^* is the structure defined as follows:

1. The domain $|M(\Gamma^*)|$ is the set of all closed terms of $\mathscr{L}$.

2. The interpretation of a constant symbol c is c itself: $c^{M(\Gamma^*)} = c$.

3. The function symbol f is assigned the function which, given as arguments the closed terms t_1, ..., t_n, has as value the closed term $f(t_1, \ldots, t_n)$:

$$f^{M(\Gamma^*)}(t_1, \ldots, t_n) = f(t_1, \ldots, t_n)$$

4. If R is an n-place predicate symbol, then

$$\langle t_1, \ldots, t_n \rangle \in R^{M(\Gamma^*)} \text{ iff } R(t_1, \ldots, t_n) \in \Gamma^*.$$

A structure M may make an existentially quantified sentence $\exists x\, A(x)$ true without there being an instance $A(t)$ that it makes true. A structure M may make all instances $A(t)$ of a universally quantified sentence $\forall x\, A(x)$ true, without making $\forall x\, A(x)$ true. This is because in general not every element of $|M|$ is the value of a closed term (M may not be covered). This is the reason the satisfaction relation is defined via variable assignments. However, for our term model $M(\Gamma^*)$ this wouldn't be necessary—because it is covered. This is the content of the next result.

12.6. CONSTRUCTION OF A MODEL

Proposition 12.10. *Let $M(\Gamma^*)$ be the term model of Definition 12.9.*

1. $M(\Gamma^*) \vDash \exists x\, A(x)$ *iff* $M(\Gamma^*) \vDash A(t)$ *for at least one term t.*

2. $M(\Gamma^*) \vDash \forall x\, A(x)$ *iff* $M(\Gamma^*) \vDash A(t)$ *for all terms t.*

Proof. 1. By Proposition 7.18, $M(\Gamma^*) \vDash \exists x\, A(x)$ iff for at least one variable assignment s, $M(\Gamma^*), s \vDash A(x)$. As $|M(\Gamma^*)|$ consists of the closed terms of $\mathcal{L}$, this is the case iff there is at least one closed term t such that $s(x) = t$ and $M(\Gamma^*), s \vDash A(x)$. By Proposition 7.22, $M(\Gamma^*), s \vDash A(x)$ iff $M(\Gamma^*), s \vDash A(t)$, where $s(x) = t$. By Proposition 7.17, $M(\Gamma^*), s \vDash A(t)$ iff $M(\Gamma^*) \vDash A(t)$, since $A(t)$ is a sentence.

2. Exercise. □

Lemma 12.11 (Truth Lemma). *Suppose A does not contain =. Then $M(\Gamma^*) \vDash A$ iff $A \in \Gamma^*$.*

Proof. We prove both directions simultaneously, and by induction on A.

1. $A \equiv \bot$: $M(\Gamma^*) \nvDash \bot$ by definition of satisfaction. On the other hand, $\bot \notin \Gamma^*$ since Γ^* is consistent.

2. $A \equiv R(t_1, \ldots, t_n)$: $M(\Gamma^*) \vDash R(t_1, \ldots, t_n)$ iff $\langle t_1, \ldots, t_n \rangle \in R^{M(\Gamma^*)}$ (by the definition of satisfaction) iff $R(t_1, \ldots, t_n) \in \Gamma^*$ (by the construction of $M(\Gamma^*)$).

3. $A \equiv \neg B$: $M(\Gamma^*) \vDash A$ iff $M(\Gamma^*) \nvDash B$ (by definition of satisfaction). By induction hypothesis, $M(\Gamma^*) \nvDash B$ iff $B \notin \Gamma^*$. Since Γ^* is consistent and complete, $B \notin \Gamma^*$ iff $\neg B \in \Gamma^*$.

4. $A \equiv B \wedge C$: exercise.

5. $A \equiv B \vee C$: $M(\Gamma^*) \vDash A$ iff $M(\Gamma^*) \vDash B$ or $M(\Gamma^*) \vDash C$ (by definition of satisfaction) iff $B \in \Gamma^*$ or $C \in \Gamma^*$ (by induction hypothesis). This is the case iff $(B \vee C) \in \Gamma^*$ (by Proposition 12.2(3)).

6. $A \equiv B \to C$: exercise.

7. $A \equiv \forall x\, B(x)$: exercise.

8. $A \equiv \exists x\, B(x)$: $M(\Gamma^*) \vDash A$ iff $M(\Gamma^*) \vDash B(t)$ for at least one term t (Proposition 12.10). By induction hypothesis, this is the case iff $B(t) \in \Gamma^*$ for at least one term t. By Proposition 12.7, this in turn is the case iff $\exists x\, B(x) \in \Gamma^*$.
□

12.7 Identity

The construction of the term model given in the preceding section is enough to establish completeness for first-order logic for sets Γ that do not contain =. The term model satisfies every $A \in \Gamma^*$ which does not contain = (and hence all $A \in \Gamma$). It does not work, however, if = is present. The reason is that Γ^* then may contain a sentence $t = t'$, but in the term model the value of any term is that term itself. Hence, if t and t' are different terms, their values in the term model—i.e., t and t', respectively—are different, and so $t = t'$ is false. We can fix this, however, using a construction known as "factoring."

Definition 12.12. Let Γ^* be a consistent and complete set of sentences in $\mathcal{L}$. We define the relation $\approx$ on the set of closed terms of $\mathcal{L}$ by

$$t \approx t' \quad \text{iff} \quad t = t' \in \Gamma^*$$

Proposition 12.13. *The relation $\approx$ has the following properties:*

1. *$\approx$ is reflexive.*

2. *$\approx$ is symmetric.*

3. *$\approx$ is transitive.*

12.7. IDENTITY

4. If $t \approx t'$, f is a function symbol, and $t_1, \ldots, t_{i-1}, t_{i+1}, \ldots, t_n$ are terms, then

$$f(t_1,\ldots,t_{i-1},t,t_{i+1},\ldots,t_n) \approx f(t_1,\ldots,t_{i-1},t',t_{i+1},\ldots,t_n).$$

5. If $t \approx t'$, R is a predicate symbol, and $t_1, \ldots, t_{i-1}, t_{i+1}, \ldots, t_n$ are terms, then

$$R(t_1,\ldots,t_{i-1},t,t_{i+1},\ldots,t_n) \in \Gamma^* \text{ iff}$$
$$R(t_1,\ldots,t_{i-1},t',t_{i+1},\ldots,t_n) \in \Gamma^*.$$

Proof. Since Γ^* is consistent and complete, $t = t' \in \Gamma^*$ iff $\Gamma^* \vdash t = t'$. Thus it is enough to show the following:

1. $\Gamma^* \vdash t = t$ for all terms t.

2. If $\Gamma^* \vdash t = t'$ then $\Gamma^* \vdash t' = t$.

3. If $\Gamma^* \vdash t = t'$ and $\Gamma^* \vdash t' = t''$, then $\Gamma^* \vdash t = t''$.

4. If $\Gamma^* \vdash t = t'$, then

$$\Gamma^* \vdash f(t_1,\ldots,t_{i-1},t,t_{i+1,},\ldots,t_n) = f(t_1,\ldots,t_{i-1},t',t_{i+1},\ldots,t_n)$$

for every n-place function symbol f and terms $t_1, \ldots, t_{i-1}, t_{i+1}, \ldots, t_n$.

5. If $\Gamma^* \vdash t = t'$ and $\Gamma^* \vdash R(t_1,\ldots,t_{i-1},t,t_{i+1},\ldots,t_n)$, then $\Gamma^* \vdash R(t_1,\ldots,t_{i-1},t',t_{i+1},\ldots,t_n)$ for every n-place predicate symbol R and terms $t_1, \ldots, t_{i-1}, t_{i+1}, \ldots, t_n$. □

Definition 12.14. Suppose Γ^* is a consistent and complete set in a language $\mathscr{L}$, t is a term, and $\approx$ as in the previous definition. Then:

$$[t]_\approx = \{t' : t' \in \text{Trm}(\mathscr{L}), t \approx t'\}$$

and $\mathrm{Trm}(\mathcal{L})/_\approx = \{[t]_\approx : t \in \mathrm{Trm}(\mathcal{L})\}$.

Definition 12.15. Let $M = M(\Gamma^*)$ be the term model for Γ^*. Then $M/_\approx$ is the following structure:

1. $|M/_\approx| = \mathrm{Trm}(\mathcal{L})/_\approx$.

2. $c^{M/_\approx} = [c]_\approx$

3. $f^{M/_\approx}([t_1]_\approx, \ldots, [t_n]_\approx) = [f(t_1, \ldots, t_n)]_\approx$

4. $\langle [t_1]_\approx, \ldots, [t_n]_\approx \rangle \in R^{M/_\approx}$ iff $M \vDash R(t_1, \ldots, t_n)$.

Note that we have defined $f^{M/_\approx}$ and $R^{M/_\approx}$ for elements of $\mathrm{Trm}(\mathcal{L})/_\approx$ by referring to them as $[t]_\approx$, i.e., via *representatives* $t \in [t]_\approx$. We have to make sure that these definitions do not depend on the choice of these representatives, i.e., that for some other choices t' which determine the same equivalence classes ($[t]_\approx = [t']_\approx$), the definitions yield the same result. For instance, if R is a one-place predicate symbol, the last clause of the definition says that $[t]_\approx \in R^{M/_\approx}$ iff $M \vDash R(t)$. If for some other term t' with $t \approx t'$, $M \nvDash R(t)$, then the definition would require $[t']_\approx \notin R^{M/_\approx}$. If $t \approx t'$, then $[t]_\approx = [t']_\approx$, but we can't have both $[t]_\approx \in R^{M/_\approx}$ and $[t]_\approx \notin R^{M/_\approx}$. However, Proposition 12.13 guarantees that this cannot happen.

Proposition 12.16. $M/_\approx$ *is well defined, i.e., if* $t_1, \ldots, t_n, t'_1, \ldots, t'_n$ *are terms, and* $t_i \approx t'_i$ *then*

1. $[f(t_1, \ldots, t_n)]_\approx = [f(t'_1, \ldots, t'_n)]_\approx$, *i.e.,*

$$f(t_1, \ldots, t_n) \approx f(t'_1, \ldots, t'_n)$$

and

2. $M \vDash R(t_1, \ldots, t_n)$ iff $M \vDash R(t'_1, \ldots, t'_n)$, i.e.,

$$R(t_1, \ldots, t_n) \in \Gamma^* \text{ iff } R(t'_1, \ldots, t'_n) \in \Gamma^*.$$

Proof. Follows from Proposition 12.13 by induction on n. □

Lemma 12.17. $M/_\approx \vDash A$ *iff* $A \in \Gamma^*$ *for all sentences* A.

Proof. By induction on A, just as in the proof of Lemma 12.11. The only case that needs additional attention is when $A \equiv t = t'$.

$M/_\approx \vDash t = t'$ iff $[t]_\approx = [t']_\approx$ (by definition of $M/_\approx$)
  iff $t \approx t'$ (by definition of $[t]_\approx$)
  iff $t = t' \in \Gamma^*$ (by definition of $\approx$). □

Note that while $M(\Gamma^*)$ is always countable and infinite, $M/_\approx$ may be finite, since it may turn out that there are only finitely many classes $[t]_\approx$. This is to be expected, since Γ may contain sentences which require any structure in which they are true to be finite. For instance, $\forall x \, \forall y \, x = y$ is a consistent sentence, but is satisfied only in structures with a domain that contains exactly one element.

12.8 The Completeness Theorem

Let's combine our results: we arrive at the completeness theorem.

Theorem 12.18 (Completeness Theorem). *Let Γ be a set of sentences. If Γ is consistent, it is satisfiable.*

Proof. Suppose Γ is consistent. By Lemma 12.6, there is a saturated consistent set $\Gamma' \supseteq \Gamma$. By Lemma 12.8, there is a $\Gamma^* \supseteq \Gamma'$ which is consistent and complete. Since $\Gamma' \subseteq \Gamma^*$, for each formula $A(x)$, Γ^* contains a sentence of the form $\exists x \, A(x) \rightarrow A(c)$ and so Γ^* is saturated. If Γ does not contain =, then by Lemma 12.11, $M(\Gamma^*) \vDash A$ iff $A \in \Gamma^*$. From this it follows in particular that for all $A \in \Gamma$, $M(\Gamma^*) \vDash A$, so Γ is satisfiable. If Γ does contain =, then by Lemma 12.17, for all sentences A, $M/_\approx \vDash A$ iff $A \in \Gamma^*$. In particular, $M/_\approx \vDash A$ for all $A \in \Gamma$, so Γ is satisfiable. □

Corollary 12.19 (Completeness Theorem, Second Version). *For all Γ and sentences A: if $\Gamma \vDash A$ then $\Gamma \vdash A$.*

Proof. Note that the Γ's in Corollary 12.19 and Theorem 12.18 are universally quantified. To make sure we do not confuse ourselves, let us restate Theorem 12.18 using a different variable: for any set of sentences Δ, if Δ is consistent, it is satisfiable. By contraposition, if Δ is not satisfiable, then Δ is inconsistent. We will use this to prove the corollary.

Suppose that $\Gamma \vDash A$. Then $\Gamma \cup \{\neg A\}$ is unsatisfiable by Proposition 7.27. Taking $\Gamma \cup \{\neg A\}$ as our Δ, the previous version of Theorem 12.18 gives us that $\Gamma \cup \{\neg A\}$ is inconsistent. By Propositions 10.19 and 11.19, $\Gamma \vdash A$. □

12.9 The Compactness Theorem

One important consequence of the completeness theorem is the compactness theorem. The compactness theorem states that if each *finite* subset of a set of sentences is satisfiable, the entire set is satisfiable—even if the set itself is infinite. This is far from obvious. There is nothing that seems to rule out, at first glance at least, the possibility of there being infinite sets of sentences which are contradictory, but the contradiction only arises, so to speak, from the infinite number. The compactness theorem says that such a scenario can be ruled out: there are no unsatisfiable infinite sets of sentences each finite subset of which is satisfiable. Like the completeness theorem, it has a version related to entailment: if an infinite set of sentences entails something, already a finite subset does.

12.9. THE COMPACTNESS THEOREM

Definition 12.20. A set Γ of formulas is *finitely satisfiable* iff every finite $\Gamma_0 \subseteq \Gamma$ is satisfiable.

Theorem 12.21 (Compactness Theorem). *The following hold for any sentences Γ and A:*

1. *$\Gamma \vDash A$ iff there is a finite $\Gamma_0 \subseteq \Gamma$ such that $\Gamma_0 \vDash A$.*

2. *Γ is satisfiable iff it is finitely satisfiable.*

Proof. We prove (2). If Γ is satisfiable, then there is a structure M such that $M \vDash A$ for all $A \in \Gamma$. Of course, this M also satisfies every finite subset of Γ, so Γ is finitely satisfiable.

Now suppose that Γ is finitely satisfiable. Then every finite subset $\Gamma_0 \subseteq \Gamma$ is satisfiable. By soundness (Corollaries 11.29 and 10.31), every finite subset is consistent. Then Γ itself must be consistent by Propositions 10.17 and 11.17. By completeness (Theorem 12.18), since Γ is consistent, it is satisfiable. □

Example 12.22. In every model M of a theory Γ, each term t of course picks out an element of $|M|$. Can we guarantee that it is also true that every element of $|M|$ is picked out by some term or other? In other words, are there theories Γ all models of which are covered? The compactness theorem shows that this is not the case if Γ has infinite models. Here's how to see this: Let M be an infinite model of Γ, and let c be a constant symbol not in the language of Γ. Let Δ be the set of all sentences $c \neq t$ for t a term in the language $\mathcal{L}$ of Γ, i.e.,

$$\Delta = \{c \neq t : t \in \mathrm{Trm}(\mathcal{L})\}.$$

A finite subset of $\Gamma \cup \Delta$ can be written as $\Gamma' \cup \Delta'$, with $\Gamma' \subseteq \Gamma$ and $\Delta' \subseteq \Delta$. Since Δ' is finite, it can contain only finitely many terms. Let $a \in |M|$ be an element of $|M|$ not picked out by any of them, and let M' be the structure that is just like M, but also $c^{M'} = a$. Since $a \neq \mathrm{Val}^M(t)$ for all t occuring in Δ', $M' \vDash \Delta'$.

Since $M \vDash \Gamma$, $\Gamma' \subseteq \Gamma$, and c does not occur in Γ, also $M' \vDash \Gamma'$. Together, $M' \vDash \Gamma' \cup \Delta'$ for every finite subset $\Gamma' \cup \Delta'$ of $\Gamma \cup \Delta$. So every finite subset of $\Gamma \cup \Delta$ is satisfiable. By compactness, $\Gamma \cup \Delta$ itself is satisfiable. So there are models $M \vDash \Gamma \cup \Delta$. Every such M is a model of Γ, but is not covered, since $\mathrm{Val}^M(c) \neq \mathrm{Val}^M(t)$ for all terms t of $\mathcal{L}$.

Example 12.23. Consider a language $\mathcal{L}$ containing the predicate symbol $<$, constant symbols 0, 1, and function symbols $+$, $\times$, $-$, $\div$. Let Γ be the set of all sentences in this language true in Q with domain $\mathbb{Q}$ and the obvious interpretations. Γ is the set of all sentences of $\mathcal{L}$ true about the rational numbers. Of course, in $\mathbb{Q}$ (and even in $\mathbb{R}$), there are no numbers which are greater than 0 but less than $1/k$ for all $k \in \mathbb{Z}^+$. Such a number, if it existed, would be an *infinitesimal*: non-zero, but infinitely small. The compactness theorem shows that there are models of Γ in which infinitesimals exist: Let Δ be $\{0 < c\} \cup \{c < (1 \div \overline{k}) : k \in \mathbb{Z}^+\}$ (where $\overline{k} = (1 + (1 + \cdots + (1 + 1)\ldots))$ with k 1's). For any finite subset Δ_0 of Δ there is a K such that all the sentences $c < (1 \div \overline{k})$ in Δ_0 have $k < K$. If we expand Q to Q' with $c^{Q'} = 1/K$ we have that $Q' \vDash \Gamma \cup \Delta_0$, and so $\Gamma \cup \Delta$ is finitely satisfiable (Exercise: prove this in detail). By compactness, $\Gamma \cup \Delta$ is satisfiable. Any model S of $\Gamma \cup \Delta$ contains an infinitesimal, namely c^S.

Example 12.24. We know that first-order logic with identity predicate can express that the size of the domain must have some minimal size: The sentence $A_{\geq n}$ (which says "there are at least n distinct objects") is true only in structures where $|M|$ has at least n objects. So if we take

$$\Delta = \{A_{\geq n} : n \geq 1\}$$

then any model of Δ must be infinite. Thus, we can guarantee that a theory only has infinite models by adding Δ to it: the models of $\Gamma \cup \Delta$ are all and only the infinite models of Γ.

So first-order logic can express infinitude. The compactness theorem shows that it cannot express finitude, however. For suppose some set of sentences Λ were satisfied in all and only finite structures. Then $\Delta \cup \Lambda$ is finitely satisfiable. Why? Suppose $\Delta' \cup \Lambda' \subseteq \Delta \cup \Lambda$ is finite with $\Delta' \subseteq \Delta$ and $\Lambda' \subseteq \Lambda$. Let n be the largest number such that $A_{\geq n} \in \Delta'$. Λ, being satisfied in all finite structures, has a model M with finitely many but $\geq n$ elements. But then $M \vDash \Delta' \cup \Lambda'$. By compactness, $\Delta \cup \Lambda$ has an infinite model, contradicting the assumption that Λ is satisfied only in finite structures.

12.10 A Direct Proof of the Compactness Theorem

We can prove the Compactness Theorem directly, without appealing to the Completeness Theorem, using the same ideas as in the proof of the completeness theorem. In the proof of the Completeness Theorem we started with a consistent set Γ of sentences, expanded it to a consistent, saturated, and complete set Γ^* of sentences, and then showed that in the term model $M(\Gamma^*)$ constructed from Γ^*, all sentences of Γ are true, so Γ is satisfiable.

We can use the same method to show that a finitely satisfiable set of sentences is satisfiable. We just have to prove the corresponding versions of the results leading to the truth lemma where we replace "consistent" with "finitely satisfiable."

Proposition 12.25. *Suppose Γ is complete and finitely satisfiable. Then:*

1. $(A \wedge B) \in \Gamma$ *iff both* $A \in \Gamma$ *and* $B \in \Gamma$.

2. $(A \vee B) \in \Gamma$ *iff either* $A \in \Gamma$ *or* $B \in \Gamma$.

3. $(A \to B) \in \Gamma$ *iff either* $A \notin \Gamma$ *or* $B \in \Gamma$.

Lemma 12.26. *Every finitely satisfiable set Γ can be extended to a saturated finitely satisfiable set Γ'.*

Proposition 12.27. *Suppose Γ is complete, finitely satisfiable, and saturated.*

1. *$\exists x\, A(x) \in \Gamma$ iff $A(t) \in \Gamma$ for at least one closed term t.*

2. *$\forall x\, A(x) \in \Gamma$ iff $A(t) \in \Gamma$ for all closed terms t.*

Lemma 12.28. *Every finitely satisfiable set Γ can be extended to a complete and finitely satisfiable set Γ^*.*

Theorem 12.29 (Compactness). *Γ is satisfiable if and only if it is finitely satisfiable.*

Proof. If Γ is satisfiable, then there is a structure M such that $M \vDash A$ for all $A \in \Gamma$. Of course, this M also satisfies every finite subset of Γ, so Γ is finitely satisfiable.

Now suppose that Γ is finitely satisfiable. By Lemma 12.26, there is a finitely satisfiable, saturated set $\Gamma' \supseteq \Gamma$. By Lemma 12.28, Γ' can be extended to a complete and finitely satisfiable set Γ^*, and Γ^* is still saturated. Construct the term model $M(\Gamma^*)$ as in Definition 12.9. Note that Proposition 12.10 did not rely on the fact that Γ^* is consistent (or complete or saturated, for that matter), but just on the fact that $M(\Gamma^*)$ is covered. The proof of the Truth Lemma (Lemma 12.11) goes through if we replace references to Proposition 12.2 and Proposition 12.7 by references to Proposition 12.25 and Proposition 12.27 □

12.11 The Löwenheim-Skolem Theorem

The Löwenheim-Skolem Theorem says that if a theory has an infinite model, then it also has a model that is at most countably

12.11. THE LÖWENHEIM-SKOLEM THEOREM

infinite. An immediate consequence of this fact is that first-order logic cannot express that the size of a structure is uncountable: any sentence or set of sentences satisfied in all uncountable structures is also satisfied in some countable structure.

Theorem 12.30. *If Γ is consistent then it has a countable model, i.e., it is satisfiable in a structure whose domain is either finite or countably infinite.*

Proof. If Γ is consistent, the structure M delivered by the proof of the completeness theorem has a domain $|M|$ that is no larger than the set of the terms of the language $\mathscr{L}$. So M is at most countably infinite. □

Theorem 12.31. *If Γ is a consistent set of sentences in the language of first-order logic without identity, then it has a countably infinite model, i.e., it is satisfiable in a structure whose domain is infinite and countable.*

Proof. If Γ is consistent and contains no sentences in which identity appears, then the structure M delivered by the proof of the completness theorem has a domain $|M|$ identical to the set of terms of the language $\mathscr{L}'$. So M is countably infinite, since $\mathrm{Trm}(\mathscr{L}')$ is. □

Example 12.32 (Skolem's Paradox). Zermelo-Fraenkel set theory **ZFC** is a very powerful framework in which practically all mathematical statements can be expressed, including facts about the sizes of sets. So for instance, **ZFC** can prove that the set $\mathbb{R}$ of real numbers is uncountable, it can prove Cantor's Theorem that the power set of any set is larger than the set itself, etc. If **ZFC** is consistent, its models are all infinite, and moreover, they all contain elements about which the theory says that they are uncountable, such as the element that makes true the theorem of **ZFC** that the power set of the natural numbers exists. By the Löwenheim-Skolem Theorem, **ZFC** also has countable models—models that contain "uncountable" sets but which themselves are countable.

Summary

The **completeness theorem** is the converse of the **soundness theorem**. In one form it states that if $\Gamma \vDash A$ then $\Gamma \vdash A$, in another that if Γ is consistent then it is satisfiable. We proved the second form (and derived the first from the second). The proof is involved and requires a number of steps. We start with a consistent set Γ. First we add infinitely many new constant symbols c_i as well as formulas of the form $\exists x\, A(x) \to A(c)$ where each formula $A(x)$ with a free variable in the expanded language is paired with one of the new constants. This results in a **saturated** consistent set of sentences containing Γ. It is still consistent. Now we take that set and extend it to a **complete consistent set**. A complete consistent set has the nice property that for any sentence A, either A or $\neg A$ is in the set (but never both). Since we started from a saturated set, we now have a saturated, complete, consistent set of sentences Γ^* that includes Γ. From this set it is now possible to define a structure M such that $M(\Gamma^*) \vDash A$ iff $A \in \Gamma^*$. In particular, $M(\Gamma^*) \vDash \Gamma$, i.e., Γ is satisfiable. If $=$ is present, the construction is slightly more complex.

Two important corollaries follow from the completeness theorem. The **compactness theorem** states that $\Gamma \vDash A$ iff $\Gamma_0 \vDash A$ for some finite $\Gamma_0 \subseteq \Gamma$. An equivalent formulation is that Γ is satisfiable iff every finite $\Gamma_0 \subseteq \Gamma$ is satisfiable. The compactness theorem is useful to prove the existence of structures with certain properties. For instance, we can use it to show that there are infinite models for every theory which has arbitrarily large finite models. This means in particular that finitude cannot be expressed in first-order logic. The second corollary, the **Löwenheim-Skolem Theorem**, states that every satisfiable Γ has a countable model. It in turn shows that uncountability cannot be expressed in first-order logic.

Problems

Problem 12.1. Complete the proof of Proposition 12.2.

Problem 12.2. Complete the proof of Proposition 12.10.

Problem 12.3. Complete the proof of Lemma 12.11.

Problem 12.4. Complete the proof of Proposition 12.13.

Problem 12.5. Use Corollary 12.19 to prove Theorem 12.18, thus showing that the two formulations of the completeness theorem are equivalent.

Problem 12.6. In order for a derivation system to be complete, its rules must be strong enough to prove every unsatisfiable set inconsistent. Which of the rules of derivation were necessary to prove completeness? Are any of these rules not used anywhere in the proof? In order to answer these questions, make a list or diagram that shows which of the rules of derivation were used in which results that lead up to the proof of Theorem 12.18. Be sure to note any tacit uses of rules in these proofs.

Problem 12.7. Prove (1) of Theorem 12.21.

Problem 12.8. In the standard model of arithmetic N, there is no element $k \in |N|$ which satisfies every formula $\bar{n} < x$ (where $\bar{n}$ is $0'^{\cdots'}$ with n $'$'s). Use the compactness theorem to show that the set of sentences in the language of arithmetic which are true in the standard model of arithmetic N are also true in a structure N' that contains an element which *does* satisfy every formula $\bar{n} < x$.

Problem 12.9. Prove Proposition 12.25. Avoid the use of $\vdash$.

Problem 12.10. Prove Lemma 12.26. (Hint: The crucial step is to show that if Γ_n is finitely satisfiable, so is $\Gamma_n \cup \{D_n\}$, without any appeal to derivations or consistency.)

Problem 12.11. Prove Proposition 12.27.

Problem 12.12. Prove Lemma 12.28. (Hint: the crucial step is to show that if Γ_n is finitely satisfiable, then either $\Gamma_n \cup \{A_n\}$ or $\Gamma_n \cup \{\neg A_n\}$ is finitely satisfiable.)

Problem 12.13. Write out the complete proof of the Truth Lemma (Lemma 12.11) in the version required for the proof of Theorem 12.29.

CHAPTER 13
Beyond First-order Logic

13.1 Overview

First-order logic is not the only system of logic of interest: there are many extensions and variations of first-order logic. A logic typically consists of the formal specification of a language, usually, but not always, a deductive system, and usually, but not always, an intended semantics. But the technical use of the term raises an obvious question: what do logics that are not first-order logic have to do with the word "logic," used in the intuitive or philosophical sense? All of the systems described below are designed to model reasoning of some form or another; can we say what makes them logical?

No easy answers are forthcoming. The word "logic" is used in different ways and in different contexts, and the notion, like that of "truth," has been analyzed from numerous philosophical stances. For example, one might take the goal of logical reasoning to be the determination of which statements are necessarily true, true a priori, true independent of the interpretation of the nonlogical terms, true by virtue of their form, or true by linguistic convention; and each of these conceptions requires a good deal of clarification. Even if one restricts one's attention to the kind of logic used in mathematics, there is little agreement as to its scope. For example, in the *Principia Mathematica*, Russell and Whitehead tried to develop mathematics on the basis of logic, in the *logicist* tradition begun by Frege. Their system of logic was a form of higher-type logic similar to the one described below. In the end they were forced to introduce axioms which, by most standards, do not seem purely logical (notably, the axiom of infinity, and the axiom of reducibility), but one might nonetheless hold that some forms of higher-order reasoning should be accepted as logical. In contrast, Quine, whose ontology does not admit "propositions" as legitimate objects of discourse, argues that second-order and higher-order logic are really manifestations of set theory in sheep's clothing; in other words, systems involving quantification over predicates are not purely logical.

For now, it is best to leave such philosophical issues for a rainy day, and simply think of the systems below as formal idealizations of various kinds of reasoning, logical or otherwise.

13.2 Many-Sorted Logic

In first-order logic, variables and quantifiers range over a single domain. But it is often useful to have multiple (disjoint) domains: for example, you might want to have a domain of numbers, a domain of geometric objects, a domain of functions from numbers to numbers, a domain of abelian groups, and so on.

Many-sorted logic provides this kind of framework. One starts with a list of "sorts"—the "sort" of an object indicates the "domain" it is supposed to inhabit. One then has variables and quantifiers for each sort, and (usually) an identity predicate for each sort. Functions and relations are also "typed" by the sorts of objects they can take as arguments. Otherwise, one keeps the

13.2. MANY-SORTED LOGIC

usual rules of first-order logic, with versions of the quantifier-rules repeated for each sort.

For example, to study international relations we might choose a language with two sorts of objects, French citizens and German citizens. We might have a unary relation, "drinks wine," for objects of the first sort; another unary relation, "eats wurst," for objects of the second sort; and a binary relation, "forms a multinational married couple," which takes two arguments, where the first argument is of the first sort and the second argument is of the second sort. If we use variables a, b, c to range over French citizens and x, y, z to range over German citizens, then

$$\forall a\, \forall x [(MarriedTo(a,x) \rightarrow (DrinksWine(a) \vee \neg EatsWurst(x))]]$$

asserts that if any French person is married to a German, either the French person drinks wine or the German doesn't eat wurst.

Many-sorted logic can be embedded in first-order logic in a natural way, by lumping all the objects of the many-sorted domains together into one first-order domain, using unary predicate symbols to keep track of the sorts, and relativizing quantifiers. For example, the first-order language corresponding to the example above would have unary predicate symbols "$German$" and "$French$," in addition to the other relations described, with the sort requirements erased. A sorted quantifier $\forall x\, A$, where x is a variable of the German sort, translates to

$$\forall x\, (German(x) \rightarrow A).$$

We need to add axioms that insure that the sorts are separate— e.g., $\forall x\, \neg (German(x) \wedge French(x))$—as well as axioms that guarantee that "drinks wine" only holds of objects satisfying the predicate $French(x)$, etc. With these conventions and axioms, it is not difficult to show that many-sorted sentences translate to first-order sentences, and many-sorted derivations translate to first-order derivations. Also, many-sorted structures "translate" to corresponding first-order structures and vice-versa, so we also have a completeness theorem for many-sorted logic.

13.3 Second-Order logic

The language of second-order logic allows one to quantify not just over a domain of individuals, but over relations on that domain as well. Given a first-order language $\mathcal{L}$, for each k one adds variables R which range over k-ary relations, and allows quantification over those variables. If R is a variable for a k-ary relation, and $t_1, \ldots, t_k$ are ordinary (first-order) terms, $R(t_1, \ldots, t_k)$ is an atomic formula. Otherwise, the set of formulas is defined just as in the case of first-order logic, with additional clauses for second-order quantification. Note that we only have the identity predicate for first-order terms: if R and S are relation variables of the same arity k, we can define $R = S$ to be an abbreviation for
$$\forall x_1 \ldots \forall x_k \, (R(x_1, \ldots, x_k) \leftrightarrow S(x_1, \ldots, x_k)).$$

The rules for second-order logic simply extend the quantifier rules to the new second order variables. Here, however, one has to be a little bit careful to explain how these variables interact with the predicate symbols of $\mathcal{L}$, and with formulas of $\mathcal{L}$ more generally. At the bare minimum, relation variables count as terms, so one has inferences of the form
$$A(R) \vdash \exists R \, A(R)$$

But if $\mathcal{L}$ is the language of arithmetic with a constant relation symbol $<$, one would also expect the following inference to be valid:
$$x < y \vdash \exists R \, R(x, y)$$
or for a given formula A,
$$A(x_1, \ldots, x_k) \vdash \exists R \, R(x_1, \ldots, x_k)$$

More generally, we might want to allow inferences of the form
$$A[\lambda \vec{x}. \, B(\vec{x})/R] \vdash \exists R \, A$$

where $A[\lambda \vec{x}. \, B(\vec{x})/R]$ denotes the result of replacing every atomic formula of the form $Rt_1, \ldots, t_k$ in A by $B(t_1, \ldots, t_k)$. This last rule

13.3. SECOND-ORDER LOGIC

is equivalent to having a *comprehension schema*, i.e., an axiom of the form

$$\exists R \, \forall x_1, \ldots, x_k \, (A(x_1, \ldots, x_k) \leftrightarrow R(x_1, \ldots, x_k)),$$

one for each formula A in the second-order language, in which R is not a free variable. (Exercise: show that if R is allowed to occur in A, this schema is inconsistent!)

When logicians refer to the "axioms of second-order logic" they usually mean the minimal extension of first-order logic by second-order quantifier rules together with the comprehension schema. But it is often interesting to study weaker subsystems of these axioms and rules. For example, note that in its full generality the axiom schema of comprehension is *impredicative*: it allows one to assert the existence of a relation $R(x_1, \ldots, x_k)$ that is "defined" by a formula with second-order quantifiers; and these quantifiers range over the set of all such relations—a set which includes R itself! Around the turn of the twentieth century, a common reaction to Russell's paradox was to lay the blame on such definitions, and to avoid them in developing the foundations of mathematics. If one prohibits the use of second-order quantifiers in the formula A, one has a *predicative* form of comprehension, which is somewhat weaker.

From the semantic point of view, one can think of a second-order structure as consisting of a first-order structure for the language, coupled with a set of relations on the domain over which the second-order quantifiers range (more precisely, for each k there is a set of relations of arity k). Of course, if comprehension is included in the derivation system, then we have the added requirement that there are enough relations in the "second-order part" to satisfy the comprehension axioms—otherwise the derivation system is not sound! One easy way to insure that there are enough relations around is to take the second-order part to consist of *all* the relations on the first-order part. Such a structure is called *full*, and, in a sense, is really the "intended structure" for the language. If we restrict our attention to full structures we have what is known as the *full* second-order semantics. In that case, specifying a structure boils down to specifying the first-order part, since the contents of the second-order part follow from that implicitly.

To summarize, there is some ambiguity when talking about second-order logic. In terms of the derivation system, one might have in mind either

1. A "minimal" second-order derivation system, together with some comprehension axioms.

2. The "standard" second-order derivation system, with full comprehension.

In terms of the semantics, one might be interested in either

1. The "weak" semantics, where a structure consists of a first-order part, together with a second-order part big enough to satisfy the comprehension axioms.

2. The "standard" second-order semantics, in which one considers full structures only.

When logicians do not specify the derivation system or the semantics they have in mind, they are usually refering to the second item on each list. The advantage to using this semantics is that, as we will see, it gives us categorical descriptions of many natural mathematical structures; at the same time, the derivation system is quite strong, and sound for this semantics. The drawback is that the derivation system is *not* complete for the semantics; in fact, *no* effectively given derivation system is complete for the full second-order semantics. On the other hand, we will see that the derivation system *is* complete for the weakened semantics; this implies that if a sentence is not provable, then there is *some* structure, not necessarily the full one, in which it is false.

The language of second-order logic is quite rich. One can identify unary relations with subsets of the domain, and so in particular you can quantify over these sets; for example, one can express induction for the natural numbers with a single axiom

$$\forall R\,((R(0) \wedge \forall x\,(R(x) \to R(x'))) \to \forall x\,R(x)).$$

If one takes the language of arithmetic to have symbols $0, ', +, \times$ and $<$, one can add the following axioms to describe their behavior:

1. $\forall x\,\neg x' = 0$
2. $\forall x\,\forall y\,(s(x) = s(y) \to x = y)$
3. $\forall x\,(x + 0) = x$
4. $\forall x\,\forall y\,(x + y') = (x + y)'$
5. $\forall x\,(x \times 0) = 0$
6. $\forall x\,\forall y\,(x \times y') = ((x \times y) + x)$
7. $\forall x\,\forall y\,(x < y \leftrightarrow \exists z\,y = (x + z'))$

It is not difficult to show that these axioms, together with the axiom of induction above, provide a categorical description of the structure N, the standard model of arithmetic, provided we are using the full second-order semantics. Given any structure M in which these axioms are true, define a function f from $\mathbb{N}$ to the domain of M using ordinary recursion on $\mathbb{N}$, so that $f(0) = 0^M$ and $f(x+1) = '^M(f(x))$. Using ordinary induction on $\mathbb{N}$ and the fact that axioms (1) and (2) hold in M, we see that f is injective. To see that f is surjective, let P be the set of elements of $|M|$ that are in the range of f. Since M is full, P is in the second-order domain. By the construction of f, we know that 0^M is in P, and that P is closed under $'^M$. The fact that the induction axiom holds in M (in particular, for P) guarantees that P is equal to the entire first-order domain of M. This shows that f is a bijection. Showing that f is a homomorphism is no more difficult, using ordinary induction on $\mathbb{N}$ repeatedly.

In set-theoretic terms, a function is just a special kind of relation; for example, a unary function f can be identified with a binary relation R satisfying $\forall x\, \exists! y\, R(x,y)$. As a result, one can quantify over functions too. Using the full semantics, one can then define the class of infinite structures to be the class of structures M for which there is an injective function from the domain of M to a proper subset of itself:

$$\exists f\, (\forall x\, \forall y\, (f(x) = f(y) \to x = y) \land \exists y\, \forall x\, f(x) \neq y).$$

The negation of this sentence then defines the class of finite structures.

In addition, one can define the class of well-orderings, by adding the following to the definition of a linear ordering:

$$\forall P\, (\exists x\, P(x) \to \exists x\, (P(x) \land \forall y\, (y < x \to \neg P(y)))).$$

This asserts that every non-empty set has a least element, modulo the identification of "set" with "one-place relation". For another example, one can express the notion of connectedness for graphs, by saying that there is no nontrivial separation of the vertices into disconnected parts:

$$\neg \exists A\, (\exists x\, A(x) \land \exists y\, \neg A(y) \land \forall w\, \forall z\, ((A(w) \land \neg A(z)) \to \neg R(w,z))).$$

For yet another example, you might try as an exercise to define the class of finite structures whose domain has even size. More strikingly, one can provide a categorical description of the real numbers as a complete ordered field containing the rationals.

In short, second-order logic is much more expressive than first-order logic. That's the good news; now for the bad. We have already mentioned that there is no effective derivation system that is complete for the full second-order semantics. For better or for worse, many of the properties of first-order logic are absent, including compactness and the Löwenheim-Skolem theorems.

On the other hand, if one is willing to give up the full second-order semantics in terms of the weaker one, then the minimal second-order derivation system is complete for this semantics. In other words, if we read ⊢ as "proves in the minimal system" and ⊨ as "logically implies in the weaker semantics", we can show that whenever $\Gamma \vDash A$ then $\Gamma \vdash A$. If one wants to include specific comprehension axioms in the derivation system, one has to restrict the semantics to second-order structures that satisfy these axioms: for example, if Δ consists of a set of comprehension axioms (possibly all of them), we have that if $\Gamma \cup \Delta \vDash A$, then $\Gamma \cup \Delta \vdash A$. In particular, if A is not provable using the comprehension axioms we are considering, then there is a model of $\neg A$ in which these comprehension axioms nonetheless hold.

The easiest way to see that the completeness theorem holds for the weaker semantics is to think of second-order logic as a many-sorted logic, as follows. One sort is interpreted as the ordinary "first-order" domain, and then for each k we have a domain of "relations of arity k." We take the language to have built-in relation symbols "$true_k(R, x_1, \ldots, x_k)$" which is meant to assert that R holds of $x_1, \ldots, x_k$, where R is a variable of the sort "k-ary relation" and $x_1, \ldots, x_k$ are objects of the first-order sort.

With this identification, the weak second-order semantics is essentially the usual semantics for many-sorted logic; and we have already observed that many-sorted logic can be embedded in first-order logic. Modulo the translations back and forth, then, the weaker conception of second-order logic is really a form of first-order logic in disguise, where the domain contains both "objects" and "relations" governed by the appropriate axioms.

13.4 Higher-Order logic

Passing from first-order logic to second-order logic enabled us to talk about sets of objects in the first-order domain, within the formal language. Why stop there? For example, third-order logic should enable us to deal with sets of sets of objects, or perhaps even sets which contain both objects and sets of objects. And fourth-order logic will let us talk about sets of objects of that kind.

As you may have guessed, one can iterate this idea arbitrarily.

In practice, higher-order logic is often formulated in terms of functions instead of relations. (Modulo the natural identifications, this difference is inessential.) Given some basic "sorts" A, B, C, ... (which we will now call "types"), we can create new ones by stipulating

If σ and τ are finite types then so is $\sigma \to \tau$.

Think of types as syntactic "labels," which classify the objects we want in our domain; $\sigma \to \tau$ describes those objects that are functions which take objects of type σ to objects of type τ. For example, we might want to have a type Ω of truth values, "true" and "false," and a type $\mathbb{N}$ of natural numbers. In that case, you can think of objects of type $\mathbb{N} \to \Omega$ as unary relations, or subsets of $\mathbb{N}$; objects of type $\mathbb{N} \to \mathbb{N}$ are functions from natural numers to natural numbers; and objects of type $(\mathbb{N} \to \mathbb{N}) \to \mathbb{N}$ are "functionals," that is, higher-type functions that take functions to numbers.

As in the case of second-order logic, one can think of higher-order logic as a kind of many-sorted logic, where there is a sort for each type of object we want to consider. But it is usually clearer just to define the syntax of higher-type logic from the ground up. For example, we can define a set of finite types inductively, as follows:

1. $\mathbb{N}$ is a finite type.

2. If σ and τ are finite types, then so is $\sigma \to \tau$.

3. If σ and τ are finite types, so is $\sigma \times \tau$.

Intuitively, $\mathbb{N}$ denotes the type of the natural numbers, $\sigma \to \tau$ denotes the type of functions from σ to τ, and $\sigma \times \tau$ denotes the type of pairs of objects, one from σ and one from τ. We can then define a set of terms inductively, as follows:

1. For each type σ, there is a stock of variables x, y, z, ... of type σ

13.4. HIGHER-ORDER LOGIC

2. 0 is a term of type $\mathbb{N}$

3. S (successor) is a term of type $\mathbb{N} \to \mathbb{N}$

4. If s is a term of type σ, and t is a term of type $\mathbb{N} \to (\sigma \to \sigma)$, then R_{st} is a term of type $\mathbb{N} \to \sigma$

5. If s is a term of type $\tau \to \sigma$ and t is a term of type τ, then $s(t)$ is a term of type σ

6. If s is a term of type σ and x is a variable of type τ, then $\lambda x.\, s$ is a term of type $\tau \to \sigma$.

7. If s is a term of type σ and t is a term of type τ, then $\langle s, t \rangle$ is a term of type $\sigma \times \tau$.

8. If s is a term of type $\sigma \times \tau$ then $p_1(s)$ is a term of type σ and $p_2(s)$ is a term of type τ.

Intuitively, R_{st} denotes the function defined recursively by

$$R_{st}(0) = s$$
$$R_{st}(x+1) = t(x, R_{st}(x)),$$

$\langle s, t \rangle$ denotes the pair whose first component is s and whose second component is t, and $p_1(s)$ and $p_2(s)$ denote the first and second elements ("projections") of s. Finally, $\lambda x.\, s$ denotes the function f defined by

$$f(x) = s$$

for any x of type σ; so item (6) gives us a form of comprehension, enabling us to define functions using terms. Formulas are built up from identity predicate statements $s = t$ between terms of the same type, the usual propositional connectives, and higher-type quantification. One can then take the axioms of the system to be the basic equations governing the terms defined above, together with the usual rules of logic with quantifiers and identity predicate.

If one augments the finite type system with a type Ω of truth values, one has to include axioms which govern its use as well. In fact, if one is clever, one can get rid of complex formulas entirely, replacing them with terms of type Ω! The proof system can then be modified accordingly. The result is essentially the *simple theory of types* set forth by Alonzo Church in the 1930s.

As in the case of second-order logic, there are different versions of higher-type semantics that one might want to use. In the full version, variables of type $\sigma \to \tau$ range over the set of *all* functions from the objects of type σ to objects of type τ. As you might expect, this semantics is too strong to admit a complete, effective derivation system. But one can consider a weaker semantics, in which a structure consists of sets of elements T_τ for each type τ, together with appropriate operations for application, projection, etc. If the details are carried out correctly, one can obtain completeness theorems for the kinds of derivation systems described above.

Higher-type logic is attractive because it provides a framework in which we can embed a good deal of mathematics in a natural way: starting with $\mathbb{N}$, one can define real numbers, continuous functions, and so on. It is also particularly attractive in the context of intuitionistic logic, since the types have clear "constructive" intepretations. In fact, one can develop constructive versions of higher-type semantics (based on intuitionistic, rather than classical logic) that clarify these constructive interpretations quite nicely, and are, in many ways, more interesting than the classical counterparts.

13.5 Intuitionistic Logic

In constrast to second-order and higher-order logic, intuitionistic first-order logic represents a restriction of the classical version, intended to model a more "constructive" kind of reasoning. The following examples may serve to illustrate some of the underlying motivations.

13.5. INTUITIONISTIC LOGIC

Suppose someone came up to you one day and announced that they had determined a natural number x, with the property that if x is prime, the Riemann hypothesis is true, and if x is composite, the Riemann hypothesis is false. Great news! Whether the Riemann hypothesis is true or not is one of the big open questions of mathematics, and here they seem to have reduced the problem to one of calculation, that is, to the determination of whether a specific number is prime or not.

What is the magic value of x? They describe it as follows: x is the natural number that is equal to 7 if the Riemann hypothesis is true, and 9 otherwise.

Angrily, you demand your money back. From a classical point of view, the description above does in fact determine a unique value of x; but what you really want is a value of x that is given *explicitly*.

To take another, perhaps less contrived example, consider the following question. We know that it is possible to raise an irrational number to a rational power, and get a rational result. For example, $\sqrt{2}^2 = 2$. What is less clear is whether or not it is possible to raise an irrational number to an *irrational* power, and get a rational result. The following theorem answers this in the affirmative:

Theorem 13.1. *There are irrational numbers a and b such that a^b is rational.*

Proof. Consider $\sqrt{2}^{\sqrt{2}}$. If this is rational, we are done: we can let $a = b = \sqrt{2}$. Otherwise, it is irrational. Then we have

$$(\sqrt{2}^{\sqrt{2}})^{\sqrt{2}} = \sqrt{2}^{\sqrt{2} \cdot \sqrt{2}} = \sqrt{2}^2 = 2,$$

which is certainly rational. So, in this case, let a be $\sqrt{2}^{\sqrt{2}}$, and let b be $\sqrt{2}$. $\square$

Does this constitute a valid proof? Most mathematicians feel that it does. But again, there is something a little bit unsatisfying here: we have proved the existence of a pair of real numbers with a certain property, without being able to say *which* pair of numbers it is. It is possible to prove the same result, but in such a way that the pair a, b *is* given in the proof: take $a = \sqrt{3}$ and $b = \log_3 4$. Then

$$a^b = \sqrt{3}^{\log_3 4} = 3^{1/2 \cdot \log_3 4} = (3^{\log_3 4})^{1/2} = 4^{1/2} = 2,$$

since $3^{\log_3 x} = x$.

Intuitionistic logic is designed to model a kind of reasoning where moves like the one in the first proof are disallowed. Proving the existence of an x satisfying $A(x)$ means that you have to give a specific x, and a proof that it satisfies A, like in the second proof. Proving that A or B holds requires that you can prove one or the other.

Formally speaking, intuitionistic first-order logic is what you get if you omit restrict a derivation system for first-order logic in a certain way. Similarly, there are intuitionistic versions of second-order or higher-order logic. From the mathematical point of view, these are just formal deductive systems, but, as already noted, they are intended to model a kind of mathematical reasoning. One can take this to be the kind of reasoning that is justified on a certain philosophical view of mathematics (such as Brouwer's intuitionism); one can take it to be a kind of mathematical reasoning which is more "concrete" and satisfying (along the lines of Bishop's constructivism); and one can argue about whether or not the formal description captures the informal motivation. But whatever philosophical positions we may hold, we can study intuitionistic logic as a formally presented logic; and for whatever reasons, many mathematical logicians find it interesting to do so.

There is an informal constructive interpretation of the intuitionist connectives, usually known as the BHK interpretation (named after Brouwer, Heyting, and Kolmogorov). It runs as follows: a proof of $A \wedge B$ consists of a proof of A paired with a proof of B; a proof of $A \vee B$ consists of either a proof of A, or a proof of B, where we have explicit information as to which is the case; a proof of $A \to B$ consists of a procedure, which transforms a proof of A to a proof of B; a proof of $\forall x\, A(x)$ consists of a procedure which returns a proof of $A(x)$ for any value of x; and a proof of $\exists x\, A(x)$ consists of a value of x, together with a proof that this value satisfies A. One can describe the interpretation in computational terms known as the "Curry-Howard isomorphism" or the "formulas-as-types paradigm": think of a formula as specifying a certain kind of data type, and proofs as computational objects of these data types that enable us to see that the corresponding formula is true.

Intuitionistic logic is often thought of as being classical logic "minus" the law of the excluded middle. This following theorem makes this more precise.

Theorem 13.2. *Intuitionistically, the following axiom schemata are equivalent:*

1. $(A \to \bot) \to \neg A$.

2. $A \vee \neg A$

3. $\neg\neg A \to A$

Obtaining instances of one schema from either of the others is a good exercise in intuitionistic logic.

The first deductive systems for intuitionistic propositional logic, put forth as formalizations of Brouwer's intuitionism, are due, independently, to Kolmogorov, Glivenko, and Heyting. The first formalization of intuitionistic first-order logic (and parts of intuitionist mathematics) is due to Heyting. Though a number of classically valid schemata are not intuitionistically valid, many are.

The *double-negation translation* describes an important relationship between classical and intuitionist logic. It is defined inductively follows (think of A^N as the "intuitionist" translation of the classical formula A):

$$A^N \equiv \neg\neg A \quad \text{for atomic formulas } A$$
$$(A \wedge B)^N \equiv (A^N \wedge B^N)$$
$$(A \vee B)^N \equiv \neg\neg(A^N \vee B^N)$$
$$(A \to B)^N \equiv (A^N \to B^N)$$
$$(\forall x\, A)^N \equiv \forall x\, A^N$$
$$(\exists x\, A)^N \equiv \neg\neg \exists x\, A^N$$

Kolmogorov and Glivenko had versions of this translation for propositional logic; for predicate logic, it is due to Gödel and Gentzen, independently. We have

Theorem 13.3. *1. $A \leftrightarrow A^N$ is provable classically*

2. If A is provable classically, then A^N is provable intuitionistically.

We can now envision the following dialogue. Classical mathematician: "I've proved A!" Intuitionist mathematician: "Your proof isn't valid. What you've really proved is A^N." Classical mathematician: "Fine by me!" As far as the classical mathematician is concerned, the intuitionist is just splitting hairs, since the two are equivalent. But the intuitionist insists there is a difference.

Note that the above translation concerns pure logic only; it does not address the question as to what the appropriate *nonlogical* axioms are for classical and intuitionistic mathematics, or what the relationship is between them. But the following slight extension of the theorem above provides some useful information:

Theorem 13.4. *If Γ proves A classically, Γ^N proves A^N intuitionistically.*

In other words, if A is provable from some hypotheses classically, then A^N is provable from their double-negation translations.

13.5. INTUITIONISTIC LOGIC

To show that a sentence or propositional formula is intuitionistically valid, all you have to do is provide a proof. But how can you show that it is not valid? For that purpose, we need a semantics that is sound, and preferably complete. A semantics due to Kripke nicely fits the bill.

We can play the same game we did for classical logic: define the semantics, and prove soundness and completeness. It is worthwhile, however, to note the following distinction. In the case of classical logic, the semantics was the "obvious" one, in a sense implicit in the meaning of the connectives. Though one can provide some intuitive motivation for Kripke semantics, the latter does not offer the same feeling of inevitability. In addition, the notion of a classical structure is a natural mathematical one, so we can either take the notion of a structure to be a tool for studying classical first-order logic, or take classical first-order logic to be a tool for studying mathematical structures. In contrast, Kripke structures can only be viewed as a logical construct; they don't seem to have independent mathematical interest.

A Kripke structure $\mathfrak{M} = \langle W, R, V \rangle$ for a propositional language consists of a set W, partial order R on W with a least element, and an "monotone" assignment of propositional variables to the elements of W. The intuition is that the elements of W represent "worlds," or "states of knowledge"; an element $v \geq u$ represents a "possible future state" of u; and the propositional variables assigned to u are the propositions that are known to be true in state u. The forcing relation $\mathfrak{M}, w \Vdash A$ then extends this relationship to arbitrary formulas in the language; read $\mathfrak{M}, w \Vdash A$ as "A is true in state w." The relationship is defined inductively, as follows:

1. $\mathfrak{M}, w \Vdash p_i$ iff p_i is one of the propositional variables assigned to w.

2. $\mathfrak{M}, w \not\Vdash \bot$.

3. $\mathfrak{M}, w \Vdash (A \wedge B)$ iff $\mathfrak{M}, w \Vdash A$ and $\mathfrak{M}, w \Vdash B$.

4. $\mathfrak{M}, w \Vdash (A \vee B)$ iff $\mathfrak{M}, w \Vdash A$ or $\mathfrak{M}, w \Vdash B$.

5. $\mathfrak{M}, w \Vdash (A \rightarrow B)$ iff, whenever $w' \geq w$ and $\mathfrak{M}, w' \Vdash A$, then $\mathfrak{M}, w' \Vdash B$.

It is a good exercise to try to show that $\neg(p \wedge q) \rightarrow (\neg p \vee \neg q)$ is not intuitionistically valid, by cooking up a Kripke structure that provides a counterexample.

13.6 Modal Logics

Consider the following example of a conditional sentence:

> If Jeremy is alone in that room, then he is drunk and naked and dancing on the chairs.

This is an example of a conditional assertion that may be materially true but nonetheless misleading, since it seems to suggest that there is a stronger link between the antecedent and conclusion other than simply that either the antecedent is false or the consequent true. That is, the wording suggests that the claim is not only true in this particular world (where it may be trivially true, because Jeremy is not alone in the room), but that, moreover, the conclusion *would have* been true *had* the antecedent been true. In other words, one can take the assertion to mean that the claim is true not just in this world, but in any "possible" world; or that it is *necessarily* true, as opposed to just true in this particular world.

Modal logic was designed to make sense of this kind of necessity. One obtains modal propositional logic from ordinary propositional logic by adding a box operator; which is to say, if A is a formula, so is $\Box A$. Intuitively, $\Box A$ asserts that A is *necessarily* true, or true in any possible world. $\Diamond A$ is usually taken to be an abbreviation for $\neg \Box \neg A$, and can be read as asserting that A is *possibly* true. Of course, modality can be added to predicate logic as well.

13.6. MODAL LOGICS

Kripke structures can be used to provide a semantics for modal logic; in fact, Kripke first designed this semantics with modal logic in mind. Rather than restricting to partial orders, more generally one has a set of "possible worlds," P, and a binary "accessibility" relation $R(x,y)$ between worlds. Intuitively, $R(p,q)$ asserts that the world q is compatible with p; i.e., if we are "in" world p, we have to entertain the possibility that the world could have been like q.

Modal logic is sometimes called an "intensional" logic, as opposed to an "extensional" one. The intended semantics for an extensional logic, like classical logic, will only refer to a single world, the "actual" one; while the semantics for an "intensional" logic relies on a more elaborate ontology. In addition to structureing necessity, one can use modality to structure other linguistic constructions, reinterpreting $\square$ and $\diamond$ according to the application. For example:

1. In provability logic, $\square A$ is read "A is provable" and $\diamond A$ is read "A is consistent."

2. In epistemic logic, one might read $\square A$ as "I know A" or "I believe A."

3. In temporal logic, one can read $\square A$ as "A is always true" and $\diamond A$ as "A is sometimes true."

One would like to augment logic with rules and axioms dealing with modality. For example, the system **S4** consists of the ordinary axioms and rules of propositional logic, together with the following axioms:

$$\square(A \to B) \to (\square A \to \square B)$$
$$\square A \to A$$
$$\square A \to \square\square A$$

as well as a rule, "from A conclude $\Box A$." **S5** adds the following axiom:

$$\Diamond A \to \Box \Diamond A$$

Variations of these axioms may be suitable for different applications; for example, **S5** is usually taken to characterize the notion of logical necessity. And the nice thing is that one can usually find a semantics for which the derivation system is sound and complete by restricting the accessibility relation in the Kripke structures in natural ways. For example, **S4** corresponds to the class of Kripke structures in which the accessibility relation is reflexive and transitive. **S5** corresponds to the class of Kripke structures in which the accessibility relation is *universal*, which is to say that every world is accessible from every other; so $\Box A$ holds if and only if A holds in every world.

13.7 Other Logics

As you may have gathered by now, it is not hard to design a new logic. You too can create your own a syntax, make up a deductive system, and fashion a semantics to go with it. You might have to be a bit clever if you want the derivation system to be complete for the semantics, and it might take some effort to convince the world at large that your logic is truly interesting. But, in return, you can enjoy hours of good, clean fun, exploring your logic's mathematical and computational properties.

Recent decades have witnessed a veritable explosion of formal logics. Fuzzy logic is designed to model reasoning about vague properties. Probabilistic logic is designed to model reasoning about uncertainty. Default logics and nonmonotonic logics are designed to model defeasible forms of reasoning, which is to say, "reasonable" inferences that can later be overturned in the face of new information. There are epistemic logics, designed to model reasoning about knowledge; causal logics, designed to model reasoning about causal relationships; and even "deontic" logics, which are designed to model reasoning about moral and ethical obligations. Depending on whether the primary motivation for introducing these systems is philosophical, mathematical, or computational, you may find such creatures studies under the rubric of mathematical logic, philosophical logic, artificial intelligence, cognitive science, or elsewhere.

The list goes on and on, and the possibilities seem endless. We may never attain Leibniz' dream of reducing all of human reason to calculation—but that can't stop us from trying.

PART III

Turing Machines

CHAPTER 14

Turing Machine Computations

14.1 Introduction

What does it mean for a function, say, from $\mathbb{N}$ to $\mathbb{N}$ to be *computable*? Among the first answers, and the most well known one, is that a function is computable if it can be computed by a Turing machine. This notion was set out by Alan Turing in 1936. Turing machines are an example of *a model of computation*—they are a mathematically precise way of defining the idea of a "computational procedure." What exactly that means is debated, but it is widely agreed that Turing machines are one way of specifying computational procedures. Even though the term "Turing machine" evokes the image of a physical machine with moving parts, strictly speaking a Turing machine is a purely mathematical construct, and as such it idealizes the idea of a computational procedure. For instance, we place no restriction on either the time or memory requirements of a Turing machine: Turing machines can compute something even if the computation would

14.1. INTRODUCTION

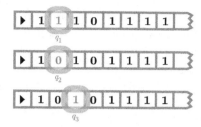

Figure 14.1: A Turing machine executing its program.

require more storage space or more steps than there are atoms in the universe.

It is perhaps best to think of a Turing machine as a program for a special kind of imaginary mechanism. This mechanism consists of a *tape* and a *read-write head*. In our version of Turing machines, the tape is infinite in one direction (to the right), and it is divided into *squares*, each of which may contain a symbol from a finite *alphabet*. Such alphabets can contain any number of different symbols, but we will mainly make do with three: ▷, ␣, and I. When the mechanism is started, the tape is empty (i.e., each square contains the symbol ␣) except for the leftmost square, which contains ▷, and a finite number of squares which contain the *input*. At any time, the mechanism is in one of a finite number of *states*. At the outset, the head scans the leftmost square and in a specified *initial state*. At each step of the mechanism's run, the content of the square currently scanned together with the state the mechanism is in and the Turing machine program determine what happens next. The Turing machine program is given by a partial function which takes as input a state q and a symbol σ and outputs a triple $\langle q', \sigma', D \rangle$. Whenever the mechanism is in state q and reads symbol σ, it replaces the symbol on the current square with σ', the head moves left, right, or stays put according to whether D is L, R, or N, and the mechanism goes into state q'.

For instance, consider the situation in Figure 14.1. The visible part of the tape of the Turing machine contains the end-of-tape symbol ▷ on the leftmost square, followed by three 1's, a 0, and four more 1's. The head is reading the third square from the left, which contains a 1, and is in state q_1—we say "the machine is reading a 1 in state q_1." If the program of the Turing machine returns, for input $\langle q_1, 1 \rangle$, the triple $\langle q_2, 0, N \rangle$, the machine would now replace the 1 on the third square with a 0, leave the read/write head where it is, and switch to state q_2. If then the program returns $\langle q_3, 0, R \rangle$ for input $\langle q_2, 0 \rangle$, the machine would now overwrite the 0 with another 0 (effectively, leaving the content of the tape under the read/write head unchanged), move one square to the right, and enter state q_3. And so on.

We say that the machine *halts* when it encounters some state, q_n, and symbol, σ such that there is no instruction for $\langle q_n, \sigma \rangle$, i.e., the transition function for input $\langle q_n, \sigma \rangle$ is undefined. In other words, the machine has no instruction to carry out, and at that point, it ceases operation. Halting is sometimes represented by a specific halt state h. This will be demonstrated in more detail later on.

The beauty of Turing's paper, "On computable numbers," is that he presents not only a formal definition, but also an argument that the definition captures the intuitive notion of computability. From the definition, it should be clear that any function computable by a Turing machine is computable in the intuitive sense. Turing offers three types of argument that the converse is true, i.e., that any function that we would naturally regard as computable is computable by such a machine. They are (in Turing's words):

1. A direct appeal to intuition.

2. A proof of the equivalence of two definitions (in case the new definition has a greater intuitive appeal).

3. Giving examples of large classes of numbers which are computable.

Our goal is to try to define the notion of computability "in principle," i.e., without taking into account practical limitations of time and space. Of course, with the broadest definition of computability in place, one can then go on to consider computation with bounded resources; this forms the heart of the subject known as "computational complexity."

Historical Remarks Alan Turing invented Turing machines in 1936. While his interest at the time was the decidability of first-order logic, the paper has been described as a definitive paper on the foundations of computer design. In the paper, Turing focuses on computable real numbers, i.e., real numbers whose decimal expansions are computable; but he notes that it is not hard to adapt his notions to computable functions on the natural numbers, and so on. Notice that this was a full five years before the first working general purpose computer was built in 1941 (by the German Konrad Zuse in his parent's living room), seven years before Turing and his colleagues at Bletchley Park built the code-breaking Colossus (1943), nine years before the American ENIAC (1945), twelve years before the first British general purpose computer—the Manchester Small-Scale Experimental Machine—was built in Manchester (1948), and thirteen years before the Americans first tested the BINAC (1949). The Manchester SSEM has the distinction of being the first stored-program computer—previous machines had to be rewired by hand for each new task.

14.2 Representing Turing Machines

Turing machines can be represented visually by *state diagrams*. The diagrams are composed of state cells connected by arrows. Unsurprisingly, each state cell represents a state of the machine. Each arrow represents an instruction that can be carried out from that state, with the specifics of the instruction written above or below the appropriate arrow. Consider the following machine, which has only two internal states, q_0 and q_1, and one instruction:

Recall that the Turing machine has a read/write head and a tape with the input written on it. The instruction can be read as *if reading a ⊔ in state q_0, write a I, move right, and move to state q_1.* This is equivalent to the transition function mapping $\langle q_0, ⊔ \rangle$ to $\langle q_1, I, R \rangle$.

Example 14.1. *Even Machine*: The following Turing machine halts if, and only if, there are an even number of I's on the tape (under the assumption that all I's come before the first ⊔ on the tape).

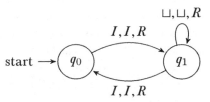

The state diagram corresponds to the following transition function:

$$\delta(q_0, I) = \langle q_1, I, R \rangle,$$
$$\delta(q_1, I) = \langle q_0, I, R \rangle,$$
$$\delta(q_1, ⊔) = \langle q_1, ⊔, R \rangle$$

The above machine halts only when the input is an even number of strokes. Otherwise, the machine (theoretically) continues to operate indefinitely. For any machine and input, it is possible to trace through the *configurations* of the machine in order to determine the output. We will give a formal definition of configurations later. For now, we can intuitively think of configurations as a series of diagrams showing the state of the machine at any

14.2. REPRESENTING TURING MACHINES

point in time during operation. Configurations show the content of the tape, the state of the machine and the location of the read/write head.

Let us trace through the configurations of the even machine if it is started with an input of four I's. In this case, we expect that the machine will halt. We will then run the machine on an input of three I's, where the machine will run forever.

The machine starts in state q_0, scanning the leftmost I. We can represent the initial state of the machine as follows:

$$\triangleright I_0 I I I \sqcup \ldots$$

The above configuration is straightforward. As can be seen, the machine starts in state one, scanning the leftmost I. This is represented by a subscript of the state name on the first I. The applicable instruction at this point is $\delta(q_0, I) = \langle q_1, I, R \rangle$, and so the machine moves right on the tape and changes to state q_1.

$$\triangleright I I_1 I I \sqcup \ldots$$

Since the machine is now in state q_1 scanning a I, we have to "follow" the instruction $\delta(q_1, I) = \langle q_0, I, R \rangle$. This results in the configuration

$$\triangleright I I I_0 I \sqcup \ldots$$

As the machine continues, the rules are applied again in the same order, resulting in the following two configurations:

$$\triangleright I I I I_1 \sqcup \ldots$$

$$\triangleright I I I I \sqcup_0 \ldots$$

The machine is now in state q_0 scanning a $\sqcup$. Based on the transition diagram, we can easily see that there is no instruction to be carried out, and thus the machine has halted. This means that the input has been accepted.

Suppose next we start the machine with an input of three I's. The first few configurations are similar, as the same instructions are carried out, with only a small difference of the tape input:

$$\triangleright I_0 I I \sqcup \ldots$$

$\triangleright I I_1 I \sqcup \ldots$

$\triangleright I I I_0 \sqcup \ldots$

$\triangleright I I I \sqcup_1 \ldots$

The machine has now traversed past all the I's, and is reading a $\sqcup$ in state q_1. As shown in the diagram, there is an instruction of the form $\delta(q_1, \sqcup) = \langle q_1, \sqcup, R \rangle$. Since the tape is filled with $\sqcup$ indefinitely to the right, the machine will continue to execute this instruction *forever*, staying in state q_1 and moving ever further to the right. The machine will never halt, and does not accept the input.

It is important to note that not all machines will halt. If halting means that the machine runs out of instructions to execute, then we can create a machine that never halts simply by ensuring that there is an outgoing arrow for each symbol at each state. The even machine can be modified to run indefinitely by adding an instruction for scanning a $\sqcup$ at q_0.

Example 14.2.

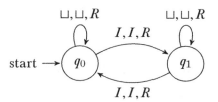

Machine tables are another way of representing Turing machines. Machine tables have the tape alphabet displayed on the *x*-axis, and the set of machine states across the *y*-axis. Inside the table, at the intersection of each state and symbol, is written the rest of the instruction—the new state, new symbol, and direction of movement. Machine tables make it easy to determine in what state, and for what symbol, the machine halts. Whenever there is a gap in the table is a possible point for the machine to halt. Unlike state diagrams and instruction sets, where the points at which the machine halts are not always immediately obvious, any halting points are quickly identified by finding the gaps in the machine table.

Example 14.3. The machine table for the even machine is:

	⊔	I	▷
q_0		I, q_1, R	
q_1	$⊔, q_1, R$	I, q_0, R	

As we can see, the machine halts when scanning a ⊔ in state q_0.

So far we have only considered machines that read and accept input. However, Turing machines have the capacity to both read and write. An example of such a machine (although there are many, many examples) is a *doubler*. A doubler, when started with a block of n I's on the tape, outputs a block of $2n$ I's.

Example 14.4. Before building a doubler machine, it is important to come up with a *strategy* for solving the problem. Since the machine (as we have formulated it) cannot remember how many I's it has read, we need to come up with a way to keep track of all the I's on the tape. One such way is to separate the output from the input with a ⊔. The machine can then erase the first I from the input, traverse over the rest of the input, leave a ⊔, and write two new I's. The machine will then go back and find the second I in the input, and double that one as well. For each one I of input, it will write two I's of output. By erasing the input as the machine goes, we can guarantee that no I is missed or doubled twice. When the entire input is erased, there will be $2n$ I's left on the tape. The state diagram of the resulting Turing machine is depicted in Figure 14.2.

14.3 Turing Machines

The formal definition of what constitutes a Turing machine looks abstract, but is actually simple: it merely packs into one mathe-

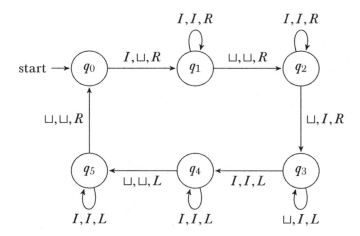

Figure 14.2: A doubler machine

matical structure all the information needed to specify the workings of a Turing machine. This includes (1) which states the machine can be in, (2) which symbols are allowed to be on the tape, (3) which state the machine should start in, and (4) what the instruction set of the machine is.

Definition 14.5 (Turing machine). A *Turing machine* M is a tuple $\langle Q, \Sigma, q_0, \delta \rangle$ consisting of

1. a finite set of *states* Q,

2. a finite *alphabet* Σ which includes ▷ and ␣,

3. an *initial state* $q_0 \in Q$,

4. a finite *instruction set* $\delta \colon Q \times \Sigma \nrightarrow Q \times \Sigma \times \{L, R, N\}$.

The partial function δ is also called the *transition function* of M.

We assume that the tape is infinite in one direction only. For this reason it is useful to designate a special symbol ▷ as a marker for the left end of the tape. This makes it easier for Turing machine programs to tell when they're "in danger" of running off the tape. We could assume that this symbol is never overwritten, i.e., that $\delta(q,\triangleright) = \langle q',\triangleright,x \rangle$ if $\delta(q,\triangleright)$ is defined. Some textbooks do this, we do not. You can simply be careful when constructing your Turing machine that it never overwrites ▷. Moreover, there are cases where allowing such overwriting provides some convenient flexibility.

Example 14.6. *Even Machine:* The even machine is formally the quadruple $\langle Q, \Sigma, q_0, \delta \rangle$ where

$$Q = \{q_0, q_1\}$$
$$\Sigma = \{\triangleright, \sqcup, I\},$$
$$\delta(q_0, I) = \langle q_1, I, R \rangle,$$
$$\delta(q_1, I) = \langle q_0, I, R \rangle,$$
$$\delta(q_1, \sqcup) = \langle q_1, \sqcup, R \rangle.$$

14.4 Configurations and Computations

Recall tracing through the configurations of the even machine earlier. The imaginary mechanism consisting of tape, read/write head, and Turing machine program is really just an intuitive way of visualizing what a Turing machine computation is. Formally, we can define the computation of a Turing machine on a given input as a sequence of *configurations*—and a configuration in turn is a sequence of symbols (corresponding to the contents of the tape at a given point in the computation), a number indicating the position of the read/write head, and a state. Using these, we can define what the Turing machine M computes on a given input.

Definition 14.7 (Configuration). A *configuration* of Turing machine $M = \langle Q, \Sigma, q_0, \delta \rangle$ is a triple $\langle C, m, q \rangle$ where

1. $C \in \Sigma^*$ is a finite sequence of symbols from Σ,

2. $m \in \mathbb{N}$ is a number $< \operatorname{len}(C)$, and

3. $q \in Q$

Intuitively, the sequence C is the content of the tape (symbols of all squares from the leftmost square to the last non-blank or previously visited square), m is the number of the square the read/write head is scanning (beginning with 0 being the number of the leftmost square), and q is the current state of the machine.

The potential input for a Turing machine is a sequence of symbols, usually a sequence that encodes a number in some form. The initial configuration of the Turing machine is that configuration in which we start the Turing machine to work on that input: the tape contains the tape end marker immediately followed by the input written on the squares to the right, the read/write head is scanning the leftmost square of the input (i.e., the square to the right of the left end marker), and the mechanism is in the designated start state q_0.

Definition 14.8 (Initial configuration). The *initial configuration* of M for input $I \in \Sigma^*$ is

$$\langle \triangleright \frown I, 1, q_0 \rangle.$$

The $\frown$ symbol is for *concatenation*—the input string begins immediately to the left end marker.

Definition 14.9. We say that a configuration $\langle C, m, q \rangle$ *yields the configuration* $\langle C', m', q' \rangle$ *in one step* (according to M), iff

1. the m-th symbol of C is σ,

2. the instruction set of M specifies $\delta(q,\sigma) = \langle q',\sigma',D\rangle$,

3. the m-th symbol of C' is σ', and

4.
   a) $D = L$ and $m' = m - 1$ if $m > 0$, otherwise $m' = 0$, or

   b) $D = R$ and $m' = m + 1$, or

   c) $D = N$ and $m' = m$,

5. if $m' = \text{len}(C)$, then $\text{len}(C') = \text{len}(C) + 1$ and the m'-th symbol of C' is ⊔. Otherwise $\text{len}(C') = \text{len}(C)$.

6. for all i such that $i < \text{len}(C)$ and $i \neq m$, $C'(i) = C(i)$,

Definition 14.10. A *run of M on input I* is a sequence C_i of configurations of M, where C_0 is the initial configuration of M for input I, and each C_i yields C_{i+1} in one step.

We say that *M halts on input I after k steps* if $C_k = \langle C, m, q\rangle$, the mth symbol of C is σ, and $\delta(q,\sigma)$ is undefined. In that case, the *output* of M for input I is O, where O is a string of symbols not ending in ⊔ such that $C = \triangleright \frown O \frown ⊔^j$ for some $i, j \in \mathbb{N}$.

According to this definition, the output O of M always ends in a symbol other than ⊔, or, if at time k the entire tape is filled with ⊔ (except for the leftmost ▷), O is the empty string.

14.5 Unary Representation of Numbers

Turing machines work on sequences of symbols written on their tape. Depending on the alphabet a Turing machine uses, these sequences of symbols can represent various inputs and outputs. Of particular interest, of course, are Turing machines which compute *arithmetical* functions, i.e., functions of natural numbers. A simple way to represent positive integers is by coding them as sequences of a single symbol I. If $n \in \mathbb{N}$, let I^n be the empty se-

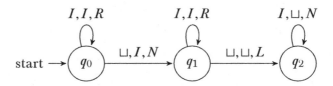

Figure 14.3: A machine computing $f(x,y) = x + y$

quence if $n = 0$, and otherwise the sequence consisting of exactly n I's.

Definition 14.11 (Computation). A Turing machine M *computes* the function $f : \mathbb{N}^k \to \mathbb{N}$ iff M halts on input

$$I^{n_1} \sqcup I^{n_2} \sqcup \ldots \sqcup I^{n_k}$$

with output $I^{f(n_1,\ldots,n_k)}$.

Example 14.12. *Addition:* Let's build a machine that computes the function $f(n,m) = n + m$. This requires a machine that starts with two blocks of I's of length n and m on the tape, and halts with one block consisting of $n+m$ I's. The two input blocks of I's are separated by a $\sqcup$, so one method would be to write a stroke on the square containing the $\sqcup$, and erase the last I.

In Example 14.4, we gave an example of a Turing machine that takes as input a sequence of I's and halts with a sequence of twice as many I's on the tape—the doubler machine. However, because the output contains $\sqcup$'s to the left of the doubled block of I's, it does not actually compute the function $f(x) = 2x$, as you might have assumed. We'll describe two ways of fixing that.

Example 14.13. The machine in Figure 14.4 computes the function $f(x) = 2x$. Instead of erasing the input and writing two I's at the far right for every I in the input as the machine from Example 14.4 does, this machine adds a single I to the right for

14.5. UNARY REPRESENTATION OF NUMBERS

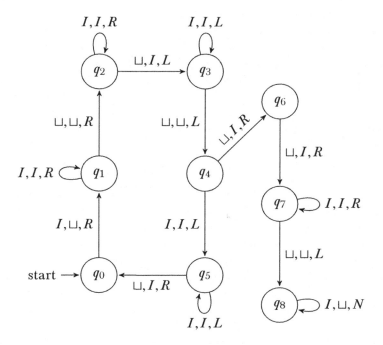

Figure 14.4: A machine computing $f(x) = 2x$

every I in the input. It has to keep track of where the input ends, so it leaves a ⊔ between the input and the added strokes, which it fills with a I at the very end. And we have to "remember" where we are in the input, so we temporarily replace a I in the input block by a ⊔.

Example 14.14. A second possibility for computing $f(x) = 2x$ is to keep the original doubler machine, but add states and instructions at the end which move the doubled block of strokes to the far left of the tape. The machine in Figure 14.5 does just this last part: started on a tape consisting of a block of ⊔'s followed by a block of I's (and the head positioned anywhere in the block

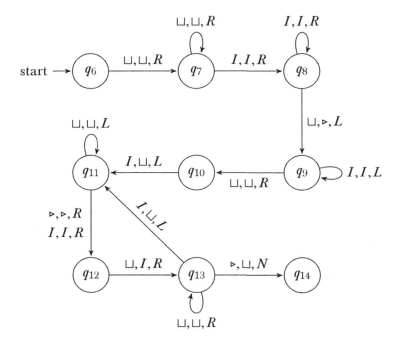

Figure 14.5: Moving a block of I's to the left

of $\sqcup$'s), it erases the I's one at a time and writes them at the beginning of the tape. In order to be able to tell when it is done, it first marks the end of the block of I's with a ▷ symbol, which gets deleted at the end. We've started numbering the states at q_6, so they can be added to the doubler machine. All you'll need is an additional instruction $\delta(q_5, \sqcup) = \langle q_6, \sqcup, N \rangle$, i.e., an arrow from q_5 to q_6 labelled $\sqcup, \sqcup, N$.

Definition 14.15. A Turing machine M computes the partial function $f \colon \mathbb{N}^k \to \mathbb{N}$ iff,

1. M halts on input $I^{n_1} \frown \sqcup \frown \ldots \frown \sqcup \frown I^{n_k}$ with output I^m if $f(n_1, \ldots, n_k) = m$.

2. M does not halt at all, or with an output that is not a single block of I's if $f(n_1, \ldots, n_k)$ is undefined.

14.6 Halting States

Although we have defined our machines to halt only when there is no instruction to carry out, common representations of Turing machines have a dedicated *halting state h*, such that $h \in Q$.

The idea behind a halting state is simple: when the machine has finished operation (it is ready to accept input, or has finished writing the output), it goes into a state h where it halts. Some machines have two halting states, one that accepts input and one that rejects input.

Example 14.16. *Halting States.* To elucidate this concept, let us begin with an alteration of the even machine. Instead of having the machine halt in state q_0 if the input is even, we can add an instruction to send the machine into a halting state.

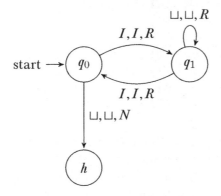

Let us further expand the example. When the machine determines that the input is odd, it never halts. We can alter the machine to include a *reject* state by replacing the looping instruction with an instruction to go to a reject state r.

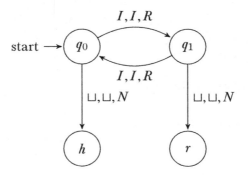

Adding a dedicated halting state can be advantageous in cases like this, where it makes explicit when the machine accepts/rejects certain inputs. However, it is important to note that no computing power is gained by adding a dedicated halting state. Similarly, a less formal notion of halting has its own advantages. The definition of halting used so far in this chapter makes the proof of the *Halting Problem* intuitive and easy to demonstrate. For this reason, we continue with our original definition.

14.7 Disciplined Machines

In section section 14.6, we considered Turing machines that have a single, designated halting state h—such machines are guaranteed to halt, if they halt at all, in state h. In this way, machines with a single halting state are more "disciplined" than we allow Turing machines in general to be. There are other restrictions we might impose on the behavior of Turing machines. For instance, we also have not prohibited Turing machines from ever erasing the tape-end marker on square 0, or to attempt to move left from square 0. (Our definition states that the head simply stays on square 0 in this case; other definitions have the machine halt.) It is likewise sometimes desirable to be able to assume that a Turing machine, if it halts at all, halts on square 1.

Definition 14.17. A Turing machine M is *disciplined* iff

1. it has a designated single halting state h,

2. it halts, if it halts at all, while scanning square 1,

3. it never erases the ▷ symbol on square 0, and

4. it never attempts to move left from square 0.

We have already discussed that any Turing machine can be changed into one with the same behavior but with a designated halting state. This is done simply by adding a new state h, and adding an instruction $\delta(q,\sigma) = \langle h, \sigma, N \rangle$ for any pair $\langle q, \sigma \rangle$ where the original δ is undefined. It is true, although tedious to prove, that any Turing machine M can be turned into a disciplined Turing machine M' which halts on the same inputs and produces the same output. For instance, if the Turing machine halts and is not on square 1, we can add some instructions to make the head move left until it finds the tape-end marker, then move one square to the right, then halt. We'll leave you to think about how the other conditions can be dealt with.

Example 14.18. In Figure 14.6, we turn the addition machine from Example 14.12 into a disciplined machine.

Proposition 14.19. *For every Turing machine M, there is a disciplined Turing machine M' which halts with output O if M halts with output O, and does not halt if M does not halt. In particular, any function $f: \mathbb{N}^n \to \mathbb{N}$ computable by a Turing machine is also computable by a disciplined Turing machine.*

14.8 Combining Turing Machines

The examples of Turing machines we have seen so far have been fairly simple in nature. But in fact, any problem that can be solved

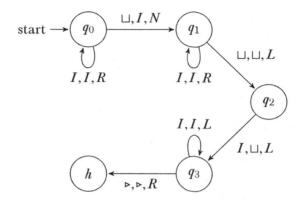

Figure 14.6: A disciplined addition machine

with any modern programming language can also be solved with Turing machines. To build more complex Turing machines, it is important to convince ourselves that we can combine them, so we can build machines to solve more complex problems by breaking the procedure into simpler parts. If we can find a natural way to break a complex problem down into constituent parts, we can tackle the problem in several stages, creating several simple Turing machines and combining them into one machine that can solve the problem. This point is especially important when tackling the Halting Problem in the next section.

How do we combine Turing machines $M = \langle Q, \Sigma, q_0, \delta \rangle$ and $M' = \langle Q', \Sigma', q_0', \delta' \rangle$? We now use the configuration of the tape after M has halted as the input configuration of a run of machine M'. To get a single Turing machine $M \frown M'$ that does this, do the following:

1. Renumber (or relabel) all the states Q' of M' so that M and M' have no states in common ($Q \cap Q' = \emptyset$).

2. The states of $M \frown M'$ are $Q \cup Q'$.

14.8. COMBINING TURING MACHINES

3. The tape alphabet is $\Sigma \cup \Sigma'$.

4. The start state is q_0.

5. The transition function is the function δ'' given by:

$$\delta''(q,\sigma) = \begin{cases} \delta(q,\sigma) & \text{if } q \in Q \\ \delta'(q,\sigma) & \text{if } q \in Q' \\ \langle q'_0, \sigma, N \rangle & \text{if } q \in Q \text{ and } \delta(q,\sigma) \text{ is undefined} \end{cases}$$

The resulting machine uses the instructions of M when it is in a state $q \in Q$, the instructions of M' when it is in a state $q \in Q'$. When it is in a state $q \in Q$ and is scanning a symbol σ for which M has no transition (i.e., M would have halted), it enters the start state of M' (and leaves the tape contents and head position as it is).

Note that unless the machine M is disciplined, we don't know where the tape head is when M halts, so the halting configuration of M need not have the head scanning square 1. When combining machines, it's important to keep this in mind.

Example 14.20. *Combining Machines:* We'll design a machine which, when started on input consisting of two blocks of I's of length n and m, halts with a single block of $2(m+n)$ I's on the tape. In order to build this machine, we can combine two machines we are already familiar with: the addition machine, and the doubler. We begin by drawing a state diagram for the addition machine.

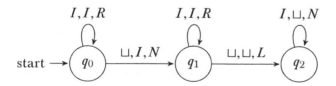

Instead of halting in state q_2, we want to continue operation in order to double the output. Recall that the doubler machine erases

the first stroke in the input and writes two strokes in a separate output. Let's add an instruction to make sure the tape head is reading the first stroke of the output of the addition machine.

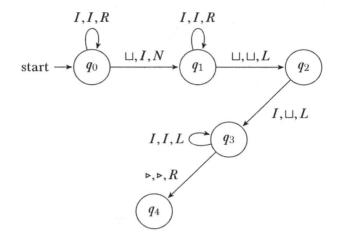

It is now easy to double the input—all we have to do is connect the doubler machine onto state q_4. This requires renaming the states of the doubler machine so that they start at q_4 instead of q_0—this way we don't end up with two starting states. The final diagram should look as in Figure 14.7.

Proposition 14.21. *If M and M' are disciplined and compute the functions $f: \mathbb{N}^k \to \mathbb{N}$ and $f': \mathbb{N} \to \mathbb{N}$, respectively, then $M \frown M'$ is disciplined and computes $f' \circ f$.*

Proof. Since M is disciplined, when it halts with output $f(n_1, \ldots, n_k) = m$, the head is scanning square 1. If we now enter the start state of M', the machine will halt with output $f'(m)$, again scanning square 1. The other conditions of Definition 14.17 are also satisfied. □

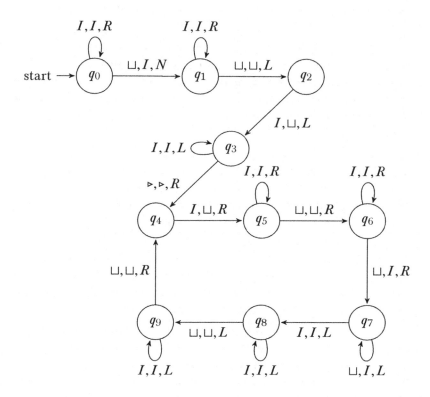

Figure 14.7: Combining adder and doubler machines

14.9 Variants of Turing Machines

There are in fact many possible ways to define Turing machines, of which ours is only one. In some ways, our definition is more liberal than others. We allow arbitrary finite alphabets, a more restricted definition might allow only two tape symbols, I and ⊔. We allow the machine to write a symbol to the tape and move at the same time, other definitions allow either writing or moving. We allow the possibility of writing without moving the tape head, other definitions leave out the N "instruction." In other ways, our definition is more restrictive. We assumed that the tape is infinite in one direction only, other definitions allow the tape to be infinite both to the left and the right. In fact, one can even allow any number of separate tapes, or even an infinite grid of squares. We represent the instruction set of the Turing machine by a transition function; other definitions use a transition relation where the machine has more than one possible instruction in any given situation.

This last relaxation of the definition is particularly interesting. In our definition, when the machine is in state q reading symbol σ, $\delta(q,\sigma)$ determines what the new symbol, state, and tape head position is. But if we allow the instruction set to be a relation between current state-symbol pairs $\langle q,\sigma \rangle$ and new state-symbol-direction triples $\langle q',\sigma',D \rangle$, the action of the Turing machine may not be uniquely determined—the instruction relation may contain both $\langle q,\sigma,q',\sigma',D \rangle$ and $\langle q,\sigma,q'',\sigma'',D' \rangle$. In this case we have a *non-deterministic* Turing machine. These play an important role in computational complexity theory.

There are also different conventions for when a Turing machine halts: we say it halts when the transition function is undefined, other definitions require the machine to be in a special designated halting state. We have explained in section 14.6 why requiring a designated halting state is not a restriction which impacts what Turing machines can compute. Since the tapes of our Turing machines are infinite in one direction only, there are cases where a Turing machine can't properly carry out an instruction: if it reads the leftmost square and is supposed to move left. According to our definition, it just stays put instead of "falling off", but we could have defined it so that it halts when that happens. This definition is also equivalent: we could simulate the behavior of a Turing machine that halts when it attempts to move left from square 0 by deleting every transition $\delta(q,\triangleright) = \langle q',\sigma,L \rangle$—then instead of attempting to move left on $\triangleright$ the machine halts.[1]

---

[1] This doesn't *quite* work, since nothing prevents us from writing and reading $\triangleright$ on squares other than square 0 (see Example 14.14). We can get around

There are also different ways of representing numbers (and hence the input-output function computed by a Turing machine): we use unary representation, but you can also use binary representation. This requires two symbols in addition to ⊔ and ▷.

Now here is an interesting fact: none of these variations matters as to which functions are Turing computable. *If a function is Turing computable according to one definition, it is Turing computable according to all of them.*

We won't go into the details of verifying this. Here's just one example: we gain no additional computing power by allowing a tape that is infinite in both directions, or multiple tapes. The reason is, roughly, that a Turing machine with a single one-way infinite tape can simulate multiple or two-way infinite tapes. E.g., using additional states and instructions, we can "translate" a program for a machine with multiple tapes or two-way infinite tape into one with a single one-way infinite tape. The translated machine can use the even squares for the squares of tape 1 (or the "positive" squares of a two-way infinite tape) and the odd squares for the squares of tape 2 (or the "negative" squares).

14.10 The Church-Turing Thesis

Turing machines are supposed to be a precise replacement for the concept of an effective procedure. Turing thought that anyone who grasped both the concept of an effective procedure and the concept of a Turing machine would have the intuition that anything that could be done via an effective procedure could be done by Turing machine. This claim is given support by the fact that all the other proposed precise replacements for the concept of an effective procedure turn out to be extensionally equivalent to the concept of a Turing machine —that is, they can compute exactly the same set of functions. This claim is called the *Church-Turing thesis.*

---

that by adding a second ▷′ symbol to use instead for such a purpose.

Definition 14.22 (Church-Turing thesis). The *Church-Turing Thesis* states that anything computable via an effective procedure is Turing computable.

The Church-Turing thesis is appealed to in two ways. The first kind of use of the Church-Turing thesis is an excuse for laziness. Suppose we have a description of an effective procedure to compute something, say, in "pseudo-code." Then we can invoke the Church-Turing thesis to justify the claim that the same function is computed by some Turing machine, even if we have not in fact constructed it.

The other use of the Church-Turing thesis is more philosophically interesting. It can be shown that there are functions which cannot be computed by Turing machines. From this, using the Church-Turing thesis, one can conclude that it cannot be effectively computed, using any procedure whatsoever. For if there were such a procedure, by the Church-Turing thesis, it would follow that there would be a Turing machine for it. So if we can prove that there is no Turing machine that computes it, there also can't be an effective procedure. In particular, the Church-Turing thesis is invoked to claim that the so-called halting problem not only cannot be solved by Turing machines, it cannot be effectively solved at all.

Summary

A **Turing machine** is a kind of idealized computation mechanism. It consists of a one-way infinite **tape**, divided into squares, each of which can contain a **symbol** from a pre-determined alphabet. The machine operates by moving a **read-write head** along the tape. It may also be in one of a pre-determined number of **states**. The actions of the read-write head are determined by a set of instructions; each instruction is conditional on the machine being in a certain state and reading a certain symbol, and specifies which symbol the machine will write onto the current square, whether it will move the read-write head one square left, right, or stay put, and which state it will switch to. If the tape contains a certain **input**, represented as a sequence of symbols on the tape, and the machine is put into the designated start state with the read-write head reading the leftmost square of the input, the instruction set will step-wise determine a sequence of **configurations** of the machine: content of tape, position of read-write head, and state of the machine. Should the machine encounter a configuration in which the instruction set does not contain an instruction for the current symbol read/state combination, the machine **halts**, and the content of the tape is the **output**.

Numbers can very easily be represented as sequences of strokes on the Tape of a Turing machine. We say a function $\mathbb{N} \to \mathbb{N}$ is **Turing computable** if there is a Turing machine which, whenever it is started on the unary representation of n as input, eventually halts with its tape containing the unary representation of $f(n)$ as output. Many familiar arithmetical functions are easily (or not-so-easily) shown to be Turing computable. Many other models of computation other than Turing machines have been proposed; and it has always turned out that the arithmetical functions computable there are also Turing computable. This is seen as support for the **Church-Turing Thesis**, that every arithmetical function that can effectively be computed is Turing computable.

Problems

Problem 14.1. Choose an arbitary input and trace through the configurations of the doubler machine in Example 14.4.

Problem 14.2. Design a Turing-machine with alphabet $\{\triangleright, \sqcup, A, B\}$ that accepts, i.e., halts on, any string of A's and B's where the number of A's is the same as the number of B's *and* all the A's precede all the B's, and rejects, i.e., does not halt on, any string where the number of A's is not equal to the number of B's or the A's do not precede all the B's. (E.g., the machine should accept *AABB*, and *AAABBB*, but reject both *AAB* and *AABBAABB*.)

Problem 14.3. Design a Turing-machine with alphabet $\{\triangleright, \sqcup, A, B\}$ that takes as input any string α of A's and B's and duplicates them to produce an output of the form $\alpha\alpha$. (E.g. input *ABBA* should result in output *ABBAABBA*).

Problem 14.4. *Alphabetical?:* Design a Turing-machine with alphabet $\{\triangleright, \sqcup, A, B\}$ that when given as input a finite sequence of A's and B's checks to see if all the A's appear to the left of all the B's or not. The machine should leave the input string on the tape, and either halt if the string is "alphabetical", or loop forever if the string is not.

Problem 14.5. *Alphabetizer:* Design a Turing-machine with alphabet $\{\triangleright, \sqcup, A, B\}$ that takes as input a finite sequence of A's and B's rearranges them so that all the A's are to the left of all the B's. (e.g., the sequence *BABAA* should become the sequence *AAABB*, and the sequence *ABBABB* should become the sequence *AABBBB*).

Problem 14.6. Give a definition for when a Turing machine M computes the function $f: \mathbb{N}^k \to \mathbb{N}^m$.

Problem 14.7. Trace through the configurations of the machine from Example 14.12 for input $\langle 3, 2 \rangle$. What happens if the machine computes $0 + 0$?

Problem 14.8. *Subtraction:* Design a Turing machine that when given an input of two non-empty strings of strokes of length n and m, where $n > m$, computes the function $f(n, m) = n - m$.

Problem 14.9. *Equality:* Design a Turing machine to compute the following function:

$$\text{equality}(n, m) = \begin{cases} 1 & \text{if } n = m \\ 0 & \text{if } n \neq m \end{cases}$$

where n and $m \in \mathbb{Z}^+$.

Problem 14.10. Design a Turing machine to compute the function $\min(x, y)$ where x and y are positive integers represented on the tape by strings of I's separated by a ⊔. You may use additional symbols in the alphabet of the machine.

The function min selects the smallest value from its arguments, so $\min(3,5) = 3$, $\min(20,16) = 16$, and $\min(4,4) = 4$, and so on.

Problem 14.11. Give a disciplined machine that computes $f(x) = x + 1$.

Problem 14.12. Find a disciplined machine which, when started on input I^n produces output $I^n \frown ⊔ \frown I^n$.

Problem 14.13. Give a disciplined Turing machine computing $f(x) = x + 2$ by taking the machine M from problem 14.11 and construct $M \frown M$.

CHAPTER 15
Undecidability

15.1 Introduction

It might seem obvious that not every function, even every arithmetical function, can be computable. There are just too many, whose behavior is too complicated. Functions defined from the decay of radioactive particles, for instance, or other chaotic or random behavior. Suppose we start counting 1-second intervals from a given time, and define the function $f(n)$ as the number of particles in the universe that decay in the n-th 1-second interval after that initial moment. This seems like a candidate for a function we cannot ever hope to compute.

But it is one thing to not be able to imagine how one would compute such functions, and quite another to actually prove that they are uncomputable. In fact, even functions that seem hopelessly complicated may, in an abstract sense, be computable. For instance, suppose the universe is finite in time—some day, in the very distant future the universe will contract into a single point, as some cosmological theories predict. Then there is only a finite (but incredibly large) number of seconds from that initial moment for which $f(n)$ is defined. And any function which is defined for only finitely many inputs is computable: we could list the outputs in one big table, or code it in one very big Turing machine state transition diagram.

We are often interested in special cases of functions whose

15.1. INTRODUCTION

values give the answers to yes/no questions. For instance, the question "is n a prime number?" is associated with the function

$$\text{isprime}(n) = \begin{cases} 1 & \text{if } n \text{ is prime} \\ 0 & \text{otherwise.} \end{cases}$$

We say that a yes/no question can be *effectively decided*, if the associated 1/0-valued function is effectively computable.

To prove mathematically that there are functions which cannot be effectively computed, or problems that cannot effectively decided, it is essential to fix a specific model of computation, and show that there are functions it cannot compute or problems it cannot decide. We can show, for instance, that not every function can be computed by Turing machines, and not every problem can be decided by Turing machines. We can then appeal to the Church-Turing thesis to conclude that not only are Turing machines not powerful enough to compute every function, but no effective procedure can.

The key to proving such negative results is the fact that we can assign numbers to Turing machines themselves. The easiest way to do this is to enumerate them, perhaps by fixing a specific way to write down Turing machines and their programs, and then listing them in a systematic fashion. Once we see that this can be done, then the existence of Turing-uncomputable functions follows by simple cardinality considerations: the set of functions from $\mathbb{N}$ to $\mathbb{N}$ (in fact, even just from $\mathbb{N}$ to $\{0,1\}$) are uncountable, but since we can enumerate all the Turing machines, the set of Turing-computable functions is only countably infinite.

We can also define *specific* functions and problems which we can prove to be uncomputable and undecidable, respectively. One such problem is the so-called *Halting Problem*. Turing machines can be finitely described by listing their instructions. Such a description of a Turing machine, i.e., a Turing machine program, can of course be used as input to another Turing machine. So we can consider Turing machines that decide questions about other Turing machines. One particularly interesting question is this: "Does the given Turing machine eventually halt when started on input n?" It would be nice if there were a Turing machine that could decide this question: think of it as a quality-control Turing machine which ensures that Turing machines don't get caught in infinite loops and such. The interesting fact, which Turing proved, is that there cannot be such a Turing machine. There cannot be a single Turing machine which, when started on input consisting of a description of a Turing machine M and some number n, will always halt with either output 1 or 0 according to whether M machine would have halted when started on input n or not.

Once we have examples of specific undecidable problems we can use them to show that other problems are undecidable, too. For instance, one celebrated undecidable problem is the question, "Is the first-order formula A valid?". There is no Turing machine which, given as input a first-order formula A, is guaranteed to halt with output 1 or 0 according to whether A is valid or not. Historically, the question of finding a procedure to effectively solve this problem was called simply "the" decision problem; and so we say that the decision problem is unsolvable. Turing and Church proved this result independently at around the same time, so it is also called the Church-Turing Theorem.

15.2 Enumerating Turing Machines

We can show that the set of all Turing machines is countable. This follows from the fact that each Turing machine can be finitely described. The set of states and the tape vocabulary are finite sets. The transition function is a partial function from $Q \times \Sigma$ to $Q \times \Sigma \times \{L, R, N\}$, and so likewise can be specified by listing its values for the finitely many argument pairs for which it is defined.

This is true as far as it goes, but there is a subtle difference. The definition of Turing machines made no resriction on what elements the set of states and tape alphabet can have. So, e.g., for every real number, there technically is a Turing machine that

15.2. ENUMERATING TURING MACHINES

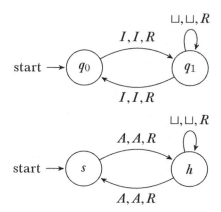

Figure 15.1: Variants of the *Even* machine

uses that number as a state. However, the *behavior* of the Turing machine is independent of which objects serve as states and vocabulary. Consider the two Turing machines in Figure 15.1. These two diagrams correspond to two machines, M with the tape alphabet $\Sigma = \{\triangleright, \sqcup, I\}$ and set of states $\{q_0, q_1\}$, and M' with alphabet $\Sigma' = \{\triangleright, \sqcup, A\}$ and states $\{s, h\}$. But their instructions are otherwise the same: M will halt on a sequence of n I's iff n is even, and M' will halt on a sequence of n A's iff n is even. All we've done is rename I to A, q_0 to s, and q_1 to h. This example generalizes: we can think of Turing machines as the same as long as one results from the other by such a renaming of symbols and states. In fact, we can simply think of the symbols and states of a Turing machine as positive integers: instead of σ_0 think 1, instead of σ_1 think 2, etc.; $\triangleright$ is 1, $\sqcup$ is 2, etc. In this way, the *Even* machine becomes the machine depicted in Figure 15.2. We might call a Turing machine with states and symbols that are positive integers a *standard* machine, and only consider standard machines from now on.[1]

---

[1] The terminology "standard machine" is not standard.

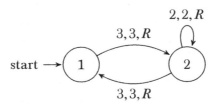

Figure 15.2: A standard *Even* machine

We wanted to show that the set of Turing machines is countable, and with the above considerations in mind, it is enough to show that the set of standard Turing machines is countable. Suppose we are given a standard Turing machine $M = \langle Q, \Sigma, q_0, \delta \rangle$. How could we describe it using a finite string of positive integers? We'll first list the number of states, the states themselves, the number of symbols, the symbols themselves, and the starting state. (Remember, all of these are positive integers, since M is a standard machine.) What about δ? The set of possible arguments, i.e., pairs $\langle q, \sigma \rangle$, is finite, since Q and Σ are finite. So the information in δ is simply the finite list of all 5-tuples $\langle q, \sigma, q', \sigma', d \rangle$ where $\delta(q, \sigma) = \langle q', \sigma', D \rangle$, and d is a number that codes the direction D (say, 1 for L, 2 for R, and 3 for N).

In this way, every standard Turing machine can be described by a finite list of positive integers, i.e., as a sequence $s_M \in (\mathbb{Z}^+)^*$. For instance, the standard *Even* machine is coded by the sequence

$$2, \underbrace{1,2}_{Q}, 3, \underbrace{1,2,3,1}_{\Sigma}, 1, \underbrace{3,2,3,2}_{\delta(1,3)=\langle 2,3,R \rangle}, \underbrace{2,2,2,2,2}_{\delta(2,2)=\langle 2,2,R \rangle}, \underbrace{2,3,1,3,2}_{\delta(2,3)=\langle 1,3,R \rangle}.$$

Theorem 15.1. *There are functions from $\mathbb{N}$ to $\mathbb{N}$ which are not Turing computable.*

Proof. We know that the set of finite sequences of positive integers $(\mathbb{Z}^+)^*$ is countable (problem 4.7). This gives us that the set of descriptions of standard Turing machines, as a subset of $(\mathbb{Z}^+)^*$, is itself enumerable. Every Turing computable function $\mathbb{N}$ to $\mathbb{N}$ is computed by some (in fact, many) Turing machines. By renaming its states and symbols to positive integers (in particular, ▷ as 1, ⊔ as 2, and I as 3) we can see that every Turing computable function is computed by a standard Turing machine. This means that the set of all Turing computable functions from $\mathbb{N}$ to $\mathbb{N}$ is also enumerable.

On the other hand, the set of all functions from $\mathbb{N}$ to $\mathbb{N}$ is not countable (problem 4.21). If all functions were computable by some Turing machine, we could enumerate the set of all functions by listing all the descriptions of Turing machines that compute them. So there are some functions that are not Turing computable. □

15.3 Universal Turing Machines

In section 15.2 we discussed how every Turing machine can be described by a finite sequence of integers. This sequence encodes the states, alphabet, start state, and instructions of the Turing machine. We also pointed out that the set of all of these descriptions is countable. Since the set of such descriptions is countably infinite, this means that there is a surjective function from $\mathbb{N}$ to these descriptions. Such a surjective function can be obtained, for instance, using Cantor's zig-zag method. It gives us a way of enumerating all (descriptions) of Turing machines. If we fix one such enumeration, it now makes sense to talk of the 1st, 2nd, ..., eth Turing machine. These numbers are called *indices*.

Definition 15.2. If M is the eth Turing machine (in our fixed enumeration), we say that e is an *index* of M. We write M_e for the eth Turing machine.

A machine may have more than one index, e.g., two descriptions of M may differ in the order in which we list its instructions, and these different descriptions will have different indices.

Importantly, it is possible to give the enumeration of Turing machine descriptions in such a way that we can effectively compute the description of M from its index, and to effectively compute an index of a machine M from its description. By the Church-Turing thesis, it is then possible to find a Turing machine which recovers the description of the Turing machine with index e and writes the corresponding description on its tape as output. The description would be a sequence of blocks of I's (representing the positive integers in the sequence describing M_e).

Given this, it now becomes natural to ask: what functions of Turing machine indices are themselves computable by Turing machines? What properties of Turing machine indices can be decided by Turing machines? An example: the function that maps an index e to the number of states the Turing machine with index e has, is computable by a Turing machine. Here's what such a Turing machine would do: started on a tape containing a single block of e I's, it would first decode e into its description. The description is now represented by a sequence of blocks of I's on the tape. Since the first element in this sequence is the number of states. So all that has to be done now is to erase everything but the first block of I's and then halt.

A remarkable result is the following:

Theorem 15.3. *There is a universal Turing machine U which, when started on input $\langle e, n \rangle$*

1. *halts iff M_e halts on input n, and*

2. *if M_e halts with output m, so does U.*

U thus computes the function $f : \mathbb{N} \times \mathbb{N} \twoheadrightarrow \mathbb{N}$ given by $f(e, n) = m$ if M_e started on input n halts with output m, and undefined otherwise.

15.3. UNIVERSAL TURING MACHINES

Proof. To actually produce U is basically impossible, since it is an extremely complicated machine. But we can describe in outline how it works, and then invoke the Church-Turing thesis. When it starts, U's tape contains a block of e I's followed by a block of n I's. It first "decodes" the index e to the right of the input n. This is produces a list of numbers (i.e., blocks of I's separated by ⊔'s) that describes the instructions of machine M_e. U then writes the number of the start state of M_e and the number 1 on the tape to the right of the description of M_e. (Again, these are represented in unary, as blocks of I's.) Next, it copies the input (block of n I's) to the right—but it replaces each I by a block of three I's (remember, the number of the I symbol is 3, 1 being the number of ▷ and 2 being the number of ⊔). At the left end of this sequence of blocks (separated by ⊔ symbols on the tape of U), it writes a single I, the code for ▷.

U now has on its tape: the index e, the number n, the code number of the start state (the "current state"), the number of the initial head position 1 (the "current head position"), and the initial contents of the "tape" (a sequence of blocks of I's representing the code numbers of the symbols of M_e—the "symbols"—separated by ⊔'s).

It now simulates what M_e would do if started on input n, by doing the following:

1. Find the number k of the "current head position" (at the beginning, that's 1),

2. Move to the kth block in the "tape" to see what the "symbol" there is,

3. Find the instruction matching the current "state" and "symbol,"

4. Move back to the kth block on the "tape" and replace the "symbol" there with the code number of the symbol M_e would write,

5. Move the head to where it records the current "state" and replace the number there with the number of the new state,

6. Move to the place where it records the "tape position" and erase a *1* or add a *1* (if the instruction says to move left or right, respectively).

7. Repeat.[2]

If M_e started on input n never halts, then U also never halts, so its output is undefined.

If in step (3) it turns out that the description of M_e contains no instruction for the current "state"/"symbol" pair, then M_e would halt. If this happens, U erases the part of its tape to the left of the "tape." For each block of three *1*'s (representing a *1* on M_e's tape), it writes a *1* on the left end of its own tape, and successively erases the "tape." When this is done, U's tape contains a single block of *1*'s of length m.

If U encounters something other than a block of three *1*'s on the "tape," it immediately halts. Since U's tape in this case does not contain a single block of *1*'s, its output is not a natural number, i.e., $f(e, n)$ is undefined in this case. □

15.4 The Halting Problem

Assume we have fixed some enumeration of Turing machine descriptions. Each Turing machine thus receives an *index*: its place in the enumeration $M_1, M_2, M_3, \ldots$ of Turing machine descriptions.

We know that there must be non-Turing-computable functions: the set of Turing machine descriptions—and hence the set of Turing machines—is countable, but the set of all functions

---

[2]We're glossing over some subtle difficulties here. E.g., U may need some extra space when it increases the counter where it keeps track of the "current head position"—in that case it will have to move the entire "tape" to the right.

15.4. THE HALTING PROBLEM

from $\mathbb{N}$ to $\mathbb{N}$ is not. But we can find specific examples of noncomputable functions as well. One such function is the halting function.

Definition 15.4 (Halting function). The *halting function h* is defined as

$$h(e, n) = \begin{cases} 0 & \text{if machine } M_e \text{ does not halt for input } n \\ 1 & \text{if machine } M_e \text{ halts for input } n \end{cases}$$

Definition 15.5 (Halting problem). The *Halting Problem* is the problem of determining (for any e, n) whether the Turing machine M_e halts for an input of n strokes.

We show that h is not Turing-computable by showing that a related function, s, is not Turing-computable. This proof relies on the fact that anything that can be computed by a Turing machine can be computed by a disciplined Turing machine (section 14.7), and the fact that two Turing machines can be hooked together to create a single machine (section 14.8).

Definition 15.6. The function s is defined as

$$s(e) = \begin{cases} 0 & \text{if machine } M_e \text{ does not halt for input } e \\ 1 & \text{if machine } M_e \text{ halts for input } e \end{cases}$$

Lemma 15.7. *The function s is not Turing computable.*

Proof. We suppose, for contradiction, that the function s is Turing computable. Then there would be a Turing machine S that computes s. We may assume, without loss of generality, that when S halts, it does so while scanning the first square (i.e., that it is disciplined). This machine can be "hooked up" to another machine J, which halts if it is started on input 0 (i.e., if it reads ⊔ in the initial state while scanning the square to the right of the end-of-tape symbol), and otherwise wanders off to the right, never halting. $S \frown J$, the machine created by hooking S to J, is a Turing machine, so it is M_e for some e (i.e., it appears somewhere in the enumeration). Start M_e on an input of e Is. There are two possibilities: either M_e halts or it does not halt.

1. Suppose M_e halts for an input of e Is. Then $s(e) = 1$. So S, when started on e, halts with a single I as output on the tape. Then J starts with a I on the tape. In that case J does not halt. But M_e is the machine $S \frown J$, so it should do exactly what S followed by J would do (i.e., in this case, wander off to the right and never halt). So M_e cannot halt for an input of e I's.

2. Now suppose M_e does not halt for an input of e Is. Then $s(e) = 0$, and S, when started on input e, halts with a blank tape. J, when started on a blank tape, immediately halts. Again, M_e does what S followed by J would do, so M_e must halt for an input of e I's.

In each case we arrive at a contradiction with our assumption. This shows there cannot be a Turing machine S: s is not Turing computable. □

Theorem 15.8 (Unsolvability of the Halting Problem). *The halting problem is unsolvable, i.e., the function h is not Turing computable.*

Proof. Suppose h were Turing computable, say, by a Turing machine H. We could use H to build a Turing machine that computes s: First, make a copy of the input (separated by a ⊔ symbol). Then move back to the beginning, and run H. We can clearly make a machine that does the former (see problem 14.12), and if H existed, we would be able to "hook it up" to such a copier machine to get a new machine which would determine if M_e halts on input e, i.e., computes s. But we've already shown that no such machine can exist. Hence, h is also not Turing computable. □

15.5 The Decision Problem

We say that first-order logic is *decidable* iff there is an effective method for determining whether or not a given sentence is valid. As it turns out, there is no such method: the problem of deciding validity of first-order sentences is unsolvable.

In order to establish this important negative result, we prove that the decision problem cannot be solved by a Turing machine. That is, we show that there is no Turing machine which, whenever it is started on a tape that contains a first-order sentence, eventually halts and outputs either 1 or 0 depending on whether the sentence is valid or not. By the Church-Turing thesis, every function which is computable is Turing computable. So if this "validity function" were effectively computable at all, it would be Turing computable. If it isn't Turing computable, then, it also cannot be effectively computable.

Our strategy for proving that the decision problem is unsolvable is to reduce the halting problem to it. This means the following: We have proved that the function $h(e,w)$ that halts with output 1 if the Turing machine described by e halts on input w and outputs 0 otherwise, is not Turing computable. We will show that if there were a Turing machine that decides validity of first-order sentences, then there is also Turing machine that computes h. Since h cannot be computed by a Turing machine, there cannot be a Turing machine that decides validity either.

The first step in this strategy is to show that for every input w and a Turing machine M, we can effectively describe a sentence $T(M,w)$ representing the instruction set of M and the input w and a sentence $E(M,w)$ expressing "M eventually halts" such that:

$\vDash T(M,w) \to E(M,w)$ iff M halts for input w.

The bulk of our proof will consist in describing these sentences $T(M,w)$ and $E(M,w)$ and in verifying that $T(M,w) \to E(M,w)$ is valid iff M halts on input w.

15.6 Representing Turing Machines

In order to represent Turing machines and their behavior by a sentence of first-order logic, we have to define a suitable language. The language consists of two parts: predicate symbols for describing configurations of the machine, and expressions for numbering execution steps ("moments") and positions on the tape.

We introduce two kinds of predicate symbols, both of them 2-place: For each state q, a predicate symbol Q_q, and for each tape symbol σ, a predicate symbol S_σ. The former allow us to describe the state of M and the position of its tape head, the latter allow us to describe the contents of the tape.

In order to express the positions of the tape head and the number of steps executed, we need a way to express numbers. This is done using a constant symbol 0, and a 1-place function ′, the successor function. By convention it is written *after* its argument (and we leave out the parentheses). So 0 names the leftmost position on the tape as well as the time before the first execution step (the initial configuration), 0′ names the square to the right of the leftmost square, and the time after the first execution step, and so on. We also introduce a predicate symbol < to express both the ordering of tape positions (when it means "to the left of") and execution steps (then it means "before").

Once we have the language in place, we list the "axioms" of $T(M,w)$, i.e., the sentences which, taken together, describe the behavior of M when run on input w. There will be sentences which lay down conditions on 0, ′, and <, sentences that describes the input configuration, and sentences that describe what the configuration of M is after it executes a particular instruction.

15.6. REPRESENTING TURING MACHINES

Definition 15.9. Given a Turing machine $M = \langle Q, \Sigma, q_0, \delta \rangle$, the language $\mathcal{L}_M$ consists of:

1. A two-place predicate symbol $Q_q(x,y)$ for every state $q \in Q$. Intuitively, $Q_q(\overline{m}, \overline{n})$ expresses "after n steps, M is in state q scanning the mth square."

2. A two-place predicate symbol $S_\sigma(x,y)$ for every symbol $\sigma \in \Sigma$. Intuitively, $S_\sigma(\overline{m}, \overline{n})$ expresses "after n steps, the mth square contains symbol σ."

3. A constant symbol o

4. A one-place function symbol $\prime$

5. A two-place predicate symbol $<$

For each number n there is a canonical term $\overline{n}$, the *numeral* for n, which represents it in $\mathcal{L}_M$. $\overline{0}$ is o, $\overline{1}$ is o$'$, $\overline{2}$ is o$''$, and so on. More formally:

$$\overline{0} = o$$
$$\overline{n+1} = \overline{n}'$$

The sentences describing the operation of the Turing machine M on input $w = \sigma_{i_1} \ldots \sigma_{i_k}$ are the following:

1. Axioms describing numbers and $<$:

   a) A sentence that says that every number is less than its successor:
   $$\forall x \, x < x'$$

   b) A sentence that ensures that $<$ is transitive:
   $$\forall x \, \forall y \, \forall z \, ((x < y \land y < z) \to x < z)$$

2. Axioms describing the input configuration:

a) After 0 steps—before the machine starts—M is in the inital state q_0, scanning square 1:

$$Q_{q_0}(\overline{1},\overline{0})$$

b) The first $k+1$ squares contain the symbols $\triangleright$, $\sigma_{i_1}, \ldots, \sigma_{i_k}$:

$$S_{\triangleright}(\overline{0},\overline{0}) \wedge S_{\sigma_{i_1}}(\overline{1},\overline{0}) \wedge \cdots \wedge S_{\sigma_{i_k}}(\overline{k},\overline{0})$$

c) Otherwise, the tape is empty:

$$\forall x\,(\overline{k} < x \to S_{\sqcup}(x,\overline{0}))$$

3. Axioms describing the transition from one configuration to the next:

For the following, let $A(x,y)$ be the conjunction of all sentences of the form

$$\forall z\,(((z < x \vee x < z) \wedge S_\sigma(z,y)) \to S_\sigma(z,y'))$$

where $\sigma \in \Sigma$. We use $A(\overline{m},\overline{n})$ to express "other than at square m, the tape after $n+1$ steps is the same as after n steps."

a) For every instruction $\delta(q_i,\sigma) = \langle q_j, \sigma', R\rangle$, the sentence:

$$\forall x\, \forall y\, ((Q_{q_i}(x,y) \wedge S_\sigma(x,y)) \to$$
$$(Q_{q_j}(x',y') \wedge S_{\sigma'}(x,y') \wedge A(x,y)))$$

This says that if, after y steps, the machine is in state q_i scanning square x which contains symbol σ, then after $y+1$ steps it is scanning square $x+1$, is in state q_j, square x now contains σ', and every square other than x contains the same symbol as it did after y steps.

15.6. REPRESENTING TURING MACHINES

b) For every instruction $\delta(q_i, \sigma) = \langle q_j, \sigma', L \rangle$, the sentence:

$$\forall x \, \forall y \, ((Q_{q_i}(x', y) \wedge S_\sigma(x', y)) \rightarrow$$
$$(Q_{q_j}(x, y') \wedge S_{\sigma'}(x', y') \wedge A(x, y))) \wedge$$
$$\forall y \, ((Q_{q_i}(0, y) \wedge S_\sigma(0, y)) \rightarrow$$
$$(Q_{q_j}(0, y') \wedge S_{\sigma'}(0, y') \wedge A(0, y)))$$

Take a moment to think about how this works: now we don't start with "if scanning square $x \ldots$" but: "if scanning square $x + 1 \ldots$" A move to the left means that in the next step the machine is scanning square x. But the square that is written on is $x+1$. We do it this way since we don't have subtraction or a predecessor function.

Note that numbers of the form $x + 1$ are $1, 2, \ldots$, i.e., this doesn't cover the case where the machine is scanning square 0 and is supposed to move left (which of course it can't—it just stays put). That special case is covered by the second conjunction: it says that if, after y steps, the machine is scanning square 0 in state q_i and square 0 contains symbol σ, then after $y + 1$ steps it's still scanning square 0, is now in state q_j, the symbol on square 0 is σ', and the squares other than square 0 contain the same symbols they contained ofter y steps.

c) For every instruction $\delta(q_i, \sigma) = \langle q_j, \sigma', N \rangle$, the sentence:

$$\forall x \, \forall y \, ((Q_{q_i}(x, y) \wedge S_\sigma(x, y)) \rightarrow$$
$$(Q_{q_j}(x, y') \wedge S_{\sigma'}(x, y') \wedge A(x, y)))$$

Let $T(M, w)$ be the conjunction of all the above sentences for Turing machine M and input w.

In order to express that M eventually halts, we have to find a sentence that says "after some number of steps, the transition function will be undefined." Let X be the set of all pairs $\langle q, \sigma \rangle$ such that $\delta(q, \sigma)$ is undefined. Let $E(M, w)$ then be the sentence

$$\exists x \, \exists y \, (\bigvee_{\langle q, \sigma \rangle \in X} (Q_q(x, y) \land S_\sigma(x, y)))$$

If we use a Turing machine with a designated halting state h, it is even easier: then the sentence $E(M, w)$

$$\exists x \, \exists y \, Q_h(x, y)$$

expresses that the machine eventually halts.

Proposition 15.10. *If $m < k$, then $T(M, w) \vDash \overline{m} < \overline{k}$*

Proof. Exercise. □

15.7 Verifying the Representation

In order to verify that our representation works, we have to prove two things. First, we have to show that if M halts on input w, then $T(M, w) \to E(M, w)$ is valid. Then, we have to show the converse, i.e., that if $T(M, w) \to E(M, w)$ is valid, then M does in fact eventually halt when run on input w.

The strategy for proving these is very different. For the first result, we have to show that a sentence of first-order logic (namely, $T(M, w) \to E(M, w)$) is valid. The easiest way to do this is to give a derivation. Our proof is supposed to work for all M and w, though, so there isn't really a single sentence for which we have to give a derivation, but infinitely many. So the best we can do is to prove by induction that, whatever M and w look like, and however many steps it takes M to halt on input w, there will be a derivation of $T(M, w) \to E(M, w)$.

Naturally, our induction will proceed on the number of steps M takes before it reaches a halting configuration. In our inductive proof, we'll establish that for each step n of the run of M on input w, $T(M, w) \vDash C(M, w, n)$, where $C(M, w, n)$ correctly

15.7. VERIFYING THE REPRESENTATION

describes the configuration of M run on w after n steps. Now if M halts on input w after, say, n steps, $C(M,w,n)$ will describe a halting configuration. We'll also show that $C(M,w,n) \vDash E(M,w)$, whenever $C(M,w,n)$ describes a halting configuration. So, if M halts on input w, then for some n, M will be in a halting configuration after n steps. Hence, $T(M,w) \vDash C(M,w,n)$ where $C(M,w,n)$ describes a halting configuration, and since in that case $C(M,w,n) \vDash E(M,w)$, we get that $T(M,w) \vDash E(M,w)$, i.e., that $\vDash T(M,w) \to E(M,w)$.

The strategy for the converse is very different. Here we assume that $\vDash T(M,w) \to E(M,w)$ and have to prove that M halts on input w. From the hypothesis we get that $T(M,w) \vDash E(M,w)$, i.e., $E(M,w)$ is true in every structure in which $T(M,w)$ is true. So we'll describe a structure M in which $T(M,w)$ is true: its domain will be $\mathbb{N}$, and the interpretation of all the Q_q and S_σ will be given by the configurations of M during a run on input w. So, e.g., $M \vDash Q_q(\overline{m}, \overline{n})$ iff T, when run on input w for n steps, is in state q and scanning square m. Now since $T(M,w) \vDash E(M,w)$ by hypothesis, and since $M \vDash T(M,w)$ by construction, $M \vDash E(M,w)$. But $M \vDash E(M,w)$ iff there is some $n \in |M| = \mathbb{N}$ so that M, run on input w, is in a halting configuration after n steps.

Definition 15.11. Let $C(M,w,n)$ be the sentence

$$Q_q(\overline{m},\overline{n}) \wedge S_{\sigma_0}(\overline{0},\overline{n}) \wedge \cdots \wedge S_{\sigma_k}(\overline{k},\overline{n}) \wedge \forall x\, (\overline{k} < x \to S_\sqcup(x,\overline{n}))$$

where q is the state of M at time n, M is scanning square m at time n, square i contains symbol σ_i at time n for $0 \le i \le k$ and k is the right-most non-blank square of the tape at time 0, or the right-most square the tape head has visited after n steps, whichever is greater.

Lemma 15.12. *If M run on input w is in a halting configuration after n steps, then $C(M,w,n) \vDash E(M,w)$.*

Proof. Suppose that M halts for input w after n steps. There is some state q, square m, and symbol σ such that:

1. After n steps, M is in state q scanning square m on which σ appears.

2. The transition function $\delta(q,\sigma)$ is undefined.

$C(M,w,n)$ is the description of this configuration and will include the clauses $Q_q(\overline{m},\overline{n})$ and $S_\sigma(\overline{m},\overline{n})$. These clauses together imply $E(M,w)$:
$$\exists x\, \exists y\, (\bigvee_{\langle q,\sigma\rangle \in X} (Q_q(x,y) \wedge S_\sigma(x,y)))$$
since $Q_{q'}(\overline{m},\overline{n}) \wedge S_{\sigma'}(\overline{m},\overline{n}) \vDash \bigvee_{\langle q,\sigma\rangle \in X}(Q_q(\overline{m},\overline{n}) \wedge S_\sigma(\overline{m},\overline{n}))$, as $\langle q',\sigma' \rangle \in X$. □

So if M halts for input w, then there is some n such that $C(M,w,n) \vDash E(M,w)$. We will now show that for any time n, $T(M,w) \vDash C(M,w,n)$.

Lemma 15.13. *For each n, if M has not halted after n steps, $T(M,w) \vDash C(M,w,n)$.*

Proof. Induction basis: If $n = 0$, then the conjuncts of $C(M,w,0)$ are also conjuncts of $T(M,w)$, so entailed by it.

Inductive hypothesis: If M has not halted before the nth step, then $T(M,w) \vDash C(M,w,n)$. We have to show that (unless $C(M,w,n)$ describes a halting configuration), $T(M,w) \vDash C(M,w,n+1)$.

Suppose $n > 0$ and after n steps, M started on w is in state q scanning square m. Since M does not halt after n steps, there must be an instruction of one of the following three forms in the program of M:

1. $\delta(q,\sigma) = \langle q',\sigma',R \rangle$

2. $\delta(q,\sigma) = \langle q',\sigma',L \rangle$

15.7. VERIFYING THE REPRESENTATION

3. $\delta(q,\sigma) = \langle q',\sigma',N\rangle$

We will consider each of these three cases in turn.

1. Suppose there is an instruction of the form (1). By Definition 15.9(3a), this means that

$$\forall x\, \forall y\, ((Q_q(x,y) \land S_\sigma(x,y)) \to \\ (Q_{q'}(x',y') \land S_{\sigma'}(x,y') \land A(x,y)))$$

is a conjunct of $T(M,w)$. This entails the following sentence (universal instantiation, $\overline{m}$ for x and $\overline{n}$ for y):

$$(Q_q(\overline{m},\overline{n}) \land S_\sigma(\overline{m},\overline{n})) \to \\ (Q_{q'}(\overline{m}',\overline{n}') \land S_{\sigma'}(\overline{m},\overline{n}') \land A(\overline{m},\overline{n})).$$

By induction hypothesis, $T(M,w) \vDash C(M,w,n)$, i.e.,

$$Q_q(\overline{m},\overline{n}) \land S_{\sigma_0}(\overline{0},\overline{n}) \land \cdots \land S_{\sigma_k}(\overline{k},\overline{n}) \land \\ \forall x\,(\overline{k} < x \to S_\sqcup(x,\overline{n}))$$

Since after n steps, tape square m contains σ, the corresponding conjunct is $S_\sigma(\overline{m},\overline{n})$, so this entails:

$$Q_q(\overline{m},\overline{n}) \land S_\sigma(\overline{m},\overline{n})$$

We now get

$$Q_{q'}(\overline{m}',\overline{n}') \land S_{\sigma'}(\overline{m},\overline{n}') \land \\ S_{\sigma_0}(\overline{0},\overline{n}') \land \cdots \land S_{\sigma_k}(\overline{k},\overline{n}') \land \\ \forall x\,(\overline{k} < x \to S_\sqcup(x,\overline{n}'))$$

as follows: The first line comes directly from the consequent of the preceding conditional, by modus ponens. Each conjunct in the middle line—which excludes $S_{\sigma_m}(\overline{m},\overline{n}')$—follows from the corresponding conjunct in $C(M,w,n)$ together with $A(\overline{m},\overline{n})$.

If $m < k$, $T(M,w) \vdash \overline{m} < \overline{k}$ (Proposition 15.10) and by transitivity of $<$, we have $\forall x\,(\overline{k} < x \to \overline{m} < x)$. If $m = k$, then $\forall x\,(\overline{k} < x \to \overline{m} < x)$ by logic alone. The last line then follows from the corresponding conjunct in $C(M,w,n)$, $\forall x\,(\overline{k} < x \to \overline{m} < x)$, and $A(\overline{m},\overline{n})$. If $m < k$, this already is $C(M,w,n+1)$.

Now suppose $m = k$. In that case, after $n+1$ steps, the tape head has also visited square $k+1$, which now is the rightmost square visited. So $C(M,w,n+1)$ has a new conjunct, $S_\sqcup(\overline{k}',\overline{n}')$, and the last conjuct is $\forall x\,(\overline{k}' < x \to S_\sqcup(x,\overline{n}'))$. We have to verify that these two sentences are also implied.

We already have $\forall x\,(\overline{k} < x \to S_\sqcup(x,\overline{n}'))$. In particular, this gives us $\overline{k} < \overline{k}' \to S_\sqcup(\overline{k}',\overline{n}')$. From the axiom $\forall x\, x < x'$ we get $\overline{k} < \overline{k}'$. By modus ponens, $S_\sqcup(\overline{k}',\overline{n}')$ follows.

Also, since $T(M,w) \vdash \overline{k} < \overline{k}'$, the axiom for transitivity of $<$ gives us $\forall x\,(\overline{k}' < x \to S_\sqcup(x,\overline{n}'))$. (We leave the verification of this as an exercise.)

2. Suppose there is an instruction of the form (2). Then, by Definition 15.9(3b),

$$\forall x\, \forall y\, ((Q_q(x',y) \wedge S_\sigma(x',y)) \to$$
$$(Q_{q'}(x,y') \wedge S_{\sigma'}(x',y') \wedge A(x,y))) \wedge$$
$$\forall y\, ((Q_{q_i}(0,y) \wedge S_\sigma(0,y)) \to$$
$$(Q_{q_j}(0,y') \wedge S_{\sigma'}(0,y') \wedge A(0,y)))$$

15.7. VERIFYING THE REPRESENTATION

is a conjunct of $T(M,w)$. If $m > 0$, then let $l = m - 1$ (i.e., $m = l + 1$). The first conjunct of the above sentence entails the following:

$$(Q_q(\overline{l}',\overline{n}) \wedge S_\sigma(\overline{l}',\overline{n})) \to$$
$$(Q_{q'}(\overline{l},\overline{n}') \wedge S_{\sigma'}(\overline{l}',\overline{n}') \wedge A(\overline{l},\overline{n}))$$

Otherwise, let $l = m = 0$ and consider the following sentence entailed by the second conjunct:

$$((Q_{q_i}(0,\overline{n}) \wedge S_\sigma(0,\overline{n})) \to$$
$$(Q_{q_j}(0,\overline{n}') \wedge S_{\sigma'}(0,\overline{n}') \wedge A(0,\overline{n})))$$

Either sentence implies

$$Q_{q'}(\overline{l},\overline{n}') \wedge S_{\sigma'}(\overline{m},\overline{n}') \wedge$$
$$S_{\sigma_0}(\overline{0},\overline{n}') \wedge \cdots \wedge S_{\sigma_k}(\overline{k},\overline{n}') \wedge$$
$$\forall x\, (\overline{k} < x \to S_\sqcup(x,\overline{n}'))$$

as before. (Note that in the first case, $\overline{l}' \equiv \overline{l+1} \equiv \overline{m}$ and in the second case $\overline{l} \equiv 0$.) But this just is $C(M,w,n+1)$.

3. Case (3) is left as an exercise.

We have shown that for any n, $T(M,w) \vDash C(M,w,n)$. □

Lemma 15.14. *If M halts on input w, then $T(M,w) \to E(M,w)$ is valid.*

Proof. By Lemma 15.13, we know that, for any time n, the description $C(M,w,n)$ of the configuration of M at time n is entailed by $T(M,w)$. Suppose M halts after k steps. At that point, it will be scanning square m, for some $m \in \mathbb{N}$. Then $C(M,w,k)$ describes a halting configuration of M, i.e., it contains as conjuncts both $Q_q(\overline{m},\overline{k})$ and $S_\sigma(\overline{m},\overline{k})$ with $\delta(q,\sigma)$ undefined. Thus, by Lemma 15.12, $C(M,w,k) \vDash E(M,w)$. But since

$T(M,w) \vDash C(M,w,k)$, we have $T(M,w) \vDash E(M,w)$ and therefore $T(M,w) \to E(M,w)$ is valid. □

To complete the verification of our claim, we also have to establish the reverse direction: if $T(M,w) \to E(M,w)$ is valid, then M does in fact halt when started on input w.

Lemma 15.15. *If $\vDash T(M,w) \to E(M,w)$, then M halts on input w.*

Proof. Consider the $\mathscr{L}_M$-structure $\mathfrak{M}$ with domain $\mathbb{N}$ which interprets 0 as 0, ′ as the successor function, and < as the less-than relation, and the predicates Q_q and S_σ as follows:

$$Q_q^{\mathfrak{M}} = \{\langle m,n\rangle : \begin{array}{l}\text{started on } w, \text{ after } n \text{ steps,}\\ M \text{ is in state } q \text{ scanning square } m\end{array}\}$$

$$S_\sigma^{\mathfrak{M}} = \{\langle m,n\rangle : \begin{array}{l}\text{started on } w, \text{ after } n \text{ steps,}\\ \text{square } m \text{ of } M \text{ contains symbol } \sigma\end{array}\}$$

In other words, we construct the structure $\mathfrak{M}$ so that it describes what M started on input w actually does, step by step. Clearly, $\mathfrak{M} \vDash T(M,w)$. If $\vDash T(M,w) \to E(M,w)$, then also $\mathfrak{M} \vDash E(M,w)$, i.e.,

$$\mathfrak{M} \vDash \exists x\, \exists y\, (\bigvee_{\langle q,\sigma\rangle \in X} (Q_q(x,y) \wedge S_\sigma(x,y))).$$

As $|\mathfrak{M}| = \mathbb{N}$, there must be $m, n \in \mathbb{N}$ so that $\mathfrak{M} \vDash Q_q(\overline{m},\overline{n}) \wedge S_\sigma(\overline{m},\overline{n})$ for some q and σ such that $\delta(q,\sigma)$ is undefined. By the definition of $\mathfrak{M}$, this means that M started on input w after n steps is in state q and reading symbol σ, and the transition function is undefined, i.e., M has halted. □

15.8 The Decision Problem is Unsolvable

Theorem 15.16. *The decision problem is unsolvable: There is no Turing machine D, which when started on a tape that contains a sentence B*

15.8. THE DECISION PROBLEM IS UNSOLVABLE

of first-order logic as input, D eventually halts, and outputs 1 *iff B is valid and* 0 *otherwise.*

Proof. Suppose the decision problem were solvable, i.e., suppose there were a Turing machine D. Then we could solve the halting problem as follows. We construct a Turing machine E that, given as input the number e of Turing machine M_e and input w, computes the corresponding sentence $T(M_e, w) \to E(M_e, w)$ and halts, scanning the leftmost square on the tape. The machine $E \frown D$ would then, given input e and w, first compute $T(M_e, w) \to E(M_e, w)$ and then run the decision problem machine D on that input. D halts with output 1 iff $T(M_e, w) \to E(M_e, w)$ is valid and outputs 0 otherwise. By Lemma 15.15 and Lemma 15.14, $T(M_e, w) \to E(M_e, w)$ is valid iff M_e halts on input w. Thus, $E \frown D$, given input e and w halts with output 1 iff M_e halts on input w and halts with output 0 otherwise. In other words, $E \frown D$ would solve the halting problem. But we know, by Theorem 15.8, that no such Turing machine can exist. □

Corollary 15.17. *It is undecidable if an arbitrary sentence of first-order logic is satisfiable.*

Proof. Suppose satisfiability were decidable by a Turing machine S. Then we could solve the decision problem as follows: Given a sentence B as input, move B to the right one square. Return to square 1 and write the symbol ¬.

Now run the Turing machine S. It eventually halts with output either 1 (if ¬B is satisfiable) or 0 (if ¬B is unsatisfiable) on the tape. If there is a I on square 1, erase it; if square 1 is empty, write a I, then halt.

This Turing machine always halts, and its output is 1 iff ¬B is unsatisfiable and 0 otherwise. Since B is valid iff ¬B is unsatisfiable, the machine outputs 1 iff B is valid, and 0 otherwise, i.e., it would solve the decision problem. □

So there is no Turing machine which always gives a correct "yes" or "no" answer to the question "Is B a valid sentence of first-order logic?" However, there *is* a Turing machine that always gives a correct "yes" answer—but simply does not halt if the answer is "no." This follows from the soundness and completeness theorem of first-order logic, and the fact that derivations can be effectively enumerated.

Theorem 15.18. *Validity of first-order sentences is semi-decidable: There is a Turing machine E, which when started on a tape that contains a sentence B of first-order logic as input, E eventually halts and outputs* 1 *iff B is valid, but does not halt otherwise.*

Proof. All possible derivations of first-order logic can be generated, one after another, by an effective algorithm. The machine E does this, and when it finds a derivation that shows that $\vdash B$, it halts with output 1. By the soundness theorem, if E halts with output 1, it's because $\vDash B$. By the completeness theorem, if $\vDash B$ there is a derivation that shows that $\vdash B$. Since E systematically generates all possible derivations, it will eventually find one that shows $\vdash B$, so will eventually halt with output 1. □

15.9 Trakthenbrot's Theorem

In section 15.6 we defined sentences $T(M,w)$ and $E(M,w)$ for a Turing machine M and input string w. Then we showed in Lemma 15.14 and Lemma 15.15 that $T(M,w) \to E(M,w)$ is valid iff M, started on input w, eventually halts. Since the Halting Problem is undecidable, this implies that validity and satisfiability of sentences of first-order logic is undecidable (Theorem 15.16 and Corollary 15.17).

But validity and satisfiability of sentences is defined for arbitrary structures, finite or infinite. You might suspect that it is easier to decide if a sentence is satisfiable in a finite structure (or valid in all finite structures). We can adapt the proof of the

15.9. TRAKTHENBROT'S THEOREM

unsolvability of the decision problem so that it shows this is not the case.

First, if you go back to the proof of Lemma 15.15, you'll see that what we did there is produce a model M of $T(M,w)$ which describes exactly what machine M does when started on input w. The domain of that model was $\mathbb{N}$, i.e., infinite. But if M actually halts on input w, we can build a finite model M' in the same way. Suppose M started on input w halts after k steps. Take as domain $|M'|$ the set $\{0,\ldots,n\}$, where n is the larger of k and the length of w, and let

$$\prime^{M'}(x) = \begin{cases} x+1 & \text{if } x < n \\ n & \text{otherwise,} \end{cases}$$

and $\langle x,y \rangle \in <^{M'}$ iff $x < y$ or $x = y = n$. Otherwise M' is defined just like M. By the definition of M', just like in the proof of Lemma 15.15, $M' \vDash T(M,w)$. And since we assumed that M halts on input w, $M' \vDash E(M,w)$. So, M' is a finite model of $T(M,w) \wedge E(M,w)$ (note that we've replaced $\rightarrow$ with $\wedge$).

We are halfway to a proof: we've shown that if M halts on input w, then $T(M,e) \wedge E(M,w)$ has a finite model. Unfortunately, the "only if" direction does not hold. For instance, if M after n steps is in state q and reads a symbol σ, and $\delta(q,\sigma) = \langle q,\sigma,N \rangle$, then the configuration after $n+1$ steps is exactly the same as the configuration after n steps (same state, same head position, same tape contents). But the machine never halts; it's in an infinite loop. The corresponding structure M' above satisfies $T(M,w)$ but not $E(M,w)$. (In it, the values of $\overline{n+l}$ are all the same, so it is finite). But by changing $T(M,w)$ in a suitable way we can rule out structures like this.

Consider the sentences describing the operation of the Turing machine M on input $w = \sigma_{i_1} \ldots \sigma_{i_k}$:

1. Axioms describing numbers and $<$ (just like in the definition of $T(M,w)$ in section 15.6).

2. Axioms describing the input configuration: just like in the definition of $T(M,w)$.

3. Axioms describing the transition from one configuration to the next:

   For the following, let $A(x,y)$ be as before, and let

   $$B(y) \equiv \forall x \, (x < y \rightarrow x \neq y).$$

   a) For every instruction $\delta(q_i, \sigma) = \langle q_j, \sigma', R \rangle$, the sentence:

   $$\forall x \, \forall y \, ((Q_{q_i}(x,y) \wedge S_\sigma(x,y)) \rightarrow$$
   $$(Q_{q_j}(x',y') \wedge S_{\sigma'}(x,y') \wedge A(x,y) \wedge B(y')))$$

   b) For every instruction $\delta(q_i, \sigma) = \langle q_j, \sigma', L \rangle$, the sentence

   $$\forall x \, \forall y \, ((Q_{q_i}(x',y) \wedge S_\sigma(x',y)) \rightarrow$$
   $$(Q_{q_j}(x,y') \wedge S_{\sigma'}(x',y') \wedge A(x,y))) \wedge$$
   $$\forall y \, ((Q_{q_i}(0,y) \wedge S_\sigma(0,y)) \rightarrow$$
   $$(Q_{q_j}(0,y') \wedge S_{\sigma'}(0,y') \wedge A(0,y) \wedge B(y')))$$

   c) For every instruction $\delta(q_i, \sigma) = \langle q_j, \sigma', N \rangle$, the sentence:

   $$\forall x \, \forall y \, ((Q_{q_i}(x,y) \wedge S_\sigma(x,y)) \rightarrow$$
   $$(Q_{q_j}(x,y') \wedge S_{\sigma'}(x,y') \wedge A(x,y) \wedge B(y')))$$

As you can see, the sentences describing the transitions of M are the same as the corresponding sentence in $T(M,w)$, except we add $B(y')$ at the end. $B(y')$ ensures that the number y' of the "next" configuration is different from all previous numbers $0, 0', \ldots$.

Let $T'(M,w)$ be the conjunction of all the above sentences for Turing machine M and input w.

Lemma 15.19. *If M started on input w halts, then $T'(M,w) \wedge E(M,w)$ has a finite model.*

Proof. Let M' be as in the proof of Lemma 15.15, except

$$|M'| = \{0,\ldots,n\},$$

$$\prime^{M'}(x) = \begin{cases} x+1 & \text{if } x < n \\ n & \text{otherwise,} \end{cases}$$

$$\langle x, y \rangle \in <^{M'} \text{ iff } x < y \text{ or } x = y = n,$$

where $n = \max(k, \text{len}(w))$ and k is the least number such that M started on input w has halted after k steps. We leave the verification that $M' \vDash T'(M,w) \wedge E(M,w)$ as an exercise. □

Lemma 15.20. *If $T'(M,w) \wedge E(M,w)$ has a finite model, then M started on input w halts.*

Proof. We show the contrapositive. Suppose that M started on w does not halt. If $T'(M,w) \wedge E(M,w)$ has no model at all, we are done. So assume M is a model of $T(M,w) \wedge E(M,w)$. We have to show that it cannot be finite.

We can prove, just like in Lemma 15.13, that if M, started on input w, has not halted after n steps, then $T'(M,w) \vDash C(M,w,n) \wedge B(\overline{n})$. Since M started on input w does not halt, $T'(M,w) \vDash C(M,w,n) \wedge B(\overline{n})$ for all $n \in \mathbb{N}$. Note that by Proposition 15.10, $T'(M,w) \vDash \overline{k} < \overline{n}$ for all $k < n$. Also $B(\overline{n}) \vDash \overline{k} < \overline{n} \to \overline{k} \neq \overline{n}$. So, $M \vDash \overline{k} \neq \overline{n}$ for all $k < n$, i.e., the infinitely many terms $\overline{k}$ must all have different values in M. But this requires that $|M|$ be infinite, so M cannot be a finite model of $T'(M,w) \wedge E(M,w)$. □

Theorem 15.21 (Trakthenbrot's Theorem). *It is undecidable if an arbitrary sentence of first-order logic has a finite model (i.e., is finitely* *satisfiable*).

Proof. Suppose there were a Turing machine F that decides the finite satisfiability problem. Then given any Turing machine M and input w, we could compute the sentence $T'(M,w) \wedge E(M,w)$, and use F to decide if it has a finite model. By Lemmas 15.19 and 15.20, it does iff M started on input w halts. So we could use F to solve the halting problem, which we know is unsolvable. □

Corollary 15.22. *There can be no derivation system that is sound and complete for* finite *validity, i.e., a derivation system which has* ⊢ B *iff* M ⊨ B *for every finite structure* M.

Proof. Exercise. □

Summary

Turing machines are determined by their instruction sets, which are finite sets of quintuples (for every state and symbol read, specify new state, symbol written, and movement of the head). The finite sets of quintuples are enumerable, so there is a way of associating a number with each Turing machine instruction set. The **index** of a Turing machine is the number associated with its instruction set under a fixed such schema. In this way we can "talk about" Turing machines indirectly—by talking about their indices.

One important problem about the behavior of Turing machines is whether they eventually halt. Let $h(e,n)$ be the function which $= 1$ if the Turing machine with index e halts when started on input n, and $= 0$ otherwise. It is called the **halting function**. The question of whether the halting function is itself Turing computable is called the **halting problem**. The answer is no: the halting problem is unsolvable. This is established using a diagonal argument.

The halting problem is only one example of a larger class of problems of the form "can X be accomplished using Turing machines." Another central problem of logic is the **decision problem for first-order logic**: is there a Turing machine that can decide if a given sentence is valid or not. This famous problem was also solved negatively: the decision problem is unsolvable. This is established by a reduction argument: we can associate with each Turing machine M and input w a first-order sentence $T(M,w) \to E(M,w)$ which is valid iff M halts when started on input w. If the decision problem were solvable, we could thus use it to solve the halting problem.

Problems

Problem 15.1. Can you think of a way to describe Turing machines that does not require that the states and alphabet symbols are explicitly listed? You may define your own notion of "standard" machine, but say something about why every Turing machine can be computed by a "standard" machine in your new sense.

Problem 15.2. The Three Halting (3-Halt) problem is the problem of giving a decision procedure to determine whether or not an arbitrarily chosen Turing Machine halts for an input of three I's on an otherwise blank tape. Prove that the 3-Halt problem is unsolvable.

Problem 15.3. Show that if the halting problem is solvable for Turing machine and input pairs M_e and n where $e \neq n$, then it is also solvable for the cases where $e = n$.

Problem 15.4. We proved that the halting problem is unsolvable if the input is a number e, which identifies a Turing machine M_e via an enumaration of all Turing machines. What if we allow the description of Turing machines from section 15.2 directly as input? Can there be a Turing machine which decides the halting problem but takes as input descriptions of Turing machines rather than indices? Explain why or why not.

Problem 15.5. Show that the *partial* function s' is defined as

$$s'(e) = \begin{cases} 1 & \text{if machine } M_e \text{ halts for input } e \\ \text{undefined} & \text{if machine } M_e \text{ does not halt for input } e \end{cases}$$

is Turing computable.

Problem 15.6. Prove Proposition 15.10. (Hint: use induction on $k - m$).

Problem 15.7. Complete case (3) of the proof of Lemma 15.13.

Problem 15.8. Give a derivation of $S_{\sigma_i}(\overline{i}, \overline{n}')$ from $S_{\sigma_i}(\overline{i}, \overline{n})$ and $A(m, n)$ (assuming $i \neq m$, i.e., either $i < m$ or $m < i$).

Problem 15.9. Give a derivation of $\forall x\, (\overline{k}' < x \to S_\sqcup(x, \overline{n}'))$ from $\forall x\, (\overline{k} < x \to S_\sqcup(x, \overline{n}'))$, $\forall x\, x < x'$, and $\forall x \forall y \forall z\, ((x < y \land y < z) \to x < z)$.)

Problem 15.10. Complete the proof of Lemma 15.19 by proving that $M' \vDash T(M, w) \land E(M, w)$.

Problem 15.11. Complete the proof of Lemma 15.20 by proving that if M, started on input w, has not halted after n steps, then $T'(M, w) \vDash B(\overline{n})$.

Problem 15.12. Prove Corollary 15.22. Observe that B is satisfied in every finite structure iff $\neg B$ is not finitely satisfiable. Explain why finite satisfiability is semi-decidable in the sense of Theorem 15.18. Use this to argue that if there were a derivation system for finite validity, then finite satisfiability would be decidable.

Emmy Noether
1882 - 1935

APPENDIX A
Proofs

A.1 Introduction

Based on your experiences in introductory logic, you might be comfortable with a derivation system—probably a natural deduction or Fitch style derivation system, or perhaps a proof-tree system. You probably remember doing proofs in these systems, either proving a formula or show that a given argument is valid. In order to do this, you applied the rules of the system until you got the desired end result. In reasoning *about* logic, we also prove things, but in most cases we are not using a derivation system. In fact, most of the proofs we consider are done in English (perhaps, with some symbolic language thrown in) rather than entirely in the language of first-order logic. When constructing such proofs, you might at first be at a loss—how do I prove something without a derivation system? How do I start? How do I know if my proof is correct?

Before attempting a proof, it's important to know what a proof is and how to construct one. As implied by the name, a *proof* is meant to show that something is true. You might think of this in terms of a dialogue—someone asks you if something is true, say, if every prime other than two is an odd number. To answer "yes" is not enough; they might want to know *why*. In this case, you'd give them a proof.

In everyday discourse, it might be enough to gesture at an

answer, or give an incomplete answer. In logic and mathematics, however, we want rigorous proof—we want to show that something is true beyond *any* doubt. This means that every step in our proof must be justified, and the justification must be cogent (i.e., the assumption you're using is actually assumed in the statement of the theorem you're proving, the definitions you apply must be correctly applied, the justifications appealed to must be correct inferences, etc.).

Usually, we're proving some statement. We call the statements we're proving by various names: propositions, theorems, lemmas, or corollaries. A proposition is a basic proof-worthy statement: important enough to record, but perhaps not particularly deep nor applied often. A theorem is a significant, important proposition. Its proof often is broken into several steps, and sometimes it is named after the person who first proved it (e.g., Cantor's Theorem, the Löwenheim-Skolem theorem) or after the fact it concerns (e.g., the completeness theorem). A lemma is a proposition or theorem that is used in the proof of a more important result. Confusingly, sometimes lemmas are important results in themselves, and also named after the person who introduced them (e.g., Zorn's Lemma). A corollary is a result that easily follows from another one.

A statement to be proved often contains assumptions that clarify which kinds of things we're proving something about. It might begin with "Let A be a formula of the form $B \to C$" or "Suppose $\Gamma \vdash A$" or something of the sort. These are *hypotheses* of the proposition, theorem, or lemma, and you may assume these to be true in your proof. They restrict what we're proving, and also introduce some names for the objects we're talking about. For instance, if your proposition begins with "Let A be a formula of the form $B \to C$," you're proving something about all formulas of a certain sort only (namely, conditionals), and it's understood that $B \to C$ is an arbitrary conditional that your proof will talk about.

A.2 Starting a Proof

But where do you even start?

You've been given something to prove, so this should be the last thing that is mentioned in the proof (you can, obviously, *announce* that you're going to prove it at the beginning, but you don't want to use it as an assumption). Write what you are trying to prove at the bottom of a fresh sheet of paper—this way you don't lose sight of your goal.

Next, you may have some assumptions that you are able to use (this will be made clearer when we talk about the *type* of proof you are doing in the next section). Write these at the top of the page and make sure to flag that they are assumptions (i.e., if you are assuming p, write "assume that p," or "suppose that p"). Finally, there might be some definitions in the question that you need to know. You might be told to use a specific definition, or there might be various definitions in the assumptions or conclusion that you are working towards. *Write these down and ensure that you understand what they mean.*

How you set up your proof will also be dependent upon the form of the question. The next section provides details on how to set up your proof based on the type of sentence.

A.3 Using Definitions

We mentioned that you must be familiar with all definitions that may be used in the proof, and that you can properly apply them. This is a really important point, and it is worth looking at in a bit more detail. Definitions are used to abbreviate properties and relations so we can talk about them more succinctly. The introduced abbreviation is called the *definiendum*, and what it abbreviates is the *definiens*. In proofs, we often have to go back to how the definiendum was introduced, because we have to exploit the logical structure of the definiens (the long version of which the defined term is the abbreviation) to get through our proof. By unpacking definitions, you're ensuring that you're getting to the heart of where the logical action is.

We'll start with an example. Suppose you want to prove the following:

Proposition A.1. *For any sets A and B, $A \cup B = B \cup A$.*

In order to even start the proof, we need to know what it means for two sets to be identical; i.e., we need to know what the "=" in that equation means for sets. Sets are defined to be identical whenever they have the same elements. So the definition we have to unpack is:

Definition A.2. Sets A and B are *identical*, $A = B$, iff every element of A is an element of B, and vice versa.

This definition uses A and B as placeholders for arbitrary sets. What it defines—the *definiendum*—is the expression "$A = B$" by giving the condition under which $A = B$ is true. This condition—"every element of A is an element of B, and vice versa"—is the *definiens*.[1] The definition specifies that $A = B$ is true if, and only if (we abbreviate this to "iff") the condition holds.

When you apply the definition, you have to match the A and B in the definition to the case you're dealing with. In our case, it means that in order for $A \cup B = B \cup A$ to be true, each $z \in A \cup B$ must also be in $B \cup A$, and vice versa. The expression $A \cup B$ in the proposition plays the role of A in the definition, and $B \cup A$ that of B. Since A and B are used both in the definition and in the statement of the proposition we're proving, but in different uses, you have to be careful to make sure you don't mix up the two. For instance, it would be a mistake to think that you could prove the proposition by showing that every element of A is an element

---

[1] In this particular case—and very confusingly!—when $A = B$, the sets A and B are just one and the same set, even though we use different letters for it on the left and the right side. But the ways in which that set is picked out may be different, and that makes the definition non-trivial.

A.3. USING DEFINITIONS

of B, and vice versa—that would show that $A = B$, not that $A \cup B = B \cup A$. (Also, since A and B may be any two sets, you won't get very far, because if nothing is assumed about A and B they may well be different sets.)

Within the proof we are dealing with set-theoretic notions such as union, and so we must also know the meanings of the symbol $\cup$ in order to understand how the proof should proceed. And sometimes, unpacking the definition gives rise to further definitions to unpack. For instance, $A \cup B$ is defined as $\{z : z \in A \text{ or } z \in B\}$. So if you want to prove that $x \in A \cup B$, unpacking the definition of $\cup$ tells you that you have to prove $x \in \{z : z \in A \text{ or } z \in B\}$. Now you also have to remember that $x \in \{z : \ldots z \ldots\}$ iff $\ldots x \ldots$. So, further unpacking the definition of the $\{z : \ldots z \ldots\}$ notation, what you have to show is: $x \in A$ or $x \in B$. So, "every element of $A \cup B$ is also an element of $B \cup A$" really means: "for every x, if $x \in A$ or $x \in B$, then $x \in B$ or $x \in A$." If we fully unpack the definitions in the proposition, we see that what we have to show is this:

Proposition A.3. *For any sets A and B: (a) for every x, if $x \in A$ or $x \in B$, then $x \in B$ or $x \in A$, and (b) for every x, if $x \in B$ or $x \in A$, then $x \in A$ or $x \in B$.*

What's important is that unpacking definitions is a necessary part of constructing a proof. Properly doing it is sometimes difficult: you must be careful to distinguish and match the variables in the definition and the terms in the claim you're proving. In order to be successful, you must know what the question is asking and what all the terms used in the question mean—you will often need to unpack more than one definition. In simple proofs such as the ones below, the solution follows almost immediately from the definitions themselves. Of course, it won't always be this simple.

A.4 Inference Patterns

Proofs are composed of individual inferences. When we make an inference, we typically indicate that by using a word like "so," "thus," or "therefore." The inference often relies on one or two facts we already have available in our proof—it may be something we have assumed, or something that we've concluded by an inference already. To be clear, we may label these things, and in the inference we indicate what other statements we're using in the inference. An inference will often also contain an explanation of *why* our new conclusion follows from the things that come before it. There are some common patterns of inference that are used very often in proofs; we'll go through some below. Some patterns of inference, like proofs by induction, are more involved (and will be discussed later).

We've already discussed one pattern of inference: unpacking, or applying, a definition. When we unpack a definition, we just restate something that involves the definiendum by using the definiens. For instance, suppose that we have already established in the course of a proof that $D = E$ (a). Then we may apply the definition of $=$ for sets and infer: "Thus, by definition from (a), every element of D is an element of E and vice versa."

Somewhat confusingly, we often do not write the justification of an inference when we actually make it, but before. Suppose we haven't already proved that $D = E$, but we want to. If $D = E$ is the conclusion we aim for, then we can restate this aim also by applying the definition: to prove $D = E$ we have to prove that every element of D is an element of E and vice versa. So our proof will have the form: (a) prove that every element of D is an element of E; (b) every element of E is an element of D; (c) therefore, from (a) and (b) by definition of $=$, $D = E$. But we would usually not write it this way. Instead we might write something like,

> We want to show $D = E$. By definition of $=$, this amounts to showing that every element of D is an el-

ement of E and vice versa.

(a) ... (a proof that every element of D is an element of E) ...

(b) ... (a proof that every element of E is an element of D) ...

Using a Conjunction

Perhaps the simplest inference pattern is that of drawing as conclusion one of the conjuncts of a conjunction. In other words: if we have assumed or already proved that p and q, then we're entitled to infer that p (and also that q). This is such a basic inference that it is often not mentioned. For instance, once we've unpacked the definition of $D = E$ we've established that every element of D is an element of E and vice versa. From this we can conclude that every element of E is an element of D (that's the "vice versa" part).

Proving a Conjunction

Sometimes what you'll be asked to prove will have the form of a conjunction; you will be asked to "prove p and q." In this case, you simply have to do two things: prove p, and then prove q. You could divide your proof into two sections, and for clarity, label them. When you're making your first notes, you might write "(1) Prove p" at the top of the page, and "(2) Prove q" in the middle of the page. (Of course, you might not be explicitly asked to prove a conjunction but find that your proof requires that you prove a conjunction. For instance, if you're asked to prove that $D = E$ you will find that, after unpacking the definition of =, you have to prove: every element of D is an element of E *and* every element of E is an element of D).

Proving a Disjunction

When what you are proving takes the form of a disjunction (i.e., it is an statement of the form "p or q"), it is enough to show that one of the disjuncts is true. However, it basically never happens that either disjunct just follows from the assumptions of your theorem. More often, the assumptions of your theorem are themselves disjunctive, or you're showing that all things of a certain kind have one of two properties, but some of the things have the one and others have the other property. This is where proof by cases is useful (see below).

Conditional Proof

Many theorems you will encounter are in conditional form (i.e., show that if p holds, then q is also true). These cases are nice and easy to set up—simply assume the antecedent of the conditional (in this case, p) and prove the conclusion q from it. So if your theorem reads, "If p then q," you start your proof with "assume p" and at the end you should have proved q.

Conditionals may be stated in different ways. So instead of "If p then q," a theorem may state that "p only if q," "q if p," or "q, provided p." These all mean the same and require assuming p and proving q from that assumption. Recall that a biconditional ("p if and only if (iff) q") is really two conditionals put together: if p then q, and if q then p. All you have to do, then, is two instances of conditional proof: one for the first conditional and another one for the second. Sometimes, however, it is possible to prove an "iff" statement by chaining together a bunch of other "iff" statements so that you start with "p" an end with "q"—but in that case you have to make sure that each step really is an "iff."

Universal Claims

Using a universal claim is simple: if something is true for anything, it's true for each particular thing. So if, say, the hypothesis of your proof is $A \subseteq B$, that means (unpacking the definition

A.4. INFERENCE PATTERNS

of $\subseteq$), that, for every $x \in A$, $x \in B$. Thus, if you already know that $z \in A$, you can conclude $z \in B$.

Proving a universal claim may seem a little bit tricky. Usually these statements take the following form: "If x has P, then it has Q" or "All Ps are Qs." Of course, it might not fit this form perfectly, and it takes a bit of practice to figure out what you're asked to prove exactly. But: we often have to prove that all objects with some property have a certain other property.

The way to prove a universal claim is to introduce names or variables, for the things that have the one property and then show that they also have the other property. We might put this by saying that to prove something for *all* Ps you have to prove it for an *arbitrary* P. And the name introduced is a name for an arbitrary P. We typically use single letters as these names for arbitrary things, and the letters usually follow conventions: e.g., we use n for natural numbers, A for formulas, A for sets, f for functions, etc.

The trick is to maintain generality throughout the proof. You start by assuming that an arbitrary object ("x") has the property P, and show (based only on definitions or what you are allowed to assume) that x has the property Q. Because you have not stipulated what x is specifically, other that it has the property P, then you can assert that all every P has the property Q. In short, x is a stand-in for *all* things with property P.

Proposition A.4. *For all sets A and B, $A \subseteq A \cup B$.*

Proof. Let A and B be arbitrary sets. We want to show that $A \subseteq A \cup B$. By definition of $\subseteq$, this amounts to: for every x, if $x \in A$ then $x \in A \cup B$. So let $x \in A$ be an arbitrary element of A. We have to show that $x \in A \cup B$. Since $x \in A$, $x \in A$ or $x \in B$. Thus, $x \in \{x : x \in A \vee x \in B\}$. But that, by definition of $\cup$, means $x \in A \cup B$. □

Proof by Cases

Suppose you have a disjunction as an assumption or as an already established conclusion—you have assumed or proved that p or q is true. You want to prove r. You do this in two steps: first you assume that p is true, and prove r, then you assume that q is true and prove r again. This works because we assume or know that one of the two alternatives holds. The two steps establish that either one is sufficient for the truth of r. (If both are true, we have not one but two reasons for why r is true. It is not necessary to separately prove that r is true assuming both p and q.) To indicate what we're doing, we announce that we "distinguish cases." For instance, suppose we know that $x \in B \cup C$. $B \cup C$ is defined as $\{x : x \in B \text{ or } x \in C\}$. In other words, by definition, $x \in B$ or $x \in C$. We would prove that $x \in A$ from this by first assuming that $x \in B$, and proving $x \in A$ from this assumption, and then assume $x \in C$, and again prove $x \in A$ from this. You would write "We distinguish cases" under the assumption, then "Case (1): $x \in B$" underneath, and "Case (2): $x \in C$" halfway down the page. Then you'd proceed to fill in the top half and the bottom half of the page.

Proof by cases is especially useful if what you're proving is itself disjunctive. Here's a simple example:

Proposition A.5. *Suppose $B \subseteq D$ and $C \subseteq E$. Then $B \cup C \subseteq D \cup E$.*

Proof. Assume (a) that $B \subseteq D$ and (b) $C \subseteq E$. By definition, any $x \in B$ is also $\in D$ (c) and any $x \in C$ is also $\in E$ (d). To show that $B \cup C \subseteq D \cup E$, we have to show that if $x \in B \cup C$ then $x \in D \cup E$ (by definition of $\subseteq$). $x \in B \cup C$ iff $x \in B$ or $x \in C$ (by definition of $\cup$). Similarly, $x \in D \cup E$ iff $x \in D$ or $x \in E$. So, we have to show: for any x, if $x \in B$ or $x \in C$, then $x \in D$ or $x \in E$.

> So far we've only unpacked definitions! We've reformulated our proposition without $\subseteq$ and $\cup$ and are left with trying to prove a universal conditional claim. By what we've discussed above, this is done by assuming

A.4. INFERENCE PATTERNS

that x is something about which we assume the "if" part is true, and we'll go on to show that the "then" part is true as well. In other words, we'll assume that $x \in B$ or $x \in C$ and show that $x \in D$ or $x \in E$.[2]

Suppose that $x \in B$ or $x \in C$. We have to show that $x \in D$ or $x \in E$. We distinguish cases.

Case 1: $x \in B$. By (c), $x \in D$. Thus, $x \in D$ or $x \in E$. (Here we've made the inference discussed in the preceding subsection!)

Case 2: $x \in C$. By (d), $x \in E$. Thus, $x \in D$ or $x \in E$. □

Proving an Existence Claim

When asked to prove an existence claim, the question will usually be of the form "prove that there is an x such that $\ldots x \ldots$", i.e., that some object that has the property described by "$\ldots x \ldots$". In this case you'll have to identify a suitable object show that is has the required property. This sounds straightforward, but a proof of this kind can be tricky. Typically it involves *constructing* or *defining* an object and proving that the object so defined has the required property. Finding the right object may be hard, proving that it has the required property may be hard, and sometimes it's even tricky to show that you've succeeded in defining an object at all!

Generally, you'd write this out by specifying the object, e.g., "let x be $\ldots$" (where $\ldots$ specifies which object you have in mind), possibly proving that $\ldots$ in fact describes an object that exists, and then go on to show that x has the property Q. Here's a simple example.

Proposition A.6. *Suppose that $x \in B$. Then there is an A such that $A \subseteq B$ and $A \neq \emptyset$.*

Proof. Assume $x \in B$. Let $A = \{x\}$.

---

[2]This paragraph just explains what we're doing—it's not part of the proof, and you don't have to go into all this detail when you write down your own proofs.

> Here we've defined the set A by enumerating its elements. Since we assume that x is an object, and we can always form a set by enumerating its elements, we don't have to show that we've succeeded in defining a set A here. However, we still have to show that A has the properties required by the proposition. The proof isn't complete without that!

Since $x \in A$, $A \neq \emptyset$.

> This relies on the definition of A as $\{x\}$ and the obvious facts that $x \in \{x\}$ and $x \notin \emptyset$.

Since x is the only element of $\{x\}$, and $x \in B$, every element of A is also an element of B. By definition of $\subseteq$, $A \subseteq B$. □

Using Existence Claims

Suppose you know that some existence claim is true (you've proved it, or it's a hypothesis you can use), say, "for some x, $x \in A$" or "there is an $x \in A$." If you want to use it in your proof, you can just pretend that you have a name for one of the things which your hypothesis says exist. Since A contains at least one thing, there are things to which that name might refer. You might of course not be able to pick one out or describe it further (other than that it is $\in A$). But for the purpose of the proof, you can pretend that you have picked it out and give a name to it. It's important to pick a name that you haven't already used (or that appears in your hypotheses), otherwise things can go wrong. In your proof, you indicate this by going from "for some x, $x \in A$" to "Let $a \in A$." Now you can reason about a, use some other hypotheses, etc., until you come to a conclusion, p. If p no longer mentions a, p is independent of the asusmption that $a \in A$, and you've shown that it follows just from the assumption "for some x, $x \in A$."

A.4. INFERENCE PATTERNS

Proposition A.7. *If $A \neq \emptyset$, then $A \cup B \neq \emptyset$.*

Proof. Suppose $A \neq \emptyset$. So for some x, $x \in A$.

> Here we first just restated the hypothesis of the proposition. This hypothesis, i.e., $A \neq \emptyset$, hides an existential claim, which you get to only by unpacking a few definitions. The definition of = tells us that $A = \emptyset$ iff every $x \in A$ is also $\in \emptyset$ and every $x \in \emptyset$ is also $\in A$. Negating both sides, we get: $A \neq \emptyset$ iff either some $x \in A$ is $\notin \emptyset$ or some $x \in \emptyset$ is $\notin A$. Since nothing is $\in \emptyset$, the second disjunct can never be true, and "$x \in A$ and $x \notin \emptyset$" reduces to just $x \in A$. So $x \neq \emptyset$ iff for some x, $x \in A$. That's an existence claim. Now we use that existence claim by introducing a name for one of the elements of A:

Let $a \in A$.

> Now we've introduced a name for one of the things $\in A$. We'll continue to argue about a, but we'll be careful to only assume that $a \in A$ and nothing else:

Since $a \in A$, $a \in A \cup B$, by definition of $\cup$. So for some x, $x \in A \cup B$, i.e., $A \cup B \neq \emptyset$.

> In that last step, we went from "$a \in A \cup B$" to "for some x, $x \in A \cup B$." That doesn't mention a anymore, so we know that "for some x, $x \in A \cup B$" follows from "for some x, $x \in A$ alone." But that means that $A \cup B \neq \emptyset$. □

It's maybe good practice to keep bound variables like "x" separate from hypothetical names like a, like we did. In practice, however, we often don't and just use x, like so:

> Suppose $A \neq \emptyset$, i.e., there is an $x \in A$. By definition of $\cup$, $x \in A \cup B$. So $A \cup B \neq \emptyset$.

However, when you do this, you have to be extra careful that you use different x's and y's for different existential claims. For instance, the following is *not* a correct proof of "If $A \neq \emptyset$ and $B \neq \emptyset$ then $A \cap B \neq \emptyset$" (which is not true).

> Suppose $A \neq \emptyset$ and $B \neq \emptyset$. So for some x, $x \in A$ and also for some x, $x \in B$. Since $x \in A$ and $x \in B$, $x \in A \cap B$, by definition of $\cap$. So $A \cap B \neq \emptyset$.

Can you spot where the incorrect step occurs and explain why the result does not hold?

A.5 An Example

Our first example is the following simple fact about unions and intersections of sets. It will illustrate unpacking definitions, proofs of conjunctions, of universal claims, and proof by cases.

Proposition A.8. *For any sets A, B, and C, $A \cup (B \cap C) = (A \cup B) \cap (A \cup C)$*

Let's prove it!

Proof. We want to show that for any sets A, B, and C, $A \cup (B \cap C) = (A \cup B) \cap (A \cup C)$

> First we unpack the definition of "=" in the statement of the proposition. Recall that proving sets identical means showing that the sets have the same elements. That is, all elements of $A \cup (B \cap C)$ are also elements of $(A \cup B) \cap (A \cup C)$, and vice versa. The "vice versa" means that also every element of $(A \cup B) \cap (A \cup C)$ must be an element of $A \cup (B \cap C)$. So in unpacking the definition, we see that we have to prove a conjunction. Let's record this:

A.5. AN EXAMPLE

By definition, $A \cup (B \cap C) = (A \cup B) \cap (A \cup C)$ iff every element of $A \cup (B \cap C)$ is also an element of $(A \cup B) \cap (A \cup C)$, and every element of $(A \cup B) \cap (A \cup C)$ is an element of $A \cup (B \cap C)$.

> Since this is a conjunction, we must prove each conjunct separately. Lets start with the first: let's prove that every element of $A \cup (B \cap C)$ is also an element of $(A \cup B) \cap (A \cup C)$.
>
> This is a universal claim, and so we consider an arbitrary element of $A \cup (B \cap C)$ and show that it must also be an element of $(A \cup B) \cap (A \cup C)$. We'll pick a variable to call this arbitrary element by, say, z. Our proof continues:

First, we prove that every element of $A \cup (B \cap C)$ is also an element of $(A \cup B) \cap (A \cup C)$. Let $z \in A \cup (B \cap C)$. We have to show that $z \in (A \cup B) \cap (A \cup C)$.

> Now it is time to unpack the definition of $\cup$ and $\cap$. For instance, the definition of $\cup$ is: $A \cup B = \{z : z \in A \text{ or } z \in B\}$. When we apply the definition to "$A \cup (B \cap C)$," the role of the "B" in the definition is now played by "$B \cap C$," so $A \cup (B \cap C) = \{z : z \in A \text{ or } z \in B \cap C\}$. So our assumption that $z \in A \cup (B \cap C)$ amounts to: $z \in \{z : z \in A \text{ or } z \in B \cap C\}$. And $z \in \{z : \ldots z \ldots\}$ iff $\ldots z \ldots$, i.e., in this case, $z \in A$ or $z \in B \cap C$.

By the definition of $\cup$, either $z \in A$ or $z \in B \cap C$.

> Since this is a disjunction, it will be useful to apply proof by cases. We take the two cases, and show that in each one, the conclusion we're aiming for (namely, "$z \in (A \cup B) \cap (A \cup C)$") obtains.

Case 1: Suppose that $z \in A$.

> There's not much more to work from based on our assumptions. So let's look at what we have to work with in the conclusion. We want to show that $z \in (A \cup B) \cap (A \cup C)$. Based on the definition of $\cap$, if we want to show that $z \in (A \cup B) \cap (A \cup C)$, we have to show that it's in both $(A \cup B)$ and $(A \cup C)$. But $z \in A \cup B$ iff $z \in A$ or $z \in B$, and we already have (as the assumption of case 1) that $z \in A$. By the same reasoning—switching C for B—$z \in A \cup C$. This argument went in the reverse direction, so let's record our reasoning in the direction needed in our proof.

Since $z \in A$, $z \in A$ or $z \in B$, and hence, by definition of $\cup$, $z \in A \cup B$. Similarly, $z \in A \cup C$. But this means that $z \in (A \cup B) \cap (A \cup C)$, by definition of $\cap$.

> This completes the first case of the proof by cases. Now we want to derive the conclusion in the second case, where $z \in B \cap C$.

Case 2: Suppose that $z \in B \cap C$.

> Again, we are working with the intersection of two sets. Let's apply the definition of $\cap$:

Since $z \in B \cap C$, z must be an element of both B and C, by definition of $\cap$.

> It's time to look at our conclusion again. We have to show that z is in both $(A \cup B)$ and $(A \cup C)$. And again, the solution is immediate.

Since $z \in B$, $z \in (A \cup B)$. Since $z \in C$, also $z \in (A \cup C)$. So, $z \in (A \cup B) \cap (A \cup C)$.

> Here we applied the definitions of $\cup$ and $\cap$ again, but since we've already recalled those definitions, and already showed that if z is in one of two sets it is in

A.5. AN EXAMPLE

their union, we don't have to be as explicit in what we've done.

We've completed the second case of the proof by cases, so now we can assert our first conclusion.

So, if $z \in A \cup (B \cap C)$ then $z \in (A \cup B) \cap (A \cup C)$.

Now we just want to show the other direction, that every element of $(A \cup B) \cap (A \cup C)$ is an element of $A \cup (B \cap C)$. As before, we prove this universal claim by assuming we have an arbitrary element of the first set and show it must be in the second set. Let's state what we're about to do.

Now, assume that $z \in (A \cup B) \cap (A \cup C)$. We want to show that $z \in A \cup (B \cap C)$.

We are now working from the hypothesis that $z \in (A \cup B) \cap (A \cup C)$. It hopefully isn't too confusing that we're using the same z here as in the first part of the proof. When we finished that part, all the assumptions we've made there are no longer in effect, so now we can make new assumptions about what z is. If that is confusing to you, just replace z with a different variable in what follows.

We know that z is in both $A \cup B$ and $A \cup C$, by definition of $\cap$. And by the definition of $\cup$, we can further unpack this to: either $z \in A$ or $z \in B$, and also either $z \in A$ or $z \in C$. This looks like a proof by cases again—except the "and" makes it confusing. You might think that this amounts to there being three possibilities: z is either in A, B or C. But that would be a mistake. We have to be careful, so let's consider each disjunction in turn.

By definition of $\cap$, $z \in A \cup B$ and $z \in A \cup C$. By definition of $\cup$, $z \in A$ or $z \in B$. We distinguish cases.

Since we're focusing on the first disjunction, we haven't gotten our second disjunction (from unpacking $A \cup C$) yet. In fact, we don't need it yet. The first case is $z \in A$, and an element of a set is also an element of the union of that set with any other. So case 1 is easy:

Case 1: Suppose that $z \in A$. It follows that $z \in A \cup (B \cap C)$.

Now for the second case, $z \in B$. Here we'll unpack the second $\cup$ and do another proof-by-cases:

Case 2: Suppose that $z \in B$. Since $z \in A \cup C$, either $z \in A$ or $z \in C$. We distinguish cases further:
Case 2a: $z \in A$. Then, again, $z \in A \cup (B \cap C)$.

Ok, this was a bit weird. We didn't actually need the assumption that $z \in B$ for this case, but that's ok.

Case 2b: $z \in C$. Then $z \in B$ and $z \in C$, so $z \in B \cap C$, and consequently, $z \in A \cup (B \cap C)$.

This concludes both proofs-by-cases and so we're done with the second half.

So, if $z \in (A \cup B) \cap (A \cup C)$ then $z \in A \cup (B \cap C)$. □

A.6 Another Example

Proposition A.9. *If $A \subseteq C$, then $A \cup (C \setminus A) = C$.*

Proof. Suppose that $A \subseteq C$. We want to show that $A \cup (C \setminus A) = C$.

We begin by observing that this is a conditional statement. It is tacitly universally quantified: the proposition holds for all sets A and C. So A and C are variables for arbitrary sets. To prove such a statement, we assume the antecedent and prove the consequent.

A.6. ANOTHER EXAMPLE

We continue by using the assumption that $A \subseteq C$. Let's unpack the definition of $\subseteq$: the assumption means that all elements of A are also elements of C. Let's write this down—it's an important fact that we'll use throughout the proof.

By the definition of $\subseteq$, since $A \subseteq C$, for all z, if $z \in A$, then $z \in C$.

We've unpacked all the definitions that are given to us in the assumption. Now we can move onto the conclusion. We want to show that $A \cup (C \setminus A) = C$, and so we set up a proof similarly to the last example: we show that every element of $A \cup (C \setminus A)$ is also an element of C and, conversely, every element of C is an element of $A \cup (C \setminus A)$. We can shorten this to: $A \cup (C \setminus A) \subseteq C$ and $C \subseteq A \cup (C \setminus A)$. (Here we're doing the opposite of unpacking a definition, but it makes the proof a bit easier to read.) Since this is a conjunction, we have to prove both parts. To show the first part, i.e., that every element of $A \cup (C \setminus A)$ is also an element of C, we assume that $z \in A \cup (C \setminus A)$ for an arbitrary z and show that $z \in C$. By the definition of $\cup$, we can conclude that $z \in A$ or $z \in C \setminus A$ from $z \in A \cup (C \setminus A)$. You should now be getting the hang of this.

$A \cup (C \setminus A) = C$ iff $A \cup (C \setminus A) \subseteq C$ and $C \subseteq (A \cup (C \setminus A))$. First we prove that $A \cup (C \setminus A) \subseteq C$. Let $z \in A \cup (C \setminus A)$. So, either $z \in A$ or $z \in (C \setminus A)$.

We've arrived at a disjunction, and from it we want to prove that $z \in C$. We do this using proof by cases.

Case 1: $z \in A$. Since for all z, if $z \in A$, $z \in C$, we have that $z \in C$.

Here we've used the fact recorded earlier which followed from the hypothesis of the proposition that $A \subseteq C$. The first case is complete, and we turn to the second case, $z \in (C \setminus A)$. Recall that $C \setminus A$ denotes the *difference* of the two sets, i.e., the set of all elements of C which are not elements of A. But any element of C not in A is in particular an element of C.

Case 2: $z \in (C \setminus A)$. This means that $z \in C$ and $z \notin A$. So, in particular, $z \in C$.

Great, we've proved the first direction. Now for the second direction. Here we prove that $C \subseteq A \cup (C \setminus A)$. So we assume that $z \in C$ and prove that $z \in A \cup (C \setminus A)$.

Now let $z \in C$. We want to show that $z \in A$ or $z \in C \setminus A$.

Since all elements of A are also elements of C, and $C \setminus A$ is the set of all things that are elements of C but not A, it follows that z is either in A or in $C \setminus A$. This may be a bit unclear if you don't already know why the result is true. It would be better to prove it step-by-step. It will help to use a simple fact which we can state without proof: $z \in A$ or $z \notin A$. This is called the "principle of excluded middle:" for any statement p, either p is true or its negation is true. (Here, p is the statement that $z \in A$.) Since this is a disjunction, we can again use proof-by-cases.

Either $z \in A$ or $z \notin A$. In the former case, $z \in A \cup (C \setminus A)$. In the latter case, $z \in C$ and $z \notin A$, so $z \in C \setminus A$. But then $z \in A \cup (C \setminus A)$.

Our proof is complete: we have shown that $A \cup (C \setminus A) = C$. □

A.7 Proof by Contradiction

In the first instance, proof by contradiction is an inference pattern that is used to prove negative claims. Suppose you want to

A.7. PROOF BY CONTRADICTION

show that some claim p is *false*, i.e., you want to show $\neg p$. The most promising strategy is to (a) suppose that p is true, and (b) show that this assumption leads to something you know to be false. "Something known to be false" may be a result that conflicts with—contradicts—p itself, or some other hypothesis of the overall claim you are considering. For instance, a proof of "if q then $\neg p$" involves assuming that q is true and proving $\neg p$ from it. If you prove $\neg p$ by contradiction, that means assuming p in addition to q. If you can prove $\neg q$ from p, you have shown that the assumption p leads to something that contradicts your other assumption q, since q and $\neg q$ cannot both be true. Of course, you have to use other inference patterns in your proof of the contradiction, as well as unpacking definitions. Let's consider an example.

Proposition A.10. *If $A \subseteq B$ and $B = \emptyset$, then A has no elements.*

Proof. Suppose $A \subseteq B$ and $B = \emptyset$. We want to show that A has no elements.

> Since this is a conditional claim, we assume the antecedent and want to prove the consequent. The consequent is: A has no elements. We can make that a bit more explicit: it's not the case that there is an $x \in A$.

A has no elements iff it's not the case that there is an x such that $x \in A$.

> So we've determined that what we want to prove is really a negative claim $\neg p$, namely: it's not the case that there is an $x \in A$. To use proof by contradiction, we have to assume the corresponding positive claim p, i.e., there is an $x \in A$, and prove a contradiction from it. We indicate that we're doing a proof by contradiction by writing "by way of contradiction, assume" or even just "suppose not," and then state the assumption p.

Suppose not: there is an $x \in A$.

> This is now the new assumption we'll use to obtain a contradiction. We have two more assumptions: that $A \subseteq B$ and that $B = \emptyset$. The first gives us that $x \in B$:

Since $A \subseteq B$, $x \in B$.

> But since $B = \emptyset$, every element of B (e.g., x) must also be an element of $\emptyset$.

Since $B = \emptyset$, $x \in \emptyset$. This is a contradiction, since by definition $\emptyset$ has no elements.

> This already completes the proof: we've arrived at what we need (a contradiction) from the assumptions we've set up, and this means that the assumptions can't all be true. Since the first two assumptions ($A \subseteq B$ and $B = \emptyset$) are not contested, it must be the last assumption introduced (there is an $x \in A$) that must be false. But if we want to be thorough, we can spell this out.

Thus, our assumption that there is an $x \in A$ must be false, hence, A has no elements by proof by contradiction. □

Every positive claim is trivially equivalent to a negative claim: p iff $\neg\neg p$. So proofs by contradiction can also be used to establish positive claims "indirectly," as follows: To prove p, read it as the negative claim $\neg\neg p$. If we can prove a contradiction from $\neg p$, we've established $\neg\neg p$ by proof by contradiction, and hence p.

In the last example, we aimed to prove a negative claim, namely that A has no elements, and so the assumption we made for the purpose of proof by contradiction (i.e., that there is an $x \in A$) was a positive claim. It gave us something to work with, namely the hypothetical $x \in A$ about which we continued to reason until we got to $x \in \emptyset$.

A.7. PROOF BY CONTRADICTION

When proving a positive claim indirectly, the assumption you'd make for the purpose of proof by contradiction would be negative. But very often you can easily reformulate a positive claim as a negative claim, and a negative claim as a positive claim. Our previous proof would have been essentially the same had we proved "$A = \emptyset$" instead of the negative consequent "A has no elements." (By definition of $=$, "$A = \emptyset$" is a general claim, since it unpacks to "every element of A is an element of $\emptyset$ and vice versa".) But it is easily seen to be equivalent to the negative claim "not: there is an $x \in A$."

So it is sometimes easier to work with $\neg p$ as an assumption than it is to prove p directly. Even when a direct proof is just as simple or even simpler (as in the next examples), some people prefer to proceed indirectly. If the double negation confuses you, think of a proof by contradiction of some claim as a proof of a contradiction from the *opposite* claim. So, a proof by contradiction of $\neg p$ is a proof of a contradiction from the assumption p; and proof by contradiction of p is a proof of a contradiction from $\neg p$.

Proposition A.11. $A \subseteq A \cup B$.

Proof. We want to show that $A \subseteq A \cup B$.

> On the face of it, this is a positive claim: every $x \in A$ is also in $A \cup B$. The negation of that is: some $x \in A$ is $\notin A \cup B$. So we can prove the claim indirectly by assuming this negated claim, and showing that it leads to a contradiction.

Suppose not, i.e., $A \nsubseteq A \cup B$.

> We have a definition of $A \subseteq A \cup B$: every $x \in A$ is also $\in A \cup B$. To understand what $A \nsubseteq A \cup B$ means, we have to use some elementary logical manipulation on the unpacked definition: it's false that every $x \in A$ is also $\in A \cup B$ iff there is *some* $x \in A$ that is $\notin C$. (This is a place where you want to be very careful:

many students' attempted proofs by contradiction fail because they analyze the negation of a claim like "all As are Bs" incorrectly.) In other words, $A \nsubseteq A \cup B$ iff there is an x such that $x \in A$ and $x \notin A \cup B$. From then on, it's easy.

So, there is an $x \in A$ such that $x \notin A \cup B$. By definition of $\cup$, $x \in A \cup B$ iff $x \in A$ or $x \in B$. Since $x \in A$, we have $x \in A \cup B$. This contradicts the assumption that $x \notin A \cup B$. □

Proposition A.12. *If $A \subseteq B$ and $B \subseteq C$ then $A \subseteq C$.*

Proof. Suppose $A \subseteq B$ and $B \subseteq C$. We want to show $A \subseteq C$.

Let's proceed indirectly: we assume the negation of what we want to etablish.

Suppose not, i.e., $A \nsubseteq C$.

As before, we reason that $A \nsubseteq C$ iff not every $x \in A$ is also $\in C$, i.e., some $x \in A$ is $\notin C$. Don't worry, with practice you won't have to think hard anymore to unpack negations like this.

In other words, there is an x such that $x \in A$ and $x \notin C$.

Now we can use this to get to our contradiction. Of course, we'll have to use the other two assumptions to do it.

Since $A \subseteq B$, $x \in B$. Since $B \subseteq C$, $x \in C$. But this contradicts $x \notin C$. □

Proposition A.13. *If $A \cup B = A \cap B$ then $A = B$.*

Proof. Suppose $A \cup B = A \cap B$. We want to show that $A = B$.

The beginning is now routine:

Assume, by way of contradiction, that $A \neq B$.

> Our assumption for the proof by contradiction is that $A \neq B$. Since $A = B$ iff $A \subseteq B$ an $B \subseteq A$, we get that $A \neq B$ iff $A \nsubseteq B$ or $B \nsubseteq A$. (Note how important it is to be careful when manipulating negations!) To prove a contradiction from this disjunction, we use a proof by cases and show that in each case, a contradiction follows.

$A \neq B$ iff $A \nsubseteq B$ or $B \nsubseteq A$. We distinguish cases.

> In the first case, we assume $A \nsubseteq B$, i.e., for some x, $x \in A$ but $\notin B$. $A \cap B$ is defined as those elements that A and B have in common, so if something isn't in one of them, it's not in the intersection. $A \cup B$ is A together with B, so anything in either is also in the union. This tells us that $x \in A \cup B$ but $x \notin A \cap B$, and hence that $A \cap B \neq A \cup B$.

Case 1: $A \nsubseteq B$. Then for some x, $x \in A$ but $x \notin B$. Since $x \notin B$, then $x \notin A \cap B$. Since $x \in A$, $x \in A \cup B$. So, $A \cap B \neq A \cup B$, contradicting the assumption that $A \cap B = A \cup B$.

Case 2: $B \nsubseteq A$. Then for some y, $y \in B$ but $y \notin A$. As before, we have $y \in A \cup B$ but $y \notin A \cap B$, and so $A \cap B \neq A \cup B$, again contradicting $A \cap B = A \cup B$. □

A.8 Reading Proofs

Proofs you find in textbooks and articles very seldom give all the details we have so far included in our examples. Authors often do not draw attention to when they distinguish cases, when they give an indirect proof, or don't mention that they use a definition. So when you read a proof in a textbook, you will often have to fill in those details for yourself in order to understand the proof. Doing this is also good practice to get the hang of the various moves you have to make in a proof. Let's look at an example.

Proposition A.14 (Absorption). *For all sets A, B,*

$$A \cap (A \cup B) = A$$

Proof. If $z \in A \cap (A \cup B)$, then $z \in A$, so $A \cap (A \cup B) \subseteq A$. Now suppose $z \in A$. Then also $z \in A \cup B$, and therefore also $z \in A \cap (A \cup B)$. □

The preceding proof of the absorption law is very condensed. There is no mention of any definitions used, no "we have to prove that" before we prove it, etc. Let's unpack it. The proposition proved is a general claim about any sets A and B, and when the proof mentions A or B, these are variables for arbitrary sets. The general claims the proof establishes is what's required to prove identity of sets, i.e., that every element of the left side of the identity is an element of the right and vice versa.

"If $z \in A \cap (A \cup B)$, then $z \in A$, so $A \cap (A \cup B) \subseteq A$."

This is the first half of the proof of the identity: it estabishes that if an arbitrary z is an element of the left side, it is also an element of the right, i.e., $A \cap (A \cup B) \subseteq A$. Assume that $z \in A \cap (A \cup B)$. Since z is an element of the intersection of two sets iff it is an element of both sets, we can conclude that $z \in A$ and also $z \in A \cup B$. In particular, $z \in A$, which is what we wanted to show. Since that's all that has to be done for the first half, we know that the rest of the proof must be a proof of the second half, i.e., a proof that $A \subseteq A \cap (A \cup B)$.

"Now suppose $z \in A$. Then also $z \in A \cup B$, and therefore also $z \in A \cap (A \cup B)$."

We start by assuming that $z \in A$, since we are showing that, for any z, if $z \in A$ then $z \in A \cap (A \cup B)$. To show that $z \in A \cap (A \cup B)$, we have to show (by definition of "$\cap$") that (i) $z \in A$ and also (ii) $z \in A \cup B$. Here (i) is just our assumption, so there is nothing further to prove, and that's why the proof does not mention it again. For (ii), recall that z is an element of a union of sets iff it is an element of at least one of those sets. Since $z \in A$, and $A \cup B$ is the union of A and B, this is the case here. So $z \in A \cup B$. We've shown both (i) $z \in A$ and (ii) $z \in A \cup B$, hence, by definition of "$\cap$," $z \in A \cap (A \cup B)$. The proof doesn't mention those definitions; it's assumed the reader has already internalized them. If you haven't, you'll have to go back and remind yourself what they are. Then you'll also have to recognize why it follows from $z \in A$ that $z \in A \cup B$, and from $z \in A$ and $z \in A \cup B$ that $z \in A \cap (A \cup B)$.

Here's another version of the proof above, with everything made explicit:

Proof. [By definition of = for sets, $A \cap (A \cup B) = A$ we have to show (a) $A \cap (A \cup B) \subseteq A$ and (b) $A \cap (A \cup B) \subseteq A$. (a): By definition of $\subseteq$, we have to show that if $z \in A \cap (A \cup B)$, then $z \in A$.] If $z \in A \cap (A \cup B)$, then $z \in A$ [since by definition of $\cap$, $z \in A \cap (A \cup B)$ iff $z \in A$ and $z \in A \cup B$], so $A \cap (A \cup B) \subseteq A$. [(b): By definition of $\subseteq$, we have to show that if $z \in A$, then $z \in A \cap (A \cup B)$.] Now suppose [(1)] $z \in A$. Then also [(2)] $z \in A \cup B$ [since by (1) $z \in A$ or $z \in B$, which by definition of $\cup$ means $z \in A \cup B$], and therefore also $z \in A \cap (A \cup B)$ [since the definition of $\cap$ requires that $z \in A$, i.e., (1), and $z \in A \cup B$), i.e., (2)]. □

A.9 I Can't Do It!

We all get to a point where we feel like giving up. But you *can* do it. Your instructor and teaching assistant, as well as your fellow students, can help. Ask them for help! Here are a few tips to help you avoid a crisis, and what to do if you feel like giving up.

To make sure you can solve problems successfully, do the following:

1. *Start as far in advance as possible.* We get busy throughout the semester and many of us struggle with procrastination, one of the best things you can do is to start your homework assignments early. That way, if you're stuck, you have time to look for a solution (that isn't crying).

2. *Talk to your classmates.* You are not alone. Others in the class may also struggle—but the may struggle with different things. Talking it out with your peers can give you a different perspective on the problem that might lead to a breakthrough. Of course, don't just copy their solution: ask them for a hint, or explain where you get stuck and ask them for the next step. And when you do get it, reciprocate. Helping someone else along, and explaining things will help you understand better, too.

3. *Ask for help.* You have many resources available to you—your instructor and teaching assistant are there for you and *want* you to succeed. They should be able to help you work out a problem and identify where in the process you're struggling.

4. *Take a break.* If you're stuck, it *might* be because you've been staring at the problem for too long. Take a short break, have a cup of tea, or work on a different problem for a while, then return to the problem with a fresh mind. Sleep on it.

Notice how these strategies require that you've started to work on the proof well in advance? If you've started the proof at 2am the day before it's due, these might not be so helpful.

This might sound like doom and gloom, but solving a proof is a challenge that pays off in the end. Some people do this as a career—so there must be something to enjoy about it. Like basically everything, solving problems and doing proofs is something that requires practice. You might see classmates who find this easy: they've probably just had lots of practice already. Try not to give in too easily.

If you do run out of time (or patience) on a particular problem: that's ok. It doesn't mean you're stupid or that you will never get it. Find out (from your instructor or another student) how it is done, and identify where you went wrong or got stuck, so you can avoid doing that the next time you encounter a similar issue. Then try to do it without looking at the solution. And next time, start (and ask for help) earlier.

A.10 Other Resources

There are many books on how to do proofs in mathematics which may be useful. Check out *How to Read and do Proofs: An Introduction to Mathematical Thought Processes* (Solow, 2013) and *How to Prove It: A Structured Approach* (Velleman, 2019) in particular. The *Book of Proof* (Hammack, 2013) and *Mathematical Reasoning* (Sandstrum, 2019) are books on proof that are freely available online. Philosophers might find *More Precisely: The Math you need to do Philosophy* (Steinhart, 2018) to be a good primer on mathematical reasoning.

There are also various shorter guides to proofs available on the internet; e.g., "Introduction to Mathematical Arguments" (Hutchings, 2003) and "How to write proofs" (Cheng, 2004).

Motivational Videos

Feel like you have no motivation to do your homework? Feeling down? These videos might help!

- https://www.youtube.com/watch?v=ZXsQAXx_ao0

- https://www.youtube.com/watch?v=BQ4yd2W50No

- https://www.youtube.com/watch?v=StTqXEQ21-Y

Problems

Problem A.1. Suppose you are asked to prove that $A \cap B \neq \emptyset$. Unpack all the definitions occuring here, i.e., restate this in a way that does not mention "$\cap$", "$=$", or "$\emptyset$".

Problem A.2. Prove *indirectly* that $A \cap B \subseteq A$.

Problem A.3. Expand the following proof of $A \cup (A \cap B) = A$, where you mention all the inference patterns used, why each step follows from assumptions or claims established before it, and where we have to appeal to which definitions.

Proof. If $z \in A \cup (A \cap B)$ then $z \in A$ or $z \in A \cap B$. If $z \in A \cap B$, $z \in A$. Any $z \in A$ is also $\in A \cup (A \cap B)$. □

APPENDIX B
Induction

B.1 Introduction

Induction is an important proof technique which is used, in different forms, in almost all areas of logic, theoretical computer science, and mathematics. It is needed to prove many of the results in logic.

Induction is often contrasted with deduction, and characterized as the inference from the particular to the general. For instance, if we observe many green emeralds, and nothing that we would call an emerald that's not green, we might conclude that all emeralds are green. This is an inductive inference, in that it proceeds from many particular cases (this emerald is green, that emerald is green, etc.) to a general claim (all emeralds are green). *Mathematical* induction is also an inference that concludes a general claim, but it is of a very different kind than this "simple induction."

Very roughly, an inductive proof in mathematics concludes that all mathematical objects of a certain sort have a certain property. In the simplest case, the mathematical objects an inductive proof is concerned with are natural numbers. In that case an inductive proof is used to establish that all natural numbers have some property, and it does this by showing that

1. 0 has the property, and

2. whenever a number k has the property, so does $k + 1$.

Induction on natural numbers can then also often be used to prove general claims about mathematical objects that can be assigned numbers. For instance, finite sets each have a finite number n of elements, and if we can use induction to show that every number n has the property "all finite sets of size n are ..." then we will have shown something about all finite sets.

Induction can also be generalized to mathematical objects that are *inductively defined*. For instance, expressions of a formal language such as those of first-order logic are defined inductively. *Structural induction* is a way to prove results about all such expressions. Structural induction, in particular, is very useful—and widely used—in logic.

B.2 Induction on $\mathbb{N}$

In its simplest form, induction is a technique used to prove results for all natural numbers. It uses the fact that by starting from 0 and repeatedly adding 1 we eventually reach every natural number. So to prove that something is true for every number, we can (1) establish that it is true for 0 and (2) show that whenever it is true for a number n, it is also true for the next number $n+1$. If we abbreviate "number n has property P" by $P(n)$ (and "number k has property P" by $P(k)$, etc.), then a proof by induction that $P(n)$ for all $n \in \mathbb{N}$ consists of:

1. a proof of $P(0)$, and

2. a proof that, for any k, if $P(k)$ then $P(k + 1)$.

To make this crystal clear, suppose we have both (1) and (2). Then (1) tells us that $P(0)$ is true. If we also have (2), we know in particular that if $P(0)$ then $P(0 + 1)$, i.e., $P(1)$. This follows from the general statement "for any k, if $P(k)$ then $P(k + 1)$" by putting 0 for k. So by modus ponens, we have that $P(1)$. From (2) again, now taking 1 for n, we have: if $P(1)$ then $P(2)$. Since we've

B.2. INDUCTION ON $\mathbb{N}$

just established $P(1)$, by modus ponens, we have $P(2)$. And so on. For any number n, after doing this n times, we eventually arrive at $P(n)$. So (1) and (2) together establish $P(n)$ for any $n \in \mathbb{N}$.

Let's look at an example. Suppose we want to find out how many different sums we can throw with n dice. Although it might seem silly, let's start with 0 dice. If you have no dice there's only one possible sum you can "throw": no dots at all, which sums to 0. So the number of different possible throws is 1. If you have only one die, i.e., $n = 1$, there are six possible values, 1 through 6. With two dice, we can throw any sum from 2 through 12, that's 11 possibilities. With three dice, we can throw any number from 3 to 18, i.e., 16 different possibilities. 1, 6, 11, 16: looks like a pattern: maybe the answer is $5n + 1$? Of course, $5n + 1$ is the maximum possible, because there are only $5n + 1$ numbers between n, the lowest value you can throw with n dice (all 1's) and $6n$, the highest you can throw (all 6's).

Theorem B.1. *With n dice one can throw all $5n + 1$ possible values between n and $6n$.*

Proof. Let $P(n)$ be the claim: "It is possible to throw any number between n and $6n$ using n dice." To use induction, we prove:

1. The *induction basis* $P(1)$, i.e., with just one die, you can throw any number between 1 and 6.

2. The *induction step*, for all k, if $P(k)$ then $P(k+1)$.

(1) Is proved by inspecting a 6-sided die. It has all 6 sides, and every number between 1 and 6 shows up one on of the sides. So it is possible to throw any number between 1 and 6 using a single die.

To prove (2), we assume the antecedent of the conditional, i.e., $P(k)$. This assumption is called the *inductive hypothesis*. We use it to prove $P(k+1)$. The hard part is to find a way of thinking about the possible values of a throw of $k + 1$ dice in terms of the possible values of throws of k dice plus of throws of the extra $k+1$-st die—this is what we have to do, though, if we want to use the inductive hypothesis.

The inductive hypothesis says we can get any number between k and $6k$ using k dice. If we throw a 1 with our $(k+1)$-st die, this adds 1 to the total. So we can throw any value between $k+1$ and $6k+1$ by throwing k dice and then rolling a 1 with the $(k+1)$-st die. What's left? The values $6k+2$ through $6k+6$. We can get these by rolling k 6s and then a number between 2 and 6 with our $(k+1)$-st die. Together, this means that with $k+1$ dice we can throw any of the numbers between $k+1$ and $6(k+1)$, i.e., we've proved $P(k+1)$ using the assumption $P(k)$, the inductive hypothesis. □

Very often we use induction when we want to prove something about a series of objects (numbers, sets, etc.) that is itself defined "inductively," i.e., by defining the $(n+1)$-st object in terms of the n-th. For instance, we can define the sum s_n of the natural numbers up to n by

$$s_0 = 0$$
$$s_{n+1} = s_n + (n+1)$$

This definition gives:

$$s_0 = 0,$$
$$s_1 = s_0 + 1 \qquad = 1,$$
$$s_2 = s_1 + 2 \qquad = 1 + 2 = 3$$
$$s_3 = s_2 + 3 \qquad = 1 + 2 + 3 = 6, \text{ etc.}$$

Now we can prove, by induction, that $s_n = n(n+1)/2$.

Proposition B.2. $s_n = n(n+1)/2$.

Proof. We have to prove (1) that $s_0 = 0 \cdot (0+1)/2$ and (2) if $s_k = k(k+1)/2$ then $s_{k+1} = (k+1)(k+2)/2$. (1) is obvious. To prove (2), we assume the inductive hypothesis: $s_k = k(k+1)/2$. Using it, we have to show that $s_{k+1} = (k+1)(k+2)/2$.

What is s_{k+1}? By the definition, $s_{k+1} = s_k + (k+1)$. By inductive hypothesis, $s_k = k(k+1)/2$. We can substitute this into the previous equation, and then just need a bit of arithmetic of fractions:

$$\begin{aligned} s_{k+1} &= \frac{k(k+1)}{2} + (k+1) = \\ &= \frac{k(k+1)}{2} + \frac{2(k+1)}{2} = \\ &= \frac{k(k+1) + 2(k+1)}{2} = \\ &= \frac{(k+2)(k+1)}{2}. \end{aligned}$$
□

The important lesson here is that if you're proving something about some inductively defined sequence a_n, induction is the obvious way to go. And even if it isn't (as in the case of the possibilities of dice throws), you can use induction if you can somehow relate the case for $k+1$ to the case for k.

B.3 Strong Induction

In the principle of induction discussed above, we prove $P(0)$ and also if $P(k)$, then $P(k+1)$. In the second part, we assume that $P(k)$ is true and use this assumption to prove $P(k+1)$. Equivalently, of course, we could assume $P(k-1)$ and use it to prove $P(k)$—the important part is that we be able to carry out the inference from any number to its successor; that we can prove the claim in question for any number under the assumption it holds for its predecessor.

There is a variant of the principle of induction in which we don't just assume that the claim holds for the predecessor $k-1$ of k, but for all numbers smaller than k, and use this assumption to establish the claim for k. This also gives us the claim $P(n)$ for all $n \in \mathbb{N}$. For once we have established $P(0)$, we have thereby established that P holds for all numbers less than 1. And if we know that if $P(l)$ for all $l < k$, then $P(k)$, we know this in particular for $k = 1$. So we can conclude $P(1)$. With this we have proved $P(0)$ and $P(1)$, i.e., $P(l)$ for all $l < 2$, and since we have also the conditional, if $P(l)$ for all $l < 2$, then $P(2)$, we can conclude $P(2)$, and so on.

In fact, if we can establish the general conditional "for all k, if $P(l)$ for all $l < k$, then $P(k)$," we do not have to establish $P(0)$ anymore, since it follows from it. For remember that a general claim like "for all $l < k$, $P(l)$" is true if there are no $l < k$. This is a case of vacuous quantification: "all As are Bs" is true if there are no As, $\forall x\,(A(x) \to B(x))$ is true if no x satisfies $A(x)$. In this case, the formalized version would be "$\forall l\,(l < k \to P(l))$"—and that is true if there are no $l < k$. And if $k = 0$ that's exactly the case: no $l < 0$, hence "for all $l < 0$, $P(0)$" is true, whatever P is. A proof of "if $P(l)$ for all $l < k$, then $P(k)$" thus automatically establishes $P(0)$.

This variant is useful if establishing the claim for k can't be made to just rely on the claim for $k - 1$ but may require the assumption that it is true for one or more $l < k$.

B.4 Inductive Definitions

In logic we very often define kinds of objects *inductively*, i.e., by specifying rules for what counts as an object of the kind to be defined which explain how to get new objects of that kind from old objects of that kind. For instance, we often define special kinds of sequences of symbols, such as the terms and formulas of a language, by induction. For a simple example, consider strings of consisting of letters a, b, c, d, the symbol ∘, and brackets [and], such as "[[c∘d][", "[a[]∘]", "a" or "[[a∘b]∘d]". You probably feel that there's something "wrong" with the first two strings: the brackets don't "balance" at all in the first, and you might feel that the "∘" should "connect" expressions that themselves make sense. The third and fourth string look better: for every "[" there's a

B.4. INDUCTIVE DEFINITIONS

closing "]" (if there are any at all), and for any ∘ we can find "nice" expressions on either side, surrounded by a pair of parentheses.

We would like to precisely specify what counts as a "nice term." First of all, every letter by itself is nice. Anything that's not just a letter by itself should be of the form "$[t \circ s]$" where s and t are themselves nice. Conversely, if t and s are nice, then we can form a new nice term by putting a ∘ between them and surround them by a pair of brackets. We might use these operations to *define* the set of nice terms. This is an *inductive definition*.

Definition B.3 (Nice terms). The set of *nice terms* is inductively defined as follows:

1. Any letter a, b, c, d is a nice term.

2. If s_1 and s_2 are nice terms, then so is $[s_1 \circ s_2]$.

3. Nothing else is a nice term.

This definition tells us that something counts as a nice term iff it can be constructed according to the two conditions (1) and (2) in some finite number of steps. In the first step, we construct all nice terms just consisting of letters by themselves, i.e.,

$$a, b, c, d$$

In the second step, we apply (2) to the terms we've constructed. We'll get

$$[a \circ a], [a \circ b], [b \circ a], \ldots, [d \circ d]$$

for all combinations of two letters. In the third step, we apply (2) again, to any two nice terms we've constructed so far. We get new nice term such as $[a \circ [a \circ a]]$—where t is a from step 1 and s is $[a \circ a]$ from step 2—and $[[b \circ c] \circ [d \circ b]]$ constructed out of the two terms $[b \circ c]$ and $[d \circ b]$ from step 2. And so on. Clause (3) rules out that anything not constructed in this way sneaks into the set of nice terms.

Note that we have not yet proved that every sequence of symbols that "feels" nice is nice according to this definition. However, it should be clear that everything we can construct does in fact "feel nice": brackets are balanced, and ∘ connects parts that are themselves nice.

The key feature of inductive definitions is that if you want to prove something about all nice terms, the definition tells you which cases you must consider. For instance, if you are told that t is a nice term, the inductive definition tells you what t can look like: t can be a letter, or it can be [s_1 ∘ s_2] for some pair of nice terms s_1 and s_2. Because of clause (3), those are the only possibilities.

When proving claims about all of an inductively defined set, the strong form of induction becomes particularly important. For instance, suppose we want to prove that for every nice term of length n, the number of [in it is $< n/2$. This can be seen as a claim about all n: for every n, the number of [in any nice term of length n is $< n/2$.

Proposition B.4. *For any n, the number of [in a nice term of length n is $< n/2$.*

Proof. To prove this result by (strong) induction, we have to show that the following conditional claim is true:

> If for every $l < k$, any nice term of length l has $< l/2$ ['s, then any nice term of length k has $< k/2$ ['s.

To show this conditional, assume that its antecedent is true, i.e., assume that for any $l < k$, nice terms of length l contain $< l/2$ ['s. We call this assumption the inductive hypothesis. We want to show the same is true for nice terms of length k.

So suppose t is a nice term of length k. Because nice terms are inductively defined, we have two cases: (1) t is a letter by itself, or (2) t is [s_1 ∘ s_2] for some nice terms s_1 and s_2.

1. t is a letter. Then $k = 1$, and the number of [in t is 0. Since $0 < 1/2$, the claim holds.

2. t is $[s_1 \circ s_2]$ for some nice terms s_1 and s_2. Let's let l_1 be the length of s_1 and l_2 be the length of s_2. Then the length k of t is $l_1 + l_2 + 3$ (the lengths of s_1 and s_2 plus three symbols [, $\circ$,]). Since $l_1 + l_2 + 3$ is always greater than l_1, $l_1 < k$. Similarly, $l_2 < k$. That means that the induction hypothesis applies to the terms s_1 and s_2: the number m_1 of [in s_1 is $< l_1/2$, and the number m_2 of [in s_2 is $< l_2/2$.

The number of [in t is the number of [in s_1, plus the number of [in s_2, plus 1, i.e., it is $m_1 + m_2 + 1$. Since $m_1 < l_1/2$ and $m_2 < l_2/2$ we have:

$$m_1 + m_2 + 1 < \frac{l_1}{2} + \frac{l_2}{2} + 1 = \frac{l_1 + l_2 + 2}{2} < \frac{l_1 + l_2 + 3}{2} = k/2.$$

In each case, we've shown that the number of [in t is $< k/2$ (on the basis of the inductive hypothesis). By strong induction, the proposition follows. □

B.5 Structural Induction

So far we have used induction to establish results about all natural numbers. But a corresponding principle can be used directly to prove results about all elements of an inductively defined set. This often called *structural* induction, because it depends on the structure of the inductively defined objects.

Generally, an inductive definition is given by (a) a list of "initial" elements of the set and (b) a list of operations which produce new elements of the set from old ones. In the case of nice terms, for instance, the initial objects are the letters. We only have one operation: the operations are

$$o(s_1, s_2) = [s_1 \circ s_2]$$

You can even think of the natural numbers $\mathbb{N}$ themselves as being given by an inductive definition: the initial object is 0, and the operation is the successor function $x + 1$.

In order to prove something about all elements of an inductively defined set, i.e., that every element of the set has a property P, we must:

1. Prove that the initial objects have P

2. Prove that for each operation o, if the arguments have P, so does the result.

For instance, in order to prove something about all nice terms, we would prove that it is true about all letters, and that it is true about $[s_1 \circ s_2]$ provided it is true of s_1 and s_2 individually.

Proposition B.5. *The number of [equals the number of] in any nice term t.*

Proof. We use structural induction. Nice terms are inductively defined, with letters as initial objects and the operation o for constructing new nice terms out of old ones.

1. The claim is true for every letter, since the number of [in a letter by itself is 0 and the number of] in it is also 0.

2. Suppose the number of [in s_1 equals the number of], and the same is true for s_2. The number of [in $o(s_1, s_2)$, i.e., in $[s_1 \circ s_2]$, is the sum of the number of [in s_1 and s_2 plus one. The number of] in $o(s_1, s_2)$ is the sum of the number of] in s_1 and s_2 plus one. Thus, the number of [in $o(s_1, s_2)$ equals the number of] in $o(s_1, s_2)$. □

Let's give another proof by structural induction: a proper initial segment of a string t of symbols is any string s that agrees with t symbol by symbol, read from the left, but t is longer. So, e.g., $[a \circ$ is a proper initial segment of $[a \circ b]$, but neither are $[b \circ$ (they disagree at the second symbol) nor $[a \circ b]$ (they are the same length).

Proposition B.6. *Every proper initial segment of a nice term t has more ['s than] 's.*

Proof. By induction on t:

1. t is a letter by itself: Then t has no proper initial segments.

2. $t = [s_1 \circ s_2]$ for some nice terms s_1 and s_2. If r is a proper initial segment of t, there are a number of possibilities:

    a) r is just [: Then r has one more [than it does].

    b) r is [r_1 where r_1 is a proper initial segment of s_1: Since s_1 is a nice term, by induction hypothesis, r_1 has more [than] and the same is true for [r_1.

    c) r is [s_1 or [s_1 ∘ : By the previous result, the number of [and] in s_1 are equal; so the number of [in [s_1 or [s_1 ∘ is one more than the number of].

    d) r is [s_1 ∘ r_2 where r_2 is a proper initial segment of s_2: By induction hypothesis, r_2 contains more [than]. By the previous result, the number of [and of] in s_1 are equal. So the number of [in [s_1 ∘ r_2 is greater than the number of].

    e) r is [s_1 ∘ s_2: By the previous result, the number of [and] in s_1 are equal, and the same for s_2. So there is one more [in [s_1 ∘ s_2 than there are]. □

B.6 Relations and Functions

When we have defined a set of objects (such as the natural numbers or the nice terms) inductively, we can also define *relations on* these objects by induction. For instance, consider the following idea: a nice term t_1 is a subterm of a nice term t_2 if it occurs as a part of it. Let's use a symbol for it: $t_1 \sqsubseteq t_2$. Every nice term is a subterm of itself, of course: $t \sqsubseteq t$. We can give an inductive definition of this relation as follows:

Definition B.7. The relation of a nice term t_1 being a subterm of t_2, $t_1 \sqsubseteq t_2$, is defined by induction on t_2 as follows:

1. If t_2 is a letter, then $t_1 \sqsubseteq t_2$ iff $t_1 = t_2$.

2. If t_2 is $[s_1 \circ s_2]$, then $t_1 \sqsubseteq t_2$ iff $t_1 = t_2$, $t_1 \sqsubseteq s_1$, or $t_1 \sqsubseteq s_2$.

This definition, for instance, will tell us that a $\sqsubseteq$ [b ∘ a]. For (2) says that a $\sqsubseteq$ [b ∘ a] iff a = [b ∘ a], or a $\sqsubseteq$ b, or a $\sqsubseteq$ a. The first two are false: a clearly isn't identical to [b ∘ a], and by (1), a $\sqsubseteq$ b iff a = b, which is also false. However, also by (1), a $\sqsubseteq$ a iff a = a, which is true.

It's important to note that the success of this definition depends on a fact that we haven't proved yet: every nice term t is either a letter by itself, or there are *uniquely determined* nice terms s_1 and s_2 such that $t = [s_1 \circ s_2]$. "Uniquely determined" here means that if $t = [s_1 \circ s_2]$ it isn't *also* $= [r_1 \circ r_2]$ with $s_1 \neq r_1$ or $s_2 \neq r_2$. If this were the case, then clause (2) may come in conflict with itself: reading t_2 as $[s_1 \circ s_2]$ we might get $t_1 \sqsubseteq t_2$, but if we read t_2 as $[r_1 \circ r_2]$ we might get not $t_1 \sqsubseteq t_2$. Before we prove that this can't happen, let's look at an example where it *can* happen.

Definition B.8. Define *bracketless terms* inductively by

1. Every letter is a bracketless term.

2. If s_1 and s_2 are bracketless terms, then $s_1 \circ s_2$ is a bracketless term.

3. Nothing else is a bracketless term.

Bracketless terms are, e.g., a, b ∘ d, b ∘ a ∘ b. Now if we defined "subterm" for bracketless terms the way we did above, the second clause would read

If $t_2 = s_1 \circ s_2$, then $t_1 \sqsubseteq t_2$ iff $t_1 = t_2$, $t_1 \sqsubseteq s_1$, or $t_1 \sqsubseteq s_2$.

B.6. RELATIONS AND FUNCTIONS

Now b ∘ a ∘ b is of the form $s_1 \circ s_2$ with

$$s_1 = \text{b} \quad \text{and} \quad s_2 = \text{a} \circ \text{b}.$$

It is also of the form $r_1 \circ r_2$ with

$$r_1 = \text{b} \circ \text{a} \quad \text{and} \quad r_2 = \text{b}.$$

Now is a ∘ b a subterm of b ∘ a ∘ b? The answer is yes if we go by the first reading, and no if we go by the second.

The property that the way a nice term is built up from other nice terms is unique is called *unique readability*. Since inductive definitions of relations for such inductively defined objects are important, we have to prove that it holds.

Proposition B.9. *Suppose t is a nice term. Then either t is a letter by itself, or there are uniquely determined nice terms s_1, s_2 such that $t = [s_1 \circ s_2]$.*

Proof. If t is a letter by itself, the condition is satisfied. So assume t isn't a letter by itself. We can tell from the inductive definition that then t must be of the form $[s_1 \circ s_2]$ for some nice terms s_1 and s_2. It remains to show that these are uniquely determined, i.e., if $t = [r_1 \circ r_2]$, then $s_1 = r_1$ and $s_2 = r_2$.

So suppose $t = [s_1 \circ s_2]$ and also $t = [r_1 \circ r_2]$ for nice terms s_1, s_2, r_1, r_2. We have to show that $s_1 = r_1$ and $s_2 = r_2$. First, s_1 and r_1 must be identical, for otherwise one is a proper initial segment of the other. But by Proposition B.6, that is impossible if s_1 and r_1 are both nice terms. But if $s_1 = r_1$, then clearly also $s_2 = r_2$. □

We can also define functions inductively: e.g., we can define the function f that maps any nice term to the maximum depth of nested [...] in it as follows:

Definition B.10. The *depth* of a nice term, $f(t)$, is defined inductively as follows:

$$f(t) = \begin{cases} 0 & \text{if } t \text{ is a letter} \\ \max(f(s_1), f(s_2)) + 1 & \text{if } t = [s_1 \circ s_2]. \end{cases}$$

For instance

$$f([a \circ b]) = \max(f(a), f(b)) + 1 =$$
$$= \max(0,0) + 1 = 1, \text{ and}$$
$$f([[a \circ b] \circ c]) = \max(f([a \circ b]), f(c)) + 1 =$$
$$= \max(1,0) + 1 = 2.$$

Here, of course, we assume that s_1 an s_2 are nice terms, and make use of the fact that every nice term is either a letter or of the form $[s_1 \circ s_2]$. It is again important that it can be of this form in only one way. To see why, consider again the bracketless terms we defined earlier. The corresponding "definition" would be:

$$g(t) = \begin{cases} 0 & \text{if } t \text{ is a letter} \\ \max(g(s_1), g(s_2)) + 1 & \text{if } t = s_1 \circ s_2. \end{cases}$$

Now consider the bracketless term a ∘ b ∘ c ∘ d. It can be read in more than one way, e.g., as $s_1 \circ s_2$ with

$$s_1 = \text{a} \quad\text{and}\quad s_2 = \text{b} \circ \text{c} \circ \text{d},$$

or as $r_1 \circ r_2$ with

$$r_1 = \text{a} \circ b \quad\text{and}\quad r_2 = \text{c} \circ \text{d}.$$

Calculating g according to the first way of reading it would give

$$g(s_1 \circ s_2) = \max(g(a), g(b \circ c \circ d)) + 1 =$$
$$= \max(0, 2) + 1 = 3$$

while according to the other reading we get

$$g(r_1 \circ r_2) = \max(g(a \circ b), g(c \circ d)) + 1 =$$
$$= \max(1,1) + 1 = 2$$

But a function must always yield a unique value; so our "definition" of g doesn't define a function at all.

Problems

Problem B.1. Define the set of supernice terms by

1. Any letter a, b, c, d is a supernice term.

2. If s is a supernice term, then so is $[s]$.

3. If s_1 and s_2 are supernice terms, then so is $[s_1 \circ s_2]$.

4. Nothing else is a supernice term.

Show that the number of [in a supernice term t of length n is $\leq n/2 + 1$.

Problem B.2. Prove by structural induction that no nice term starts with].

Problem B.3. Give an inductive definition of the function l, where $l(t)$ is the number of symbols in the nice term t.

Problem B.4. Prove by structural induction on nice terms t that $f(t) < l(t)$ (where $l(t)$ is the number of symbols in t and $f(t)$ is the depth of t as defined in Definition B.10).

APPENDIX C
Biographies

C.1 Georg Cantor

An early biography of Georg Cantor (GAY-org KAHN-tor) claimed that he was born and found on a ship that was sailing for Saint Petersburg, Russia, and that his parents were unknown. This, however, is not true; although he was born in Saint Petersburg in 1845.

Fig. C.1: Georg Cantor

Cantor received his doctorate in mathematics at the University of Berlin in 1867. He is known for his work in set theory, and is credited with founding set theory as a distinctive research discipline. He was the first to prove that there are infinite sets of different sizes. His theories, and especially his theory of infinities, caused much debate among mathematicians at the time, and his work was controversial.

Cantor's religious beliefs and his mathematical work were inextricably tied; he even claimed that the theory of transfinite numbers had been communicated to him directly by God. In later life, Cantor suffered from mental illness. Beginning in 1894, and more frequently towards his later years, Cantor was hospitalized. The heavy criticism of his work, including a falling out with the mathematician Leopold Kronecker, led to depression and a lack of interest in mathematics. During depressive episodes, Cantor would turn to philosophy and literature, and even published a theory that Francis Bacon was the author of Shakespeare's plays.

Cantor died on January 6, 1918, in a sanatorium in Halle.

Further Reading For full biographies of Cantor, see Dauben (1990) and Grattan-Guinness (1971). Cantor's radical views are also described in the BBC Radio 4 program *A Brief History of Mathematics* (du Sautoy, 2014). If you'd like to hear about Cantor's theories in rap form, see Rose (2012).

C.2 Alonzo Church

Alonzo Church was born in Washington, DC on June 14, 1903. In early childhood, an air gun incident left Church blind in one eye. He finished preparatory school in Connecticut in 1920 and began his university education at Princeton that same year. He completed his doctoral studies in 1927. After a couple years abroad, Church returned to Princeton. Church was known exceedingly polite and careful. His blackboard writing was immaculate, and he would preserve important pa-

Fig. C.2: Alonzo Church

pers by carefully covering them in Duco cement (a clear glue). Outside of his academic pursuits, he enjoyed reading science fiction magazines and was not afraid to write to the editors if he spotted any inaccuracies in the writing.

Church's academic achievements were great. Together with his students Stephen Kleene and Barkley Rosser, he developed a theory of effective calculability, the lambda calculus, independently of Alan Turing's development of the Turing machine. The two definitions of computability are equivalent, and give rise to what is now known as the *Church-Turing Thesis*, that a function of the natural numbers is effectively computable if and only if it is computable via Turing machine (or lambda calculus). He also proved what is now known as *Church's Theorem*: The decision problem for the validity of first-order formulas is unsolvable.

Church continued his work into old age. In 1967 he left Princeton for UCLA, where he was professor until his retirement in 1990. Church passed away on August 1, 1995 at the age of 92.

Further Reading For a brief biography of Church, see Enderton (2019). Church's original writings on the lambda calculus and the Entscheidungsproblem (Church's Thesis) are Church (1936a,b). Aspray (1984) records an interview with Church about the Princeton mathematics community in the 1930s. Church wrote a series of book reviews of the *Journal of Symbolic Logic* from 1936 until 1979. They are all archived on John MacFarlane's website (MacFarlane, 2015).

C.3 Gerhard Gentzen

Gerhard Gentzen is known primarily as the creator of structural proof theory, and specifically the creation of the natural deduction and sequent calculus derivation systems. He was born on November 24, 1909 in Greifswald, Germany. Gerhard was homeschooled for three years before attending preparatory school, where he was behind most of his classmates in terms of educa-

C.3. GERHARD GENTZEN

tion. Despite this, he was a brilliant student and showed a strong aptitude for mathematics. His interests were varied, and he, for instance, also write poems for his mother and plays for the school theatre.

Gentzen began his university studies at the University of Greifswald, but moved around to Göttingen, Munich, and Berlin. He received his doctorate in 1933 from the University of Göttingen under Hermann Weyl. (Paul Bernays supervised most of his work, but was dismissed from the university by the Nazis.) In 1934, Gentzen began work as an assistant to David Hilbert. That same year he developed the sequent calculus and natural deduction derivation systems, in his papers *Untersuchungen über das logische Schließen I–II [Investigations Into Logical Deduction I–II]*. He proved the consistency of the Peano axioms in 1936.

Fig. C.3: Gerhard Gentzen

Gentzen's relationship with the Nazis is complicated. At the same time his mentor Bernays was forced to leave Germany, Gentzen joined the university branch of the SA, the Nazi paramilitary organization. Like many Germans, he was a member of the Nazi party. During the war, he served as a telecommunications officer for the air intelligence unit. However, in 1942 he was released from duty due to a nervous breakdown. It is unclear whether or not Gentzen's loyalties lay with the Nazi party, or whether he joined the party in order to ensure academic success.

In 1943, Gentzen was offered an academic position at the Mathematical Institute of the German University of Prague, which he accepted. However, in 1945 the citizens of Prague revolted against German occupation. Soviet forces arrived in the city and arrested all the professors at the university. Because of his membership in Nazi organizations, Gentzen was taken to a forced labour camp. He died of malnutrition while in his cell on August 4, 1945 at the age of 35.

Further Reading For a full biography of Gentzen, see Menzler-Trott (2007). An interesting read about mathematicians under Nazi rule, which gives a brief note about Gentzen's life, is given by Segal (2014). Gentzen's papers on logical deduction are available in the original german (Gentzen, 1935a,b). English translations of Gentzen's papers have been collected in a single volume by Szabo (1969), which also includes a biographical sketch.

C.4 Kurt Gödel

Kurt Gödel (GER-dle) was born on April 28, 1906 in Brünn in the Austro-Hungarian empire (now Brno in the Czech Republic). Due to his inquisitive and bright nature, young Kurtele was often called "Der kleine Herr Warum" (Little Mr. Why) by his family. He excelled in academics from primary school onward, where he got less than the highest grade only in mathematics. Gödel was often absent from school due to poor health and was exempt from physical education. He was diagnosed with

Fig. C.4: Kurt Gödel

rheumatic fever during his childhood. Throughout his life, he believed this permanently affected his heart despite medical assessment saying otherwise.

Gödel began studying at the University of Vienna in 1924 and completed his doctoral studies in 1929. He first intended to study physics, but his interests soon moved to mathematics and especially logic, in part due to the influence of the philosopher Rudolf Carnap. His dissertation, written under the supervision of Hans Hahn, proved the completeness theorem of first-order predicate logic with identity (Gödel, 1929). Only a year later, he obtained his most famous results—the first and second incompleteness theorems (published in Gödel 1931). During his time in Vienna, Gödel was heavily involved with the Vienna Circle, a group of scientifically-minded philosophers that included Carnap, whose work was especially influenced by Gödel's results.

In 1938, Gödel married Adele Nimbursky. His parents were not pleased: not only was she six years older than him and already divorced, but she worked as a dancer in a nightclub. Social pressures did not affect Gödel, however, and they remained happily married until his death.

After Nazi Germany annexed Austria in 1938, Gödel and Adele emigrated to the United States, where he took up a position at the Institute for Advanced Study in Princeton, New Jersey. Despite his introversion and eccentric nature, Gödel's time at Princeton was collaborative and fruitful. He published essays in set theory, philosophy and physics. Notably, he struck up a particularly strong friendship with his colleague at the IAS, Albert Einstein.

In his later years, Gödel's mental health deteriorated. His wife's hospitalization in 1977 meant she was no longer able to cook his meals for him. Having suffered from mental health issues throughout his life, he succumbed to paranoia. Deathly afraid of being poisoned, Gödel refused to eat. He died of starvation on January 14, 1978, in Princeton.

Further Reading For a complete biography of Gödel's life is available, see John Dawson (1997). For further biographical pieces, as well as essays about Gödel's contributions to logic and philosophy, see Wang (1990), Baaz et al. (2011), Takeuti et al. (2003), and Sigmund et al. (2007).

Gödel's PhD thesis is available in the original German (Gödel, 1929). The original text of the incompleteness theorems is (Gödel, 1931). All of Gödel's published and unpublished writings, as well as a selection of correspondence, are available in English in his *Collected Papers* Feferman et al. (1986, 1990).

For a detailed treatment of Gödel's incompleteness theorems, see Smith (2013). For an informal, philosophical discussion of Gödel's theorems, see Mark Linsenmayer's podcast (Linsenmayer, 2014).

C.5 Emmy Noether

Emmy Noether (NER-ter) was born in Erlangen, Germany, on March 23, 1882, to an upper-middle class scholarly family. Hailed as the "mother of modern algebra," Noether made groundbreaking contributions to both mathematics and physics, despite significant barriers to women's education. In Germany at the time, young girls were meant to be educated in arts and were not allowed to attend college preparatory schools. However, after auditing classes at the Universities of Göttingen and Erlangen (where her father was professor of mathematics), Noether was eventually able to enroll as a student at Erlangen in 1904, when their policy was updated to allow female students. She re-

Fig. C.5: Emmy Noether

ceived her doctorate in mathematics in 1907.

Despite her qualifications, Noether experienced much resistance during her career. From 1908–1915, she taught at Erlangen without pay. During this time, she caught the attention of David Hilbert, one of the world's foremost mathematicians of the time, who invited her to Göttingen. However, women were prohibited from obtaining professorships, and she was only able to lecture under Hilbert's name, again without pay. During this time she proved what is now known as Noether's theorem, which is still used in theoretical physics today. Noether was finally granted the right to teach in 1919. Hilbert's response to continued resistance of his university colleagues reportedly was: "Gentlemen, the faculty senate is not a bathhouse."

In the later 1920s, she concentrated on work in abstract algebra, and her contributions revolutionized the field. In her proofs she often made use of the so-called ascending chain condition, which states that there is no infinite strictly increasing chain of certain sets. For instance, certain algebraic structures now known as Noetherian rings have the property that there are no infinite sequences of ideals $I_1 \subsetneq I_2 \subsetneq \ldots$. The condition can be generalized to any partial order (in algebra, it concerns the special case of ideals ordered by the subset relation), and we can also consider the dual descending chain condition, where every strictly *de*creasing sequence in a partial order eventually ends. If a partial order satisfies the descending chain condition, it is possible to use induction along this order in a similar way in which we can use induction along the $<$ order on $\mathbb{N}$. Such orders are called *well-founded* or *Noetherian*, and the corresponding proof principle *Noetherian induction*.

Noether was Jewish, and when the Nazis came to power in 1933, she was dismissed from her position. Luckily, Noether was able to emigrate to the United States for a temporary position at Bryn Mawr, Pennsylvania. During her time there she also lectured at Princeton, although she found the university to be unwelcoming to women (Dick, 1981, 81). In 1935, Noether underwent an operation to remove a uterine tumour. She died from an infection as a result of the surgery, and was buried at Bryn Mawr.

Further Reading   For a biography of Noether, see Dick (1981). The Perimeter Institute for Theoretical Physics has their lectures on Noether's life and influence available online (Institute, 2015). If you're tired of reading, *Stuff You Missed in History Class* has a podcast on Noether's life and influence (Frey and Wilson, 2015). The collected works of Noether are available in the original German (Jacobson, 1983).

C.6 Bertrand Russell

Bertrand Russell is hailed as one of the founders of modern analytic philosophy. Born May 18, 1872, Russell was not only known for his work in philosophy and logic, but wrote many popular books in various subject areas. He was also an ardent political activist throughout his life.

Fig. C.6: Bertrand Russell

Russell was born in Trellech, Monmouthshire, Wales. His parents were members of the British nobility. They were free-thinkers, and even made friends with the radicals in Boston at the time. Unfortunately, Russell's parents died when he was young, and Russell was sent to live with his grandparents. There, he was given a religious upbringing (something his parents had wanted to avoid at all costs). His grandmother was very strict in all matters of morality. During adolescence he was mostly homeschooled by private tutors.

Russell's influence in analytic philosophy, and especially logic, is tremendous. He studied mathematics and philosophy at Trinity College, Cambridge, where he was influenced by the mathematician and philosopher Alfred North Whitehead. In 1910, Russell and Whitehead published the first volume of *Principia Mathematica*, where they championed the view that mathematics is reducible to logic. He went on to publish hundreds of books, essays and political pamphlets. In 1950, he won the Nobel Prize for literature.

Russell's was deeply entrenched in politics and social activism. During World War I he was arrested and sent to prison for six months due to pacifist activities and protest. While in prison, he was able to write and read, and claims to have found the experience "quite agreeable." He remained a pacifist throughout his life, and was again incarcerated for attending a nuclear disarmament rally in 1961. He also survived a plane crash in 1948, where the only survivors were those sitting in the smoking section. As such, Russell claimed that he owed his life to smoking. Russell was married four times, but had a reputation for carrying on extra-marital affairs. He died on February 2, 1970 at the age of 97 in Penrhyndeudraeth, Wales.

Further Reading Russell wrote an autobiography in three parts, spanning his life from 1872–1967 (Russell, 1967, 1968, 1969). The Bertrand Russell Research Centre at McMaster University is home of the Bertrand Russell archives. See their website at Duncan (2015), for information on the volumes of his collected works (including searchable indexes), and archival projects. Russell's paper *On Denoting* (Russell, 1905) is a classic of 20th century analytic philosophy.

The Stanford Encyclopedia of Philosophy entry on Russell (Irvine, 2015) has sound clips of Russell speaking on Desire and Political theory. Many video interviews with Russell are available online. To see him talk about smoking and being involved in a plane crash, e.g., see Russell (n.d.). Some of Russell's works, including his *Introduction to Mathematical Philosophy* are available as free audiobooks on LibriVox (n.d.).

C.7 Alfred Tarski

Alfred Tarski was born on January 14, 1901 in Warsaw, Poland (then part of the Russian Empire). Described as "Napoleonic," Tarski was boisterous, talkative, and intense. His energy was often reflected in his lectures—he once set fire to a wastebasket while disposing of a cigarette during a lecture, and was forbidden from lecturing in that building again.

Fig. C.7: Alfred Tarski

Tarski had a thirst for knowledge from a young age. Although later in life he would tell students that he studied logic because it was the only class in which he got a B, his high school records show that he got A's across the board—even in logic. He studied at the University of Warsaw from 1918 to 1924. Tarski first intended to study biology, but became interested in mathematics, philosophy, and logic, as the university was the center of the Warsaw School of Logic and Philosophy. Tarski earned his doctorate in 1924 under the supervision of Stanisław Leśniewski.

Before emigrating to the United States in 1939, Tarski completed some of his most important work while working as a secondary school teacher in Warsaw. His work on logical consequence and logical truth were written during this time. In 1939, Tarski was visiting the United States for a lecture tour. During his visit, Germany invaded Poland, and because of his Jewish heritage, Tarski could not return. His wife and children remained in Poland until the end of the war, but were then able to emigrate to the United States as well. Tarski taught at Harvard, the College of the City of New York, and the Institute for Advanced Study at Princeton, and finally the University of California, Berkeley. There he founded the multidisciplinary program in Logic and the Methodology of Science. Tarski died on October 26, 1983 at the age of 82.

Further Reading For more on Tarski's life, see the biography *Alfred Tarski: Life and Logic* (Feferman and Feferman, 2004). Tarski's seminal works on logical consequence and truth are available in English in (Corcoran, 1983). All of Tarski's original works have been collected into a four volume series, (Tarski, 1981).

C.8 Alan Turing

Alan Turing was born in Maida Vale, London, on June 23, 1912. He is considered the father of theoretical computer science. Turing's interest in the physical sciences and mathematics started at a young age. However, as a boy his interests were not represented well in his schools, where emphasis was placed on literature and classics. Consequently, he did poorly in school and was reprimanded by many of his teachers.

Turing attended King's College, Cambridge as an undergraduate, where he studied mathematics. In 1936 Turing developed (what is now called) the Turing machine as an attempt to precisely define the notion of a computable function and to prove the undecidability of the decision problem. He was beaten to the result by Alonzo Church, who proved the result via his own lambda calculus. Turing's paper was still published with reference to Church's result. Church invited Turing to Princeton, where he spent 1936–1938, and obtained a doctorate under Church.

Despite his interest in logic, Turing's earlier interests in physical sciences remained prevalent. His practical skills were put to work during his service with the British cryptanalytic department at Bletchley Park during World War II. Turing was a central figure in cracking the cypher used by German Naval communications—the Enigma code. Turing's expertise in statistics and cryptography, together with the introduction of electronic machinery, gave the team the ability to crack the code by creating a decrypting machine called a "bombe." His ideas also helped in the creation of the world's first programmable electronic computer, the Colossus, also used at Bletchley park to break the German Lorenz cypher.

Fig. C.8: Alan Turing

Turing was gay. Nevertheless, in 1942 he proposed to Joan Clarke, one of his teammates at Bletchley Park, but later broke off the engagement and confessed to her that he was homosexual. He had several lovers throughout his lifetime, although homosexual acts were then criminal offences in the UK. In 1952, Turing's house was burgled by a friend of his lover at the time, and when filing a police report, Turing admitted to having a homosexual relationship, under the impression that the government was on their way to legalizing homosexual acts. This was not true, and he was charged with gross indecency. Instead of going to prison, Turing opted for a hormone treatment that reduced libido. Turing was found dead on June 8, 1954, of a cyanide overdose—most likely suicide. He was given a royal pardon by Queen Elizabeth II in 2013.

Further Reading For a comprehensive biography of Alan Turing, see Hodges (2014). Turing's life and work inspired a play, *Breaking the Code*, which was produced in 1996 for TV starring Derek Jacobi as Turing. *The Imitation Game*, an Academy Award nominated film starring Bendict Cumberbatch and Kiera Knightley, is also loosely based on Alan Turing's life and time at Bletchley Park (Tyldum, 2014).

Radiolab (2012) has several podcasts on Turing's life and work. BBC Horizon's documentary *The Strange Life and Death of Dr. Turing* is available to watch online (Sykes, 1992). (Theelen, 2012) is a short video of a working LEGO Turing Machine—made to honour Turing's centenary in 2012.

Turing's original paper on Turing machines and the decision problem is Turing (1937).

C.9 Ernst Zermelo

Ernst Zermelo was born on July 27, 1871 in Berlin, Germany. He had five sisters, though his family suffered from poor health and only three survived to adulthood. His parents also passed away when he was young, leaving him and his siblings orphans when he was seventeen. Zermelo had a deep interest in the arts, and especially in poetry. He was known for being sharp, witty, and critical. His most celebrated mathematical achievements include the introduction of the axiom of choice (in 1904), and his axiomatization of set theory (in 1908).

Fig. C.9: Ernst Zermelo

Zermelo's interests at university were varied. He took courses in physics, mathematics, and philosophy. Under the supervision of Hermann Schwarz, Zermelo completed his dissertation *Investigations in the Calculus of Variations* in 1894 at the University of Berlin. In 1897, he decided to pursue more studies at the University of Göttigen, where he was heavily influenced by the foundational work of David Hilbert. In 1899 he became eligible for professorship, but did not get one until eleven years later—possibly due to his strange demeanour and "nervous haste."

Zermelo finally received a paid professorship at the University of Zurich in 1910, but was forced to retire in 1916 due to tuberculosis. After his recovery, he was given an honourary professorship at the University of Freiburg in 1921. During this time he worked on foundational mathematics. He became irritated with the works of Thoralf Skolem and Kurt Gödel, and publicly criticized their approaches in his papers. He was dismissed from his position at Freiburg in 1935, due to his unpopularity and his opposition to Hitler's rise to power in Germany.

The later years of Zermelo's life were marked by isolation. After his dismissal in 1935, he abandoned mathematics. He moved to the country where he lived modestly. He married in 1944, and became completely dependent on his wife as he was going blind. Zermelo lost his sight completely by 1951. He passed away in Günterstal, Germany, on May 21, 1953.

Further Reading For a full biography of Zermelo, see Ebbinghaus (2015). Zermelo's seminal 1904 and 1908 papers are available to read in the original German (Zermelo, 1904, 1908). Zermelo's collected works, including his writing on physics, are available in English translation in (Ebbinghaus et al., 2010; Ebbinghaus and Kanamori, 2013).

APPENDIX D
The Greek Alphabet

Alpha	α	A	Nu	ν	N
Beta	β	B	Xi	ξ	Ξ
Gamma	γ	Γ	Omicron	o	O
Delta	δ	Δ	Pi	π	Π
Epsilon	ε	E	Rho	ρ	P
Zeta	ζ	Z	Sigma	σ	Σ
Eta	η	H	Tau	τ	T
Theta	θ	Θ	Upsilon	υ	Υ
Iota	ι	I	Phi	φ	Φ
Kappa	κ	K	Chi	χ	X
Lambda	λ	Λ	Psi	ψ	Ψ
Mu	μ	M	Omega	ω	Ω

Glossary

anti-symmetric R is anti-symmetric iff, whenever both Rxy and Ryx, then $x = y$; in other words: if $x \neq y$ then not Rxy or not Ryx (see section 2.2).

assumption A formula that stands topmost in a derivation, also called an initial formula. It may be discharged or undischarged (see section 11.1).

asymmetric R is asymmetric if for no pair $x, y \in A$ we have Rxy and Ryx (see section 2.4).

bijection A function that is both surjective and injective (see section 3.2).

binary relation A subset of A^2; we write Rxy (or xRy) for $\langle x, y \rangle \in R$ (see section 2.1).

bound Occurrence of a variable within the scope of a quantifier that uses the same variable (see section 6.7).

Cartesian product $(A \times B)$ Set of all pairs of elements of A and B; $A \times B = \{\langle x, y \rangle : x \in A \text{ and } y \in B\}$ (see section 1.5).

Church-Turing Theorem States that there is no Turing machine which decides if a given sentence of first-order logic is valid or not (see section 15.8).

Church-Turing Thesis states that anything computable via an effective procedure is Turing computable (see section 14.10).

closed A set of sentences Γ is closed iff, whenever $\Gamma \models A$ then $A \in \Gamma$. The set $\{A : \Gamma \models A\}$ is the closure of Γ (see section 8.1).

compactness theorem States that every finitely satisfiable set of sentences is satisfiable (see section 12.9).

complete consistent set A set of sentences is complete and consistent iff it is consistent, and for every sentence A either A or $\neg A$ is in the set (see section 12.3).

completeness Property of a derivation system; it is complete if, whenever Γ entails A, then there is also a derivation that establishes $\Gamma \vdash A$; equivalently, iff every consistent set of sentences is satisfiable (see section 12.1).

completeness theorem States that first-order logic is complete: every consistent set of sentences is satisfiable.

composition ($g \circ f$) The function resulting from "chaining together" f and g; $(g \circ f)(x) = g(f(x))$ (see section 3.5).

connected R is connected if for all $x, y \in A$ with $x \neq y$, then either Rxy or Ryx (see section 2.2).

consistent In the sequent calculus, a set of sentences Γ is consistent iff there is no derivation of a sequent $\Gamma_0 \Rightarrow$ with $\Gamma_0 \subseteq \Gamma$ (see section 10.8). In natural deduction, Γ is consistent iff $\Gamma \nvdash \bot$ (see section 11.7). If Γ is not consistent, it is inconsistent..

covered A structure in which every element of the domain is the value of some closed term (see section 7.2).

decision problem Problem of deciding if a given sentence of first-order logic is valid or not (see Church-Turing Theorem).

deduction theorem Relates entailment and provability of a sentence from an assumption with that of a corresponding conditional. In the semantic form (Theorem 7.29), it states that $\Gamma \cup \{A\} \models B$ iff $\Gamma \models A \to B$. In the proof-theoretic form, it states that $\Gamma \cup \{A\} \vdash B$ iff $\Gamma \vdash A \to B$.

derivability ($\Gamma \vdash A$) In the sequent calculus, A is derivable from Γ if there is a derivation of a sequent $\Gamma_0 \Rightarrow A$ where $\Gamma_0 \subseteq \Gamma$ is a finite sequence of sentences in Γ (see section 10.8). In natural deduction, A is derivable from Γ if there is a derivation with end-formula A and in which every assumption is either discharged or is in Γ (see section 11.7).

derivation In the sequent calculus, a tree of sequents in which every sequent is either an initial sequent or follows from the sequents immediately above it by a rule of inference (see section 10.1). In natural deduction, a tree of formulas in which every formula is either an assumption or follows from the formulas immediately above it by a rule of inference (see section 11.1).

difference $(A \setminus B)$ the set of all elements of A which are not also elements of B: $A \setminus B = \{x : x \in A \text{ and } x \notin B\}$ (see section 1.4).

discharged An assumption in a derivation may be discharged by an inference rule below it (the rule and the assumption are then assigned a matching label, e.g., $[A]^2$). If it is not discharged, it is called undischarged (see section 11.1).

disjoint two sets with no elements in common (see section 1.4).

domain (of a function) $(\mathrm{dom}(f))$ The set of objects for which a (partial) function is defined (see section 3.1).

domain (of a structure) $(|M|)$ Non-empty set from from which a structure takes assignments and values of variables (see section 7.2).

eigenvariable In the sequent calculus, a special constant symbol in a premise of a $\exists$L or $\forall$R inference which may not appear in the conclusion (see section 10.1). In natural deduction, a special constant symbol in a premise of a $\exists$Elim or $\forall$Intro inference which may not appear in the conclusion or any undischarged assumption (see section 11.1).

entailment $(\Gamma \vDash A)$ A set of sentences Γ entails a sentence A iff for every structure M with $M \vDash \Gamma$, $M \vDash A$ (see section 7.7).

enumeration A possibly infinite list of all elements of a set A; formally a surjective function $f: \mathbb{N} \to A$ (see section 4.2).

equinumerous A and B are equinumerous iff there is a total bijection from A to B (see section 4.8).

equivalence relation a reflexive, symmetric, and transitive relation (see section 2.2).

extensionality (of satisfaction) Whether or not a formula A is satisfied depends only on the assignments to the non-logical symbols and free variables that actually occur in A.

extensionality (of sets) Sets A and B are identical, $A = B$, iff every element of A is also an element of B, and vice versa (see section 1.1).

finitely satisfiable Γ is finitely satisfiable iff every finite $\Gamma_0 \subseteq \Gamma$ is satisfiable (see section 12.9).

formula Expressions of a first-order language $\mathscr{L}$ which express relations or properties, or are true or false (see section 6.3).

free An occurrence of a variable that is not bound (see section 6.7).

free for A term t is free for x in A if none of the free occurrences of x in A occur in the scope of a quantifier that binds a variable in t (see section 6.8).

function ($f: A \to B$) A mapping of each element of a domain (of a function) A to an element of the codomain B (see section 3.1).

graph (of a function) the relation $R_f \subseteq A \times B$ defined by $R_f = \{\langle x,y \rangle : f(x) = y\}$, if $f: A \twoheadrightarrow B$ (see section 3.3).

halting problem The problem of determining (for any e, n) whether the Turing machine M_e halts for an input of n strokes (see section 15.4).

inconsistent see consistent.

injective $f: A \to B$ is injective iff for each $y \in B$ there is at most one $x \in A$ such that $f(x) = y$; equivalently if whenever $x \neq x'$ then $f(x) \neq f(x')$ (see section 3.2).

intersection $(A \cap B)$ The set of all things which are elements of both A and B: $A \cap B = \{x : x \in A \wedge x \in B\}$ (see section 1.4).

inverse function If $f: A \to B$ is a bijection, $f^{-1}: B \to A$ is the function with $f^{-1}(y) =$ whatever unique $x \in A$ is such that $f(x) = y$ (see section 3.4).

inverse relation (R^{-1}) The relation R "turned around"; $R^{-1} = \{\langle y, x \rangle : \langle x, y \rangle \in R\}$ (see section 2.6).

irreflexive R is irreflexive if, for no $x \in A$, Rxx (see section 2.4).

Löwenheim-Skolem Theorem States that every satisfiable set of sentences has a countable model (see section 12.11).

linear order A connected partial order (see section 2.4).

model A structure in which every sentence in Γ is true is a model of Γ (see section 8.2).

partial function ($f: A \twoheadrightarrow B$) A partial function is a mapping which assigns to every element of A at most one element of B. If f assigns an element of B to $x \in A$, $f(x)$ is defined, and otherwise undefined (see section 3.6).

partial order A reflexive, anti-symmetric, transitive relation (see section 2.4).

power set ($\wp(A)$) The set consisting of all subsets of a set A, $\wp(A) = \{x : x \subseteq A\}$ (see section 1.2).

preorder A reflexive and transitive relation (see section 2.4).

range (ran(f)) the subset of the codomain that is actually output by f; ran(f) = $\{y \in B : f(x) = y$ for some $x \in A\}$ (see section 3.1).

reflexive R is reflexive iff, for every $x \in A$, Rxx (see section 2.2).

satisfiable A set of sentences Γ is satisfiable if $M \vDash \Gamma$ for some structure M, otherwise it is unsatisfiable (see section 7.7).

sentence A formula with no free variable. (see section 6.7).

sequence (finite) (A^*) A finite string of elements of A; an element of A^n for some n (see section 1.3).

sequence (infinite) (A^ω) A gapless, unending sequence of elements of A; formally, a function $s\colon \mathbb{Z}^+ \to A$ (see section 1.3).

sequent An expression of the form $\Gamma \Rightarrow \Delta$ where Γ and Δ are finite sequences of sentences (see section 10.1).

set A collection of objects, considered independently of the way it is specified, of the order of the objects in the set, and of their multiplicity (see section 1.1).

soundness Property of a derivation system: it is sound if whenever $\Gamma \vdash A$ then $\Gamma \vDash A$ (see section 10.12 and section 11.11).

strict linear order A connected strict order (see section 2.4).

strict order An irreflexive, asymmetric, and transitive relation (see section 2.4).

structure (M) An interpretation of a first-order language, consisting of a domain (of a structure) and assignments of the constant, predicate and function symbols of the language (see section 7.2).

subformula Part of a formula which is itself a formula (see section 6.6).

subset ($A \subseteq B$) A set every element of which is an element of a given set B (see section 1.2).

surjective $f\colon A \to B$ is surjective iff the range of f is all of B, i.e., for every $y \in B$ there is at least one $x \in A$ such that $f(x) = y$ (see section 3.2).

symmetric R is symmetric iff, whenever Rxy then also Ryx (see section 2.2).

theorem ($\vdash A$) In the sequent calculus, a formula A is a theorem (of logic) if there is a derivation of the sequent $\Rightarrow A$ (see section 10.8). In natural deduction, a formula A is a theorem if there is a derivation of A with all assumptions discharged (see section 11.7). We also say that A is a theorem of a theory Γ if $\Gamma \vdash A$.

total order see linear order.

transitive R is transitive iff, whenever Rxy and Ryz, then also Rxz (see section 2.2).

transitive closure (R^+) the smallest transitive relation containing R (see section 2.6).

undischarged see discharged.

union ($A \cup B$) The set of all elements of A and B together: $A \cup B = \{x : x \in A \lor x \in B\}$ (see section 1.4).

valid ($\vDash A$) A sentence A is *valid* iff $M \vDash A$ for every structure M (see section 7.7).

variable assignment A function which maps each variable to an element of $|M|$ (see section 7.4).

x-**variant** Two variable assignments are *x*-variants, $s \sim_x s'$, if they differ at most in what they assign to x (see section 7.4).

Photo Credits

Georg Cantor, p. 366: Portrait of Georg Cantor by Otto Zeth courtesy of the Universitätsarchiv, Martin-Luther Universität Halle–Wittenberg. UAHW Rep. 40-VI, Nr. 3 Bild 102.

Alonzo Church, p. 367: Portrait of Alonzo Church, undated, photographer unknown. Alonzo Church Papers; 1924–1995, (C0948) Box 60, Folder 3. Manuscripts Division, Department of Rare Books and Special Collections, Princeton University Library. © Princeton University. The Open Logic Project has obtained permission to use this image for inclusion in non-commercial OLP-derived materials. Permission from Princeton University is required for any other use.

Gerhard Gentzen, p. 369: Portrait of Gerhard Gentzen playing ping-pong courtesy of Ekhart Mentzler-Trott.

Kurt Gödel, p. 370: Portrait of Kurt Gödel, ca. 1925, photographer unknown. From the Shelby White and Leon Levy Archives Center, Institute for Advanced Study, Princeton, NJ, USA, on deposit at Princeton University Library, Manuscript Division, Department of Rare Books and Special Collections, Kurt Gödel Papers, (C0282), Box 14b, #110000. The Open Logic Project has obtained permission from the Institute's Archives Center to use this image for inclusion in non-commercial OLP-derived materials. Permission from the Archives Center is required for any other use.

Emmy Noether, p. 372: Portrait of Emmy Noether, ca. 1922, courtesy of the Abteilung für Handschriften und Seltene Drucke,

Niedersächsische Staats- und Universitätsbibliothek Göttingen, Cod. Ms. D. Hilbert 754, Bl. 14 Nr. 73. Restored from an original scan by Joel Fuller.

Bertrand Russell, p. 374: Portrait of Bertrand Russell, ca. 1907, courtesy of the William Ready Division of Archives and Research Collections, McMaster University Library. Bertrand Russell Archives, Box 2, f. 4.

Alfred Tarski, p. 376: Passport photo of Alfred Tarski, 1939. Cropped and restored from a scan of Tarski's passport by Joel Fuller. Original courtesy of Bancroft Library, University of California, Berkeley. Alfred Tarski Papers, Banc MSS 84/49. The Open Logic Project has obtained permission to use this image for inclusion in non-commercial OLP-derived materials. Permission from Bancroft Library is required for any other use.

Alan Turing, p. 378: Portrait of Alan Mathison Turing by Elliott & Fry, 29 March 1951, NPG x82217, © National Portrait Gallery, London. Used under a Creative Commons BY-NC-ND 3.0 license.

Ernst Zermelo, p. 379: Portrait of Ernst Zermelo, ca. 1922, courtesy of the Abteilung für Handschriften und Seltene Drucke, Niedersächsische Staats- und Universitätsbibliothek Göttingen, Cod. Ms. D. Hilbert 754, Bl. 6 Nr. 25.

Bibliography

Aspray, William. 1984. The Princeton mathematics community in the 1930s: Alonzo Church. URL http://www.princeton.edu/mudd/finding_aids/mathoral/pmc05.htm. Interview.

Baaz, Matthias, Christos H. Papadimitriou, Hilary W. Putnam, Dana S. Scott, and Charles L. Harper Jr. 2011. *Kurt Gödel and the Foundations of Mathematics: Horizons of Truth*. Cambridge: Cambridge University Press.

Cantor, Georg. 1892. Über eine elementare Frage der Mannigfaltigkeitslehre. *Jahresbericht der deutschen Mathematiker-Vereinigung* 1: 75–8.

Cheng, Eugenia. 2004. How to write proofs: A quick quide. URL http://http://eugeniacheng.com/wp-content/uploads/2017/02/cheng-proofguide.pdf.

Church, Alonzo. 1936a. A note on the Entscheidungsproblem. *Journal of Symbolic Logic* 1: 40–41.

Church, Alonzo. 1936b. An unsolvable problem of elementary number theory. *American Journal of Mathematics* 58: 345–363.

Corcoran, John. 1983. *Logic, Semantics, Metamathematics*. Indianapolis: Hackett, 2nd ed.

Dauben, Joseph. 1990. *Georg Cantor: His Mathematics and Philosophy of the Infinite*. Princeton: Princeton University Press.

Dick, Auguste. 1981. *Emmy Noether 1882–1935*. Boston: Birkhäuser.

du Sautoy, Marcus. 2014. A brief history of mathematics: Georg Cantor. URL http://www.bbc.co.uk/programmes/b00ss1j0. Audio Recording.

Duncan, Arlene. 2015. The Bertrand Russell Research Centre. URL http://russell.mcmaster.ca/.

Ebbinghaus, Heinz-Dieter. 2015. *Ernst Zermelo: An Approach to his Life and Work*. Berlin: Springer-Verlag.

Ebbinghaus, Heinz-Dieter, Craig G. Fraser, and Akihiro Kanamori. 2010. *Ernst Zermelo. Collected Works*, vol. 1. Berlin: Springer-Verlag.

Ebbinghaus, Heinz-Dieter and Akihiro Kanamori. 2013. *Ernst Zermelo: Collected Works*, vol. 2. Berlin: Springer-Verlag.

Enderton, Herbert B. 2019. Alonzo Church: Life and Work. In *The Collected Works of Alonzo Church*, eds. Tyler Burge and Herbert B. Enderton. Cambridge, MA: MIT Press.

Feferman, Anita and Solomon Feferman. 2004. *Alfred Tarski: Life and Logic*. Cambridge: Cambridge University Press.

Feferman, Solomon, John W. Dawson Jr., Stephen C. Kleene, Gregory H. Moore, Robert M. Solovay, and Jean van Heijenoort. 1986. *Kurt Gödel: Collected Works. Vol. 1: Publications 1929–1936*. Oxford: Oxford University Press.

Feferman, Solomon, John W. Dawson Jr., Stephen C. Kleene, Gregory H. Moore, Robert M. Solovay, and Jean van Heijenoort. 1990. *Kurt Gödel: Collected Works. Vol. 2: Publications 1938–1974*. Oxford: Oxford University Press.

Frege, Gottlob. 1884. *Die Grundlagen der Arithmetik: Eine logisch mathematische Untersuchung über den Begriff der Zahl*. Breslau: Wilhelm Koebner. Translation in Frege (1953).

Frege, Gottlob. 1953. *Foundations of Arithmetic*, ed. J. L. Austin. Oxford: Basil Blackwell & Mott, 2nd ed.

Frey, Holly and Tracy V. Wilson. 2015. Stuff you missed in history class: Emmy Noether, mathematics trailblazer. URL https://www.iheart.com/podcast/stuff-you-missed-in-history-cl-21124503/episode/emmy-noether-mathematics-trailblazer-30207491/. Podcast audio.

Gentzen, Gerhard. 1935a. Untersuchungen über das logische Schließen I. *Mathematische Zeitschrift* 39: 176–210. English translation in Szabo (1969), pp. 68–131.

Gentzen, Gerhard. 1935b. Untersuchungen über das logische Schließen II. *Mathematische Zeitschrift* 39: 176–210, 405–431. English translation in Szabo (1969), pp. 68–131.

Gödel, Kurt. 1929. Über die Vollständigkeit des Logikkalküls [On the completeness of the calculus of logic]. Dissertation, Universität Wien. Reprinted and translated in Feferman et al. (1986), pp. 60–101.

Gödel, Kurt. 1931. über formal unentscheidbare Sätze der *Principia Mathematica* und verwandter Systeme I [On formally undecidable propositions of *Principia Mathematica* and related systems I]. *Monatshefte für Mathematik und Physik* 38: 173–198. Reprinted and translated in Feferman et al. (1986), pp. 144–195.

Grattan-Guinness, Ivor. 1971. Towards a biography of Georg Cantor. *Annals of Science* 27(4): 345–391.

Hammack, Richard. 2013. *Book of Proof.* Richmond, VA: Virginia Commonwealth University. URL http://www.people.vcu.edu/~rhammack/BookOfProof/BookOfProof.pdf.

Hodges, Andrew. 2014. *Alan Turing: The Enigma*. London: Vintage.

Hutchings, Michael. 2003. Introduction to mathematical arguments. URL `https://math.berkeley.edu/~hutching/teach/proofs.pdf`.

Institute, Perimeter. 2015. Emmy Noether: Her life, work, and influence. URL `https://www.youtube.com/watch?v=tNNyAyMRsgE`. Video Lecture.

Irvine, Andrew David. 2015. Sound clips of Bertrand Russell speaking. URL `http://plato.stanford.edu/entries/russell/russell-soundclips.html`.

Jacobson, Nathan. 1983. *Emmy Noether: Gesammelte Abhandlungen—Collected Papers.* Berlin: Springer-Verlag.

John Dawson, Jr. 1997. *Logical Dilemmas: The Life and Work of Kurt Gödel.* Boca Raton: CRC Press.

LibriVox. n.d. Bertrand Russell. URL `https://librivox.org/author/1508?primary_key=1508&search_category=author&search_page=1&search_form=get_results`. Collection of public domain audiobooks.

Linsenmayer, Mark. 2014. The partially examined life: Gödel on math. URL `http://www.partiallyexaminedlife.com/2014/06/16/ep95-godel/`. Podcast audio.

MacFarlane, John. 2015. Alonzo Church's JSL reviews. URL `http://johnmacfarlane.net/church.html`.

Magnus, P. D., Tim Button, J. Robert Loftis, Aaron Thomas-Bolduc, Robert Trueman, and Richard Zach. 2021. *Forall x: Calgary. An Introduction to Formal Logic.* Calgary: Open Logic Project, f21 ed. URL `https://forallx.openlogicproject.org/`.

Menzler-Trott, Eckart. 2007. *Logic's Lost Genius: The Life of Gerhard Gentzen.* Providence: American Mathematical Society.

Potter, Michael. 2004. *Set Theory and its Philosophy.* Oxford: Oxford University Press.

Radiolab. 2012. The Turing problem. URL `http://www.radiolab.org/story/193037-turing-problem/`. Podcast audio.

Rose, Daniel. 2012. A song about Georg Cantor. URL `https://www.youtube.com/watch?v=QUP5Z4Fb5k4`. Audio Recording.

Russell, Bertrand. 1905. On denoting. *Mind* 14: 479–493.

Russell, Bertrand. 1967. *The Autobiography of Bertrand Russell*, vol. 1. London: Allen and Unwin.

Russell, Bertrand. 1968. *The Autobiography of Bertrand Russell*, vol. 2. London: Allen and Unwin.

Russell, Bertrand. 1969. *The Autobiography of Bertrand Russell*, vol. 3. London: Allen and Unwin.

Russell, Bertrand. n.d. Bertrand Russell on smoking. URL `https://www.youtube.com/watch?v=80oLTiVW_lc`. Video Interview.

Sandstrum, Ted. 2019. *Mathematical Reasoning: Writing and Proof.* Allendale, MI: Grand Valley State University. URL `https://scholarworks.gvsu.edu/books/7/`.

Segal, Sanford L. 2014. *Mathematicians under the Nazis.* Princeton: Princeton University Press.

Sigmund, Karl, John Dawson, Kurt Mühlberger, Hans Magnus Enzensberger, and Juliette Kennedy. 2007. Kurt Gödel: Das Album–The Album. *The Mathematical Intelligencer* 29(3): 73–76.

Smith, Peter. 2013. *An Introduction to Gödel's Theorems.* Cambridge: Cambridge University Press.

Solow, Daniel. 2013. *How to Read and Do Proofs.* Hoboken, NJ: Wiley.

Steinhart, Eric. 2018. *More Precisely: The Math You Need to Do Philosophy.* Peterborough, ON: Broadview, 2nd ed.

Sykes, Christopher. 1992. BBC Horizon: The strange life and death of Dr. Turing. URL https://www.youtube.com/watch?v=gyusnGbBSHE.

Szabo, Manfred E. 1969. *The Collected Papers of Gerhard Gentzen.* Amsterdam: North-Holland.

Takeuti, Gaisi, Nicholas Passell, and Mariko Yasugi. 2003. *Memoirs of a Proof Theorist: Gödel and Other Logicians.* Singapore: World Scientific.

Tarski, Alfred. 1981. *The Collected Works of Alfred Tarski*, vol. I–IV. Basel: Birkhäuser.

Theelen, Andre. 2012. Lego turing machine. URL https://www.youtube.com/watch?v=FTSAiF9AHN4.

Turing, Alan M. 1937. On computable numbers, with an application to the "Entscheidungsproblem". *Proceedings of the London Mathematical Society, 2nd Series* 42: 230–265.

Tyldum, Morten. 2014. The imitation game. Motion picture.

Velleman, Daniel J. 2019. *How to Prove It: A Structured Approach.* Cambridge: Cambridge University Press, 3rd ed.

Wang, Hao. 1990. *Reflections on Kurt Gödel.* Cambridge: MIT Press.

Zermelo, Ernst. 1904. Beweis, daß jede Menge wohlgeordnet werden kann. *Mathematische Annalen* 59: 514–516. English translation in (Ebbinghaus et al., 2010, pp. 115–119).

Zermelo, Ernst. 1908. Untersuchungen über die Grundlagen der Mengenlehre I. *Mathematische Annalen* 65(2): 261–281. English translation in (Ebbinghaus et al., 2010, pp. 189-229).

About the Open Logic Project

The *Open Logic Text* is an open-source, collaborative textbook of formal meta-logic and formal methods, starting at an intermediate level (i.e., after an introductory formal logic course). Though aimed at a non-mathematical audience (in particular, students of philosophy and computer science), it is rigorous.

Coverage of some topics currently included may not yet be complete, and many sections still require substantial revision. We plan to expand the text to cover more topics in the future. We also plan to add features to the text, such as a glossary, a list of further reading, historical notes, pictures, better explanations, sections explaining the relevance of results to philosophy, computer science, and mathematics, and more problems and examples. If you find an error, or have a suggestion, please let the project team know.

The project operates in the spirit of open source. Not only is the text freely available, we provide the LaTeX source under the Creative Commons Attribution license, which gives anyone the right to download, use, modify, re-arrange, convert, and re-distribute our work, as long as they give appropriate credit. Please see the Open Logic Project website at openlogicproject.org for additional information.

Made in United States
North Haven, CT
13 May 2023

36538575R00228